AF581669

Impacts of Anthropogenic Activities on Watersheds in a Changing Climate

Impacts of Anthropogenic Activities on Watersheds in a Changing Climate

Editors

Luís Filipe Sanches Fernandes
Fernando António Leal Pacheco

MDPI • Basel • Beijing • Wuhan • Barcelona • Belgrade • Manchester • Tokyo • Cluj • Tianjin

Editors
Luís Filipe Sanches Fernandes
Universidade de Trás-os-Montes e Alto Douro
Portugal

Fernando António Leal Pacheco
University of Trás-os-Montes e Alto Douro
Portugal

Editorial Office
MDPI
St. Alban-Anlage 66
4052 Basel, Switzerland

This is a reprint of articles from the Special Issue published online in the open access journal *Water* (ISSN 2073-4441) (available at: https://www.mdpi.com/journal/water/special_issues/Watersheds_Anthropogenic_Activities).

For citation purposes, cite each article independently as indicated on the article page online and as indicated below:

LastName, A.A.; LastName, B.B.; LastName, C.C. Article Title. *Journal Name* **Year**, *Volume Number*, Page Range.

ISBN 978-3-0365-0266-3 (Hbk)
ISBN 978-3-0365-0267-0 (PDF)

Cover image courtesy of Fernando António Leal Pacheco.

Contents

About the Editors . . . vii

Fernando António Leal Pacheco and Luís Filipe Sanches Fernandes
Watersheds, Anthropogenic Activities and the Role of Adaptation to Environmental Impacts
Reprinted from: *Water* **2020**, *12*, 3451, doi:10.3390/w12123451 . . . **1**

Andrei-Emil Briciu, Dumitru Mihăilă, Adrian Graur, Dinu Iulian Oprea, Alin Prisăcariu and Petruţ Ionel Bistricean
Changes in the Water Temperature of Rivers Impacted by the Urban Heat Island: Case Study of Suceava City
Reprinted from: *Water* **2020**, *12*, 1343, doi:10.3390/w12051343 . . . **7**

Baojia Du, Zongming Wang, Dehua Mao, Huiying Li and Hengxing Xiang
Tracking Lake and Reservoir Changes in the Nenjiang Watershed, Northeast China: Patterns, Trends, and Drivers
Reprinted from: *Water* **2020**, *12*, 1108, doi:10.3390/w12041108 . . . **31**

Donatella Pavanelli, Claudio Cavazza, Stevo Lavrnić and Attilio Toscano
The Long-Term Effects of Land Use and Climate Changes on the Hydro-Morphology of the Reno River Catchment (Northern Italy)
Reprinted from: *Water* **2019**, *11*, 1831, doi:10.3390/w11091831 . . . **47**

Regina Maria Bessa Santos, Luís Filipe Sanches Fernandes, Rui Manuel Vitor Cortes and Fernando António Leal Pacheco
Hydrologic Impacts of Land Use Changes in the Sabor River Basin: A Historical View and Future Perspectives
Reprinted from: *Water* **2019**, *11*, 1464, doi:10.3390/w11071464 . . . **63**

Xiujie Wang, Pengfei Zhang, Lüliu Liu, Dandan Li and Yanpeng Wang
Effects of Human Activities on Hydrological Components in the Yiluo River Basin in Middle Yellow River
Reprinted from: *Water* **2019**, *11*, 689, doi:10.3390/w11040689 . . . **89**

Anildo Monteiro Caldas, Teresa Cristina Tarlé Pissarra, Renata Cristina Araújo Costa, Fernando Cartaxo Rolim Neto, Marcelo Zanata, Roberto da Boa Viagem Parahyba, Luis Filipe Sanches Fernandes and Fernando António Leal Pacheco
Flood Vulnerability, Environmental Land Use Conflicts, and Conservation of Soil and Water: A Study in the Batatais SP Municipality, Brazil
Reprinted from: *Water* **2018**, *10*, 1357, doi:10.3390/w10101357 . . . **105**

António Carlos Pinheiro Fernandes, Luís Filipe Sanches Fernandes, Rui Manuel Vitor Cortes and Fernando António Leal Pacheco
The Role of Landscape Configuration, Season, and Distance from Contaminant Sources on the Degradation of Stream Water Quality in Urban Catchments
Reprinted from: *Water* **2019**, *11*, 2025, doi:10.3390/w11102025 . . . **125**

Carlos Alberto Valera, Teresa Cristina Tarlé Pissarra, Marcílio Vieira Martins Filho, Renato Farias do Valle Júnior, Caroline Fávaro Oliveira, João Paulo Moura, Luís Filipe Sanches Fernandes and Fernando António Leal Pacheco
The Buffer Capacity of Riparian Vegetation to Control Water Quality in Anthropogenic Catchments from a Legally Protected Area: A Critical View over the Brazilian New Forest Code
Reprinted from: *Water* **2019**, *11*, 549, doi:10.3390/w11030549 . . . **145**

Teresa Cristina Tarlé Pissarra, Carlos Alberto Valera, Renata Cristina Araújo Costa, Hygor Evangelista Siqueira, Marcílio Vieira Martins Filho, Renato Farias do Valle Júnior, Luís Filipe Sanches Fernandes and Fernando António Leal Pacheco
A Regression Model of Stream Water Quality Based on Interactions between Landscape Composition and Riparian Buffer Width in Small Catchments
Reprinted from: *Water* **2019**, *11*, 1757, doi:10.3390/w11091757 . . . . **161**

Andrei-Emil Briciu, Adrian Graur, Dinu Iulian Oprea and Constantin Filote
A Methodology for the Fast Comparison of Streamwater Diurnal Cycles at Two Monitoring Points
Reprinted from: *Water* **2019**, *11*, 2524, doi:10.3390/w11122524 . . . . **179**

Marcelo Alvares Tenenwurcel, Maíse Soares de Moura, Adriana Monteiro da Costa, Paula Karen Mota, João Hebert Moreira Viana, Luís Filipe Sanches Fernandes and Fernando António Leal Pacheco
An Improved Model for the Evaluation of Groundwater Recharge Based on the Concept of Conservative Use Potential: A Study in the River Pandeiros Watershed, Minas Gerais, Brazil
Reprinted from: *Water* **2020**, *12*, 1001, doi:10.3390/w12041001 . . . . **199**

Mariana Bárbara Lopes Simedo, Teresa Cristina Tarlé Pissarra, Antonio Lucio Mello Martins, Maria Conceição Lopes, Renata Cristina Araújo Costa, Marcelo Zanata, Fernando António Leal Pacheco and Luís Filipe Sanches Fernandes
The Assessment of Hydrological Availability and the Payment for Ecosystem Services: A Pilot Study in a Brazilian Headwater Catchment
Reprinted from: *Water* **2020**, *12*, 2726, doi:10.3390/w12102726 . . . . **223**

About the Editors

Luís Filipe Sanches Fernandes was born in 1968 in Portugal. He holds a Ph.D. in Civil Engineering from the University of University of Tr´as-os-Montes and Alto Douro (UTAD, Portugal). He is a Professor in the Engineering Department (Civil Engineering) in the area of hydraulics, water resources, and environment at the UTAD, where he has taught since 1995. He is a full member at the Centre for the Research and Technology of Agro-Environmental and Biological Sciences (CITAB), where he was also a member of the executive committee for approximately 5 years. Moreover, he holds a MSc in Municipal Engineering from University of Minho. His Ph.D. thesis was entitled "A model for simulating hydrographs in river basins using neural networks", which he defended at UTAD. His aggregation title was in Territorial and Environmental Engineering Sciences, specialization in Civil Engineering, which he defended at the University of Évora (Portugal). Prof. Sanches Fernandes is a member of the Portuguese Association of Water Resources and senior member of Order of Engineers. He has a 28 h-index Scopus with more than 80 articles published in the fields of hydraulics, water resources, and environment. He has written 6 book chapters and more than 100 conference proceedings and communications (oral and poster). He has participated in 10 financed projects, being the leader in 2. He has supervised or co-supervised 3 Ph.D. theses and 35 master dissertations. Responsible for several university extension projects with high funding values, he was a member of 2 terms in pedagogical councils, 2 terms in scientific councils, and 1 term in general council in University of Trás-os-Montes and Alto Douro.

Fernando António Leal Pacheco was born in 1967 in Mozambique. He holds a Ph.D. degree in Hydrogeology (2001) from the Trás-os-Montes and Alto Douro University (UTAD, Portugal) and an Aggregation title in Environmental Geochemistry (2011) from the UTAD. He has been the head of the Geology Department of UTAD since 2013. He joined the Vila Real Chemistry Research Centre of UTAD in 2005, where he still develops most of his research. He has published over 100 research papers in international journals, co-authored papers with over 100 scientists of various nationalities (e.g., Portugal, The Netherlands, Brazil, Hungary, Germany, Italy, and Spain). Prof. Pacheco is a top reviewer certified by Clarivate Analytics Publons. In 2016, he received the "Sentinel of Science Award" as the Top Overall Contributor to the Peer Review of the Field of Earth and Planetary Sciences. In 2017, he was awarded as the Top Overall Contributor to Peer Review of Journal Science of Total Environment. He is also an editor for various scientific journals (e.g., Science of the Total Environment, Water, Sustainability, International Journal of Environmental Research and Public Health, Arabian Journal of Geosciences). Prof. Pacheco is actively involved in scientific cooperation on land management and water security projects with several Brazilian institutions: Federal Institute of Triângulo Mineiro (IFTM), Regional Coordination of Environmental Justice Prosecutors in the Paranaíba and Baixo Rio Grande River Basins (MPMG), Minas Gerais Institute for Water Management (IGAM), and San Paulo State University (UNESP). Prof. Pacheco's research interests and expertises cover the topics of hydrologic models coupled with weathering algorithms, especially in areas with significant anthropogenic pressure; multivariate statistical and environmental analyses of surface and groundwater databases, with a focus on the prevention of surface and groundwater contamination; land degradation and management, as well as the negative impacts of inadequate land uses on soil erosion, surface, and groundwater quality; water security issues, such as conjunctive use of surface and groundwater sources in public water supply systems; and the attenuation of hydrologic extremes (floods, droughts) through implementation of detention basins and decentralized rainwater harvesting systems in catchments.

Editorial

Watersheds, Anthropogenic Activities and the Role of Adaptation to Environmental Impacts

Fernando António Leal Pacheco [1,*] and Luís Filipe Sanches Fernandes [2]

[1] CQVR–Centro de Química de Vila Real, Universidade de Trás-os-Montes e Alto Douro, Ap. 1013, 5001-801 Vila Real, Portugal

[2] CITAB–Centro de Investigação e Tecnologias Agroambientais e Biológicas, Universidade de Trás-os-Montes e Alto Douro, Ap. 1013, 5001–801 Vila Real, Portugal; lfilipe@utad.pt

* Correspondence: fpacheco@utad.pt

Received: 25 November 2020; Accepted: 7 December 2020; Published: 9 December 2020

Runoff has shaped the Earth into watersheds, and humans have appropriated many of them. Since then, societies have settled on and used the watersheds and their resources to develop activities aiming at welfare. A vital resource is stream water, fed by the surface and underground compartments. However, overexploitation, contamination by humanity and climate variations reduce the availability of water for the targeted uses (e.g., drinking water and irrigation), endangering expected well-being besides the surrounding ecosystems. Hydrologic and statistical modeling have been used to assess the impacts of humans on the quality of water. Some studies refer to the consequences for water quality of point-source contamination, namely, direct discharges into streams of untreated or poorly treated domestic or industrial effluents [1]. Other work relates water quality degradation with diffuse pollution sources, considering the complementary role of natural processes such as soil erosion [2,3]. There is also a great deal of research where a nexus is established between anthropogenic activities, poor water quality and ecological integrity degradation [4–9]. With the purpose of helping in developing and implementing strategies for the mitigation of negative anthropogenic impacts and conservation of water quality in catchments, some studies have explored the controlling factors for contaminant propagation, at the regional (catchment) or local (reach) scales. That has included the influences of the season, scale, landscape configuration and share of riparian vegetation along the streams, among others [10–14].

Humanity also interferes with the hydrology of streams and rivers through changes in land use, including urbanization, forest–agriculture conversions or other impacting actions. The consequences for water resources and watershed management are numerous, including changes in the share of water balance components (e.g., surface flow, infiltration/groundwater flow, and evapotranspiration), potential water scarcity problems derived therefrom, hydromorphological changes in stream banks and urban floods [15–18]. This disturbance may be amplified by climate variations, leading to the need for anticipating adaptation or mitigation measures. A diversity of approaches to adaptation to water scarcity have been proposed to date, depending on the targeted water use. In the rural environment, the alternatives comprise the storage of rainwater in small dam lakes (rainwater harvesting), to be used in the irrigation of cropland or combating wildfires [19,20]. In the urban environment, especially around the large metropolises, the conjunctive use of various sources (surface water, groundwater, desalinized water and recycled wastewater) can be a reliable path to adaptation, aided by controlled groundwater recharge where possible [21–25]. The incentive to produce clean drinking water in headwater catchments through the adaptation of land uses (e.g., a shift from extensive monocultures to agro-forestry systems), aiming at the reduction of treatment and hence final costs, is a promising adaptation measure. However, the implementation of these land use and land cover (LULC) changes is challenging because it presupposes the payment for this environmental service. The landowners should receive the incentive from the final consumers, but the value is not easy to define, and the procedure is difficult to implement [26]. The control of floods has been practiced by the use of detention basins (structural interventions) and LULC changes such as forestations (non-structural measures) [27,28]. They both have advantages and disadvantages. Some authors argue that, for larger floods, the detention of river water in a large dam is essential to attain the objective of reducing or suppressing the inundated areas and their negative environmental, economic and social consequences [29].

The consequences of anthropogenic activities for watersheds are particularly addressed in this Special Issue, "Impacts of Anthropogenic Activities on Watersheds in a Changing Climate". In that regard, this Special Issue explores the impacts of human activities on the water quality and hydrology of streams and reservoirs, on hydromorphologic changes along river channels and on urban floods. It also discusses prominent factors controlling freshwater quality at the watershed and reach scales. Finally, it proposes adaptation measures to mitigate water scarcity in catchments and water quality degradation for drinking water supplies. During our working period, we received many submissions, which presented significant contributions for the main topics of interest of our Special Issue. However, only 12 high-quality papers were accepted after several rounds of strict and rigorous review. These 12 contributions are summarized in the forthcoming paragraphs, being integrated in a coherent narrative.

Contribution 1 discussed the impact of urban heat islands on the temperatures of rivers crossing cities. The authors studied the Suceava River (Romania) water temperature and concluded about a general increase downstream of Suceava city. They also observed daily water temperature profiles with steeper slopes and earlier moments of the maximum and minimum temperatures than upstream, which were considered a consequence of urban heat islands. Contribution 2 used Landsat series images from 1980 to 2015 and Structural Equation Models to investigate the diverse impacts of climatic and anthropogenic variables on nearly two hundred lakes and reservoirs widespread in the Nenjiang watershed (China). The observations allowed documenting a 42% decline in the total area and 51% in the number of lakes from 1980 to 2010, and a slight increase in these parameters afterwards. The statistical models exposed a strong relationship between the lake changes and the mean annual precipitation falling over the watershed, not ruling out the contribution of the agricultural consumption of water. Contribution 3 described the impacts of land use and land cover (LULC) changes on river flow rates and morphologies using the Reno River as a representative example. The Reno's watershed was characterized by forest exploitation and agricultural production until World War II, but the progressive abandonment of agriculture since then converted the catchment's occupation into meadows, forests and uncultivated land. These LULC changes produced reductions in the river flow rate and suspended sediment yield (−36% and −38%, respectively). These reductions produced a 40–80% decline in the riverbed area, a development of vegetation in the riparian buffer strips and a river channel change from a braided to single channel. Contributions 4 and 5 both studied the impacts of LULC on flow components. In the first case, the work was carried out in the Sabor River Basin (Portugal), and the authors investigated the roles of afforestation as well as of wildfires. While afforestation resulted in decreases in water yield, surface flow and groundwater flow and increases in evapotranspiration and lateral flow, wildfires caused an increase in surface flow and a decrease in lateral flow. In the second case, the studied area was the Yiluo River Basin (Middle Yellow River), and the aim was to investigate the roles of LULC and river regulation actions. In regard to LULC, the results showed that increased areas of urban land and decreased areas of vegetation land resulted in a decrease in both groundwater and evapotranspiration, but an increase in average surface runoff. Besides, the expansion of water areas resulted in an increase in evapotranspiration but had little effect on groundwater and surface runoff. Contribution 6 investigated the impact of LULC change on floods, focusing the analysis on a specific type of change characterized by the conversion of a natural use to one disrespecting land capability. This type of conversion is said to trigger environmental land use conflicts that are capable of amplifying environmental impacts, namely, floods. In the studied area (Batatais town, state of São Paulo, Brazil), the authors estimated that approximately 60% of areas were affected by these conflicts. The results helped explaining the severe floods that affect the city recurrently. They also supported the proposal of preventive and recovery measures in the context of a land consolidation–water management plan.

Contributions 1 to 6 describe anthropogenic activities developed on watersheds from around the planet, and the environmental impacts derived therefrom. Following this first group of contributions, a set of three other studies discuss the drivers and factors controlling the aforementioned impacts. Contribution 7 used Partial Least Squares–Path Models (PLS–PM) to investigate the effects on stream water quality of the landscape configuration, season, and distance from contaminant emissions from diffuse and point sources. The study was carried out in the Ave River Basin (Portugal). Overall, the PLS-PM results evidenced significant cause–effect relationships between landscape metrics and stream water quality (e.g., chemical and biochemical oxygen demands, and nitrogen and phosphorus emissions from the diffuse and point sources) at 10 km or larger scales, regardless of the season. Contribution 8 analyzed the effects of 15, 30 and 50 m-wide riparian forest on the water quality (e.g., turbidity) of three headwater catchments located in the EPA-URB—Environmental Protection Area of Uberaba River Basin (state of Minas Gerais, Brazil), which was legally protected for the conservation of water resources but is extensively used for sugar cane production. The results suggest that these vegetative barriers are not wide enough to preserve the legally protected water resources. Contribution 9 was also conducted in the EPA-URB. It expanded the analysis of Contribution 8 by exploring the combined effect of landscape composition and buffer strip width (L) on stream water quality. The landscape composition was assessed by the forest (F) to agriculture (A) ratio (F/A) derived from EPA-URB's LULC map. The water quality was evaluated by an index (IWQ) expressed as a function of physico-chemical parameters. The combined effect, $F/A \times L$, was quantified by multiple regressions with interaction terms. According to the authors, the interaction between F/A and L reduced the range of L values required to sustain the IWQ at a fair level by about 40%, relative to models where the terms were ignored. Notwithstanding the interaction, the calculated L ranges (45–175 m) required for a fair level of IWQ are much larger than the maximum width imposed by the Brazilian laws (30 m).

The last group of contributions to the Special Issue addressed topics related to the monitoring and planning of anthropogenic activities, as well as their impacts and mitigation or adaptation measures. Contribution 10 was a companion paper of Contribution 1 and presented the monitoring of the water level, specific conductivity, dissolved oxygen, oxidation–reduction potential and pH of the Suceava River for 365 days (2018–2019, with an hourly sampling frequency). As mentioned above, Suceava city is an urban heat island that affects the temperature of the Suceava River. The aim of Contribution 10 was to assess the diurnal cycle of other parameters to determine if they were also impacted by Suceava city. The results made it evident that the specific conductivity is higher downstream of the city (yearly averages: 483.1 μS/cm upstream and 549 μS/cm downstream) because of both treated and untreated waters being discharged directly or indirectly into the watercourse. Contribution 11 addressed the mapping of groundwater recharge from a management standpoint, being developed in the Pandeiros River Basin (state of Minas Gerais, Brazil). The immediate goal was to delineate areas with the maximum potential for recharge that could be protected from human activities and the pollution derived therefrom, for example, through LULC conversions (e.g., an extensive monoculture to agro-forestry system), and at the same time are prepared for the implementation of artificial recharge measures (e.g., the LULC conversions plus the construction of small dams). The last study (Contribution 12) frames the improvement of groundwater recharge in the topics of water security and payment for environmental services. The work took place in the Olaria Stream Basin (state of São paulo, Brazil) and involved the monitoring of stream flow discharge in three small headwater sub-basins. During the monitoring program, one sub-basin underwent recovery from antecedent gully erosion, reforestation with native trees along the drainage network, and conversion from an agricultural area into an agro-forestry system. For that sub-basin, the monitoring results revealed an increase in the stream flow discharge, which was related to the increase in recharge and explained by the aforementioned management practices. The stream flow increase improved the security of the water supply systems of various municipalities fed by these headwater catchments. The study authors argued that consumers should be willing to pay the costs faced by the land owners who implemented the management practices, as well as those for their future maintenance, because the stream flow increase was viewed as a service producing more water for public use.

Author Contributions: Both authors participated equally in the preparation and writing of this Editorial. All authors have read and agreed to the published version of the manuscript.

Funding: The author integrated in the CITAB was funded by National Funds of FCT (Portuguese Foundation for Science and Technology) under the project UIDB/04033/2020. The author integrated in the CQ-VR was funded by National Funds of FCT under the projects UIDB/QUI/00616/2020 and UIDP/00616/2020.

Conflicts of Interest: The authors declare no conflict of interest.

List of Contributions:

1. Briciu, A.-E.; Mihăilă, D.; Graur, A.; Oprea, D.I.; Prisăcariu, A.; Bistricean, P.I. Changes in the water temperature of rivers impacted by the Urban Heat Island: Case study of Suceava City. *Water* **2020**, *12*, 1343.
2. Du, B.; Wang, Z.; Mao, D.; Li, H.; Xiang, H. Tracking lake and reservoir changes in the Nenjiang Watershed, Northeast China: Patterns, trends, and drivers. *Water* **2020**, *12*, 1108.
3. Pavanelli, D.; Cavazza, C.; Lavrnić, S.; Toscano, A. The long-term effects of land use and climate changes on the hydro-morphology of the Reno River catchment (Northern Italy). *Water* **2019**, *11*, 1831.
4. Santos, R.M.B.; Sanches Fernandes, L.F.; Vitor Cortes, R.M.; Leal Pacheco, F.A. Hydrologic impacts of land use changes in the Sabor River basin: A historical view and future perspectives. *Water* **2019**, *11*, 1464.
5. Wang, X.; Zhang, P.; Liu, L.; Li, D.; Wang, Y. Effects of human activities on hydrological components in the Yiluo River basin in Middle Yellow river. *Water* **2019**, *11*, 689.
6. Caldas, A.; Pissarra, T.; Costa, R.; Neto, F.; Zanata, M.; Parahyba, R.; Sanches Fernandes, L.; Pacheco, F. Flood vulnerability, environmental land use conflicts, and conservation of soil and water: A Study in the Batatais SP Municipality, Brazil. *Water* **2018**, *10*, 1357.
7. Fernandes, A.C.P.; Sanches Fernandes, L.F.; Cortes, R.M.V.; Leal Pacheco, F.A. The role of landscape configuration, season, and distance from contaminant sources on the degradation of stream water quality in urban catchments. *Water* **2019**, *11*, 2025.
8. Valera, C.; Pissarra, T.; Filho, M.; Valle Júnior, R.; Oliveira, C.; Moura, J.; Sanches Fernandes, L.; Pacheco, F. The buffer capacity of riparian vegetation to control water quality in anthropogenic catchments from a legally protected area: A critical view over the Brazilian new forest code. *Water* **2019**, *11*, 549.
9. Pissarra, T.C.T.; Valera, C.A.; Costa, R.C.A.; Siqueira, H.E.; Martins Filho, M.V.; Valle Júnior, R.F. do; Sanches Fernandes, L.F.; Pacheco, F.A.L. A Regression model of stream water quality based on interactions between landscape composition and riparian buffer width in small catchments. *Water* **2019**, *11*, 1757.
10. Briciu, A.-E.; Graur, A.; Oprea, D.I.; Filote, C. A Methodology for the Fast Comparison of streamwater diurnal cycles at two monitoring points. *Water* **2019**, *11*, 2524.
11. Alvares Tenenwurcel, M.; Soares de Moura, M.; Monteiro da Costa, A.; Karen Mota, P.; Moreira Viana, J.H.; Fernandes, L.F.S.; Leal Pacheco, F.A. An improved model for the evaluation of groundwater recharge based on the concept of conservative use potential: A study in the river Pandeiros watershed, Minas Gerais, Brazil. *Water* **2020**, *12*, 1001.
12. Lopes Simedo, M.B.; Pissarra, T.C.T.; Mello Martins, A.L.; Lopes, M.C.; Araújo Costa, R.C.; Zanata, M.; Pacheco, F.A.L.; Fernandes, L.F.S. The assessment of hydrological availability and the payment for ecosystem services: A pilot study in a Brazilian headwater catchment. *Water* **2020**, *12*, 2726.

References

1. Álvarez, X.; Valero, E.; Santos, R.M.B.; Varandas, S.G.P.; Sanches Fernandes, L.F.; Pacheco, F.A.L. Anthropogenic nutrients and eutrophication in multiple land use watersheds: Best management practices and policies for the protection of water resources. *Land Use Policy* **2017**, *69*, 1–11. [CrossRef]
2. Oliveira, C.F.; do Valle Junior, R.F.; Valera, C.A.; Rodrigues, V.S.; Sanches Fernandes, L.F.; Pacheco, F.A.L. The modeling of pasture conservation and of its impact on stream water quality using partial least squares-path modeling. *Sci. Total Environ.* **2019**, *697*, 134081. [CrossRef] [PubMed]
3. Rodrigues, V.S.; do Valle Júnior, R.F.; Sanches Fernandes, L.F.; Pacheco, F.A.L. The assessment of water erosion using partial least squares-path modeling: A study in a legally protected area with environmental land use conflicts. *Sci. Total Environ.* **2019**, *691*, 1225–1241. [CrossRef] [PubMed]
4. Fonseca, A.R.; Sanches Fernandes, L.F.; Fontainhas-Fernandes, A.; Monteiro, S.M.; Pacheco, F.A.L. From catchment to fish: Impact of anthropogenic pressures on gill histopathology. *Sci. Total Environ.* **2016**, *550*, 972–986. [CrossRef]
5. Ferreira, A.R.L.; Sanches Fernandes, L.F.; Cortes, R.M.V.; Pacheco, F.A.L. Assessing anthropogenic impacts on riverine ecosystems using nested partial least squares regression. *Sci. Total Environ.* **2017**, *583*, 466–477. [CrossRef]

6. Fonseca, A.R.; Sanches Fernandes, L.F.; Fontainhas-Fernandes, A.; Monteiro, S.M.; Pacheco, F.A.L. The impact of freshwater metal concentrations on the severity of histopathological changes in fish gills: A statistical perspective. *Sci. Total Environ.* **2017**, *599*, 217–226. [CrossRef]
7. Santos, R.M.B.; Sanches Fernandes, L.F.; Cortes, R.M.V.; Varandas, S.G.P.; Jesus, J.J.B.; Pacheco, F.A.L. Integrative assessment of river damming impacts on aquatic fauna in a Portuguese reservoir. *Sci. Total Environ.* **2017**, *601*, 1108–1118. [CrossRef]
8. Sanches Fernandes, L.F.; Fernandes, A.C.P.; Ferreira, A.R.L.; Cortes, R.M.V.; Pacheco, F.A.L. A partial least squares—Path modeling analysis for the understanding of biodiversity loss in rural and urban watersheds in Portugal. *Sci. Total Environ.* **2018**, *626*, 1069–1085. [CrossRef]
9. Fernandes, A.C.P.; Sanches Fernandes, L.F.; Moura, J.P.; Cortes, R.M.V.; Pacheco, F.A.L. A structural equation model to predict macroinvertebrate-based ecological status in catchments influenced by anthropogenic pressures. *Sci. Total Environ.* **2019**, *681*, 242–257. [CrossRef]
10. Pissarra, T.C.T.; Valera, C.A.; Costa, R.C.A.; Siqueira, H.E.; Martins Filho, M.V.; do Valle Júnior, R.F.; Sanches Fernandes, L.F.; Pacheco, F.A.L. A regression model of stream water quality based on interactions between landscape composition and riparian buffer width in small catchments. *Water* **2019**, *11*, 1757. [CrossRef]
11. Valera, C.; Pissarra, T.; Filho, M.; Valle Júnior, R.; Oliveira, C.; Moura, J.; Sanches Fernandes, L.; Pacheco, F. The buffer capacity of riparian vegetation to control water quality in anthropogenic catchments from a legally protected area: A critical view over the brazilian new forest code. *Water* **2019**, *11*, 549. [CrossRef]
12. Fernandes, A.C.P.; Sanches Fernandes, L.F.; Terêncio, D.P.S.; Cortes, R.M.V.; Pacheco, F.A.L. Seasonal and scale effects of anthropogenic pressures on water quality and ecological integrity: A study in the Sabor river basin (NE Portugal) using partial least squares-path modeling. *Water* **2019**, *11*, 1941. [CrossRef]
13. Fernandes, A.C.P.; Sanches Fernandes, L.F.; Cortes, R.M.V.; Leal Pacheco, F.A. The role of landscape configuration, season, and distance from contaminant sources on the degradation of stream water quality in urban catchments. *Water* **2019**, *11*, 2025. [CrossRef]
14. Cortes, R.; Peredo, A.; Terêncio, D.; Sanches Fernandes, L.; Moura, J.; Jesus, J.; Magalhães, M.; Ferreira, P.; Pacheco, F. Undamming the Douro river catchment: A stepwise approach for prioritizing dam removal. *Water* **2019**, *11*, 693. [CrossRef]
15. Fernandes, L.F.S.; Pinto, A.A.S.; Terêncio, D.P.S.; Pacheco, F.A.L.; Cortes, R.M.V. Combination of ecological engineering procedures applied to morphological stabilization of estuarine banks after dredging. *Water* **2020**, *12*, 391. [CrossRef]
16. Caldas, A.; Pissarra, T.; Costa, R.; Neto, F.; Zanata, M.; Parahyba, R.; Sanches Fernandes, L.; Pacheco, F. Flood vulnerability, environmental land use conflicts, and conservation of soil and water: A study in the Batatais SP municipality, Brazil. *Water* **2018**, *10*, 1357. [CrossRef]
17. Santos, R.M.B.; Sanches Fernandes, L.F.; Vitor Cortes, R.M.; Leal Pacheco, F.A. Hydrologic impacts of land use changes in the sabor river basin: A historical view and future perspectives. *Water* **2019**, *11*, 1464. [CrossRef]
18. Santos, R.M.B.; Fernandes, L.F.S.; Cortes, R.M.V.; Pacheco, F.A.L. Development of a hydrologic and water allocation model to assess water availability in the Sabor river basin (Portugal). *Int. J. Environ. Res. Public Health* **2019**, *16*, 2419. [CrossRef]
19. Terêncio, D.P.S.; Sanches Fernandes, L.F.; Cortes, R.M.V.; Pacheco, F.A.L. Improved framework model to allocate optimal rainwater harvesting sites in small watersheds for agro-forestry uses. *J. Hydrol.* **2017**, *550*, 318–330. [CrossRef]
20. Terêncio, D.P.S.; Sanches Fernandes, L.F.; Cortes, R.M.V.; Moura, J.P.; Pacheco, F.A.L. Rainwater harvesting in catchments for agro-forestry uses: A study focused on the balance between sustainability values and storage capacity. *Sci. Total Environ.* **2018**, *613*, 1079–1092. [CrossRef]
21. Soares, S.; Terêncio, D.; Fernandes, L.; Machado, J.; Pacheco, F. The potential of small dams for conjunctive water management in rural municipalities. *Int. J. Environ. Res. Public Health* **2019**, *16*, 1239. [CrossRef] [PubMed]
22. da Costa, A.M.; de Salis, H.H.C.; Viana, J.H.M.; Leal Pacheco, F.A. Groundwater recharge potential for sustainable water use in urban areas of the Jequitiba river basin, Brazil. *Sustainability* **2019**, *11*, 2955. [CrossRef]
23. De Salis, H.H.C.; da Costa, A.M.; Künne, A.; Sanches Fernandes, L.F.; Leal Pacheco, F.A. Conjunctive water resources management in densely urbanized karst areas: A study in the Sete Lagoas region, State of Minas Gerais, Brazil. *Sustainability* **2019**, *11*, 3944. [CrossRef]

24. Cardoso de Salis, H.H.; Monteiro da Costa, A.; Moreira Vianna, J.H.; Azeneth Schuler, M.; Künne, A.; Sanches Fernandes, L.F.; Leal Pacheco, F.A. Hydrologic modeling for sustainable water resources management in urbanized Karst areas. *Int. J. Environ. Res. Public Health* **2019**, *16*, 2542. [CrossRef] [PubMed]
25. Alvares Tenenwurcel, M.; Soares de Moura, M.; Monteiro da Costa, A.; Karen Mota, P.; Moreira Viana, J.H.; Fernandes, L.F.S.; Leal Pacheco, F.A. An improved model for the evaluation of groundwater recharge based on the concept of conservative use potential: A study in the river Pandeiros watershed, Minas Gerais, Brazil. *Water* **2020**, *12*, 1001. [CrossRef]
26. Simedo, M.B.L.; Pissarra, T.C.T.; Martins, A.L.M.; Lopes, M.C.; Costa, R.C.A.; Zanata, M.; Pacheco, F.A.L.; Fernandes, L.F.S. The assessment of hydrological availability and the payment for ecosystem services: A pilot study in a brazilian headwater catchment. *Water* **2020**, *12*, 2726. [CrossRef]
27. Bellu, A.; Sanches Fernandes, L.F.; Cortes, R.M.V.; Pacheco, F.A.L. A framework model for the dimensioning and allocation of a detention basin system: The case of a flood-prone mountainous watershed. *J. Hydrol.* **2016**, *533*, 567–580. [CrossRef]
28. Terêncio, D.P.S.; Fernandes, L.F.S.; Cortes, R.M.V.; Moura, J.P.; Pacheco, F.A.L. Flood risk attenuation in critical zones of continental Portugal using sustainable detention basins. *Sci. Total Environ.* **2020**. [CrossRef]
29. Salgado Terêncio, D.P.; Sanches Fernandes, L.F.; Vitor Cortes, R.M.; Moura, J.P.; Leal Pacheco, F.A. Can land cover changes mitigate large floods? A reflection based on partial least squares-path modeling. *Water* **2019**, *11*, 684. [CrossRef]

Publisher's Note: MDPI stays neutral with regard to jurisdictional claims in published maps and institutional affiliations.

Article

Changes in the Water Temperature of Rivers Impacted by the Urban Heat Island: Case Study of Suceava City

Andrei-Emil Briciu [1,*], Dumitru Mihăilă [1], Adrian Graur [2], Dinu Iulian Oprea [1], Alin Prisăcariu [1] and Petruţ Ionel Bistricean [3]

[1] Department of Geography, Ștefan cel Mare University of Suceava, 720229 Suceava, Romania; dumitrum@atlas.usv.ro (D.M.); dinuo@atlas.usv.ro (D.I.O.); alinprisecaru@yahoo.com (A.P.)

[2] Computers, Electronics and Automation Department, Ștefan cel Mare University of Suceava, 720229 Suceava, Romania; Adrian.Graur@usv.ro

[3] Suceava Weather Station, Regional Meteorological Centre of Moldova, National Meteorological Administration of Romania, 720237 Suceava, Romania; petricabistricean@gmail.com

* Correspondence: andreibriciu@atlas.usv.ro

Received: 26 March 2020; Accepted: 6 May 2020; Published: 9 May 2020

Abstract: Cities alter the thermal regime of urban rivers in very variable ways which are not yet deciphered for the territory of Romania. The urban heat island of Suceava city was measured in 2019 and its impact on Suceava River was assessed using hourly and daily values from a network of 12 water and air monitoring stations. In 2019, Suceava River water temperature was 11.54 °C upstream of Suceava city (Mihoveni) and 11.97 °C downstream (Tişăuţi)—a 3.7% increase in the water temperature downstream. After the stream water passes through the city, the diurnal thermal profile of Suceava River water temperature shows steeper slopes and earlier moments of the maximum and minimum temperatures than upstream because of the urban heat island. In an average day, an increase of water temperature with a maximum of 0.99 °C occurred downstream, partly explained by the 2.46 °C corresponding difference between the urban floodplain and the surrounding area. The stream water diurnal cycle has been shifted towards a variation specific to that of the local air temperature. The heat exchange between Suceava River and Suceava city is bidirectional. The stream water diurnal thermal cycle is statistically more significant downstream due to the heat transfer from the city into the river. This transfer occurs partly through urban tributaries which are 1.94 °C warmer than Suceava River upstream of Suceava city. The wavelet coherence analyses and ANCOVA (analysis of covariance) prove that there are significant (0.95 confidence level) causal relationships between the changes in Suceava River water temperature downstream and the fluctuations of the urban air temperature. The complex bidirectional heat transfer and the changes in the diurnal thermal profiles are important to be analysed in other urban systems in order to decipher in more detail the observed causal relationships.

Keywords: diurnal thermal profile; urban tributaries; wavelet; covariance

---

1. Introduction

An increase in water temperature of rivers all over the world due to global warming is depicted in numerous studies [1,2]. At the same time, the increasing urban population worldwide leads to a greater and more territorially focused impact of cities on the environment. This impact, enhanced by the current climate changes, is exerted on urban rivers through multiple anthropogenic stressors such as the increasing imperviousness of catchments or the diminishing of areas with riparian vegetation [3,4]. The land use changes associated with the urbanization of an area (e.g., widespread paving) [4,5] lead to alterations in the thermal regime of the air in cities, and urban heat islands occur [6]. The warmer urban surfaces generate temperature surges in streams after localized rainstorms [3,7]. The impervious urban areas reduce rainwater infiltration in the catchment and, as a consequence, the high waters are more intense and the baseflow is weaker [8,9]. The shallower waters during baseflow are more prone to being heated by the urban air [3]. The higher temperature of urban stream waters reduce their self-purification capacity by affecting the aquatic biota and the amount of dissolved oxygen in the water [10]. Changes in fish populations' structure and diversity are easily identifiable in rivers switched to an urban regime, and species that better tolerate warmer waters replace the previous ones [11,12].

Rivers passing through cities may have mean temperatures above or below the corresponding mean air temperature [13]. In both cases, discussions may arise concerning not only the influence of the city on the river, but also the inverse case. Colder rivers can have a cooling effect on a city, which may be enhanced by air circulation along the stream corridor [14–16]. At the opposite side, the warmer rivers have an effect of enhancing the urban heat island [17].

Due to the important role of water temperature in aquatic ecosystems, river water temperature is included in water quality indices [18]. Also, some software solutions were proposed to model the thermal impact of cities on rivers in order to find pollution causes and improve city management [19,20]. Overall, the increase in the urban stream water temperature is recognised as an essential aspect of the urban stream syndrome [21].

As an effect of climate change, river temperatures in north-eastern Romania are expected to increase by 1–2 °C at the end of this century, according to some models [13]. The predicted increases in air temperature in this part of Europe [22] will be higher in urban areas. The temporal evolution of stream water temperature of rivers in Romania was analysed only in a few studies [23–27]. Fewer are the studies about river water temperature in cities [26,27]. In this study, we aim to describe the thermal impact of Suceava city on its stream waters, especially Suceava River, by using high temporal resolution data and multiple analysis methods. The syntheses obtained from our data are relevant for a usual city–river relationship in Romania. Specific objectives of this study are (1) to find if and how the urban heat island alters the water temperature of Suceava River and (2) to identify the role played by urban tributaries and wastewaters in the observed variations of Suceava River temperature.

2. Materials and Methods

2.1. Study Area

Suceava city is the centre of a metropolitan area of approximately 150,000 people and is located in the north-eastern part of Romania, in Suceava county. It is a city divided by Suceava River, which flows from NW towards SE and crosses the industrial area in the middle of the city. Terrain elevation in the city ranges from 265 m above sea level (a.s.l.), in the floodplain/south-eastern part of the city, to 435 m a.s.l., in the hills of the northern limits. Climate is temperate continental. Winters record small amounts of precipitations and frequent negative temperatures, while summers are warm and rainy, with frequent torrential rainfalls. Mean air temperature is 8.16 °C, whilst the average annual sum of precipitation is 621 mm.

Suceava River has an average flow rate of 16.87 m^3/s at Iţcani station, located into the city [27]. It also has an annual flow regime imposed by the regional rainfalls and records frequent high waters (over 50 m^3/s) in the late spring and early to mid-summertime interval. This river has well mixed waters due to its average maximum depth of under 1 m of water and due to numerous transversal hydraulic infrastructures, that generate hydraulic jumps (e.g., downstream of the Iţcani and Burdujeni bridges). The main stream water monitoring stations used in this study are positioned on Suceava River upstream and downstream of Suceava city (Figure 1, Table 1). The other 3 monitoring stations are located on 3 small streamflows, typical for the tributaries received by Suceava River between Mihoveni (upstream) and Tişăuţi (downstream). The cumulative streamflow of Suceava River tributaries in the study area is roughly 0.5 m^3/s. The Collector streamflow is a semi-artificial streamflow which represents a mixture of urban rainwater and groundwater drainage and, in a small proportion, wastewater.

It was created in the communist era as part of the measures to drain water from Suceava floodplain. Untreated wastewaters are also present in the flow of Şcheia and Cetăţii Creek, included in our monitoring network, and in the other streamflows, as a result of illicit domestic wastewater discharges. The wastewater treatment plant of Suceava city is located on the left bank of Suceava River and, in 2019, it discharged an average of 0.34 m^3/s of water. This is also one of the 5 stations used to collect air temperature and relative humidity in Suceava floodplain (Figure 1, Table 1). These latter stations are part of a wider monitoring network (19 stations) that we used prior to our hydrological analysis in order to obtain maps of the urban heat island and its meteorological effects that help us to better understand our study area (Figure 2, Supplementary Figure S1).

Major and static sources of heat are a few and are represented by the thermoelectric plant (near the wastewater treatment plant), the cellulose factory and 5 plants that belongs to the food industry. Many (tens of thousands) minor static sources are represented by domestic water and air heaters (such as boilers). Tens of thousands of cars are a daily source of mobile thermal pollution. In the official build-up area of Suceava city, the industrial areas represent 11.16%, other buildings (mostly domestic) represent 24.37%, roads and pavements represent 6.14% and the remaining space is constituted of semi-natural spaces, such as private yards, parks or bare lands [27].

The urban heat island of Suceava is easily identifiable in summer (Figure 2b) and produces a distinct area of lower relative humidity in winter (Figure 2c).

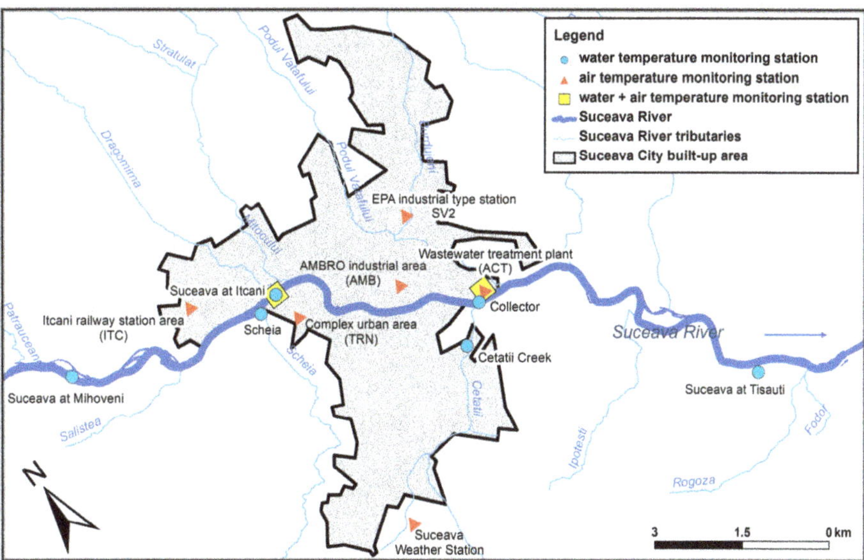

Figure 1. Map of the study area (centred on Suceava city) and location of the monitoring stations.

Table 1. Details of the monitoring stations used in this study.

Site Name	Coordinates (WGS84)	Elevation (m above Sea Level)	Riverbed-Banks/ Active Surface Type	Surrounding Land Cover	Data Type
Water monitoring stations					
Suceava at Mihoveni	47.681356° N, 26.199646° E	280	concrete	complex	own
Şcheia	47.670982° N, 26.241660° E	276	partly concrete	built area	own
Cetăţii Creek	47.647981° N, 26.274243° E	270	partly concrete	built area	own
Collector	47.651907° N, 26.282219° E	269	concrete	built area	own
Suceava at Tişăuţi	47.617615° N, 26.322629° E	260	partly natural	pasture	own
Air monitoring stations					
Suceava Weather Station	47.632349° N, 26.240747° E	359	grass	cultivated land	public
EPA industrial type SV2	47.669097° N, 26.280840° E	289	concrete	built area	public
Iţcani railway station area (TRN)	47.678598° N, 26.229750° E	280	partly grass	built area	own
AMBRO industrial area (AMB)	47.661483° N, 26.270342° E	279	grass	built area	own
Complex urban area (ITC)	47.667320° N, 26.247910° E	276	partly grass	complex	own
Water + air monitoring stations					
Suceava at Iţcani	47.671850° N, 26.246900° E	273	natural/grass	built area/pasture	public
Wastewater treatment plant (ACT)	47.653000° N, 26.284566° E	271	concrete/grass	complex	public/own

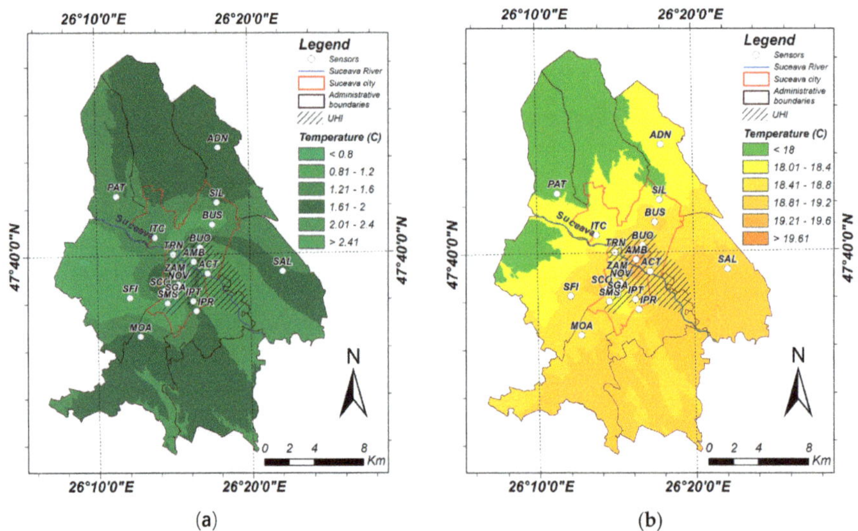

Figure 2. Urban heat island (UHI) above Suceava metropolitan area as revealed by hourly air temperature measurements in 2019 during winter—(**a**) (December, January, February)—and summer months—(**b**) (June–August)—UHI represents the area most impacted by the urban heat island.

2.2. Instruments

From June 28 until September 28, 2018, water temperature data used in this study for the analysis of Mihoveni (upstream) and Tişăuţi (downstream) monitoring stations was measured with TruBlue 585 CTD instruments (accuracy: ±0.2 °C, resolution: 0.01 °C). From September 2018 until the end of 2019, water temperature and level measurements in our monitoring stations on Suceava

River were obtained with 2 AquaTROLL500 instruments (accuracy: ±0.1 °C, resolution: 0.01 °C). The temperature measurements of the other streamflows were obtained with DS1922L-F5 iButton instruments (accuracy: ±0.5 °C, resolution: 0.0625 °C). Different stream waters, e.g., Suceava River or its tributaries, were measured with instruments with an accuracy identical for the entire streamflow type.

Air temperature and relative humidity (RH) in the stations used to create the maps in Figure 2 (with the exception of the official weather station of Suceava city) were measured with CEM DT-171 thermohygrometers (accuracy: ±1 °C and ±3.0% to ±5.0% RH, resolution: 0.1 °C and 0.1% RH) placed in miniature weather stations, made of wood, with good air circulation, preventing the direct solar radiation, with waterproof roof and fixed on wooden pillars at 2 m from the ground.

2.3. Data

We used hourly and daily time series with different lengths from 2018 and 2019 at UTC+2. The year 2019 was selected for in-depth analyses due to the broader data availability for all parameters. Data used in this study is divided into data obtained from our own instruments and data provided by various Romanian institutions. Data from our own instruments is represented by measurements of water temperature (5 sites) and, secondarily, water level (we used here, for a complementary analysis, only level data measured at Tişăuţi with AT500 instruments); also, from the air environment, we obtained hourly values of temperature and relative humidity in 18 sites, of which 4 (located in Suceava floodplain—Table 1) were selected for further integration in the meteorological-hydrological analysis and comparison. Water temperature at Mihoveni (upstream) and Tişăuţi (downstream) was recorded from 28 June 2018 until the end of 2019. Water level at Tişăuţi used here was recorded for the entire year of 2019. Water temperature of the other streamflows was recorded from 20 September 2019 until the end of 2019. The sensors were placed in areas with persistent water flow. In the case of Suceava River, sensors were permanently under at least a 0.33 m column of water; for the tributaries, the minimum water column had a height of 0.1 m (maximum possible during low flow). Our meteorological data were recorded during the entire year of 2019 and above mostly grassy active surfaces. Water temperature data of Suceava River at Mihoveni (upstream) and Tişăuţi (downstream) and air temperature and relative humidity data of Suceava floodplain are available at http://water.usv.ro/data.php.

Data from Romanian institutions is represented by data provided by ANAR (Romanian Waters National Administration—discharge and temperature data of Suceava River and atmospheric precipitation, both at Iţcani—daily values for the entire year of 2019), ACET Suceava (the company responsible for water supply and sewerage in Suceava metropolitan area—discharge and temperature data of Suceava wastewater treatment plant effluent—daily data of 2019), ANM (National Meteorological Administration—air temperature, relative humidity and precipitation at Suceava Weather Station—hourly and daily data of 2019) and ANPM (National Environment Protection Agency—air temperature, relative humidity and solar radiation at SV2 station—hourly and daily data for 2019).

In 2019, the average yearly flow rate of Suceava River was 19.5 m^3/s at Iţcani station, which is only 15.6% higher than the multiannual discharge of 16.87 m3/s (6.9 m3/s standard deviation) [27] and, therefore, 2019 can be considered an average hydrological year. But, on the other side, the atmospheric environment above the city during the same year was hotter (+2.09 °C) and drier (−86 mm; −13.8%) than the last 60 years' average, the mark of a continuous climate change and increased urbanization of Suceava city metropolitan area. We suggest that 2019 can be taken into account as a case study year, which is valuable especially because studies using hourly data on thermal pollution of urban rivers are missing in Romania.

2.4. Analysis Methods

The maps that used interpolation of stations/points with measurements of air temperature and relative humidity were created in ArcGIS with a hybrid method that combines the multiple regression method with the surface interpolation of stations/points using an inverse distance weighted (IDW) method [28]. The spatial tendency of a parameter is calculated by applying the multiple regression. The difference between the real values and those estimated by regression (residues) are then interpolated using the IDW. Finally, the raster with the interpolated residues is added to the spatial trend in order to correct the spatial representation of the analysed parameter.

In this study, boxplots, seasonal adjustment, smoothing, normalising, surface plots and scalograms are done in MATLAB. The surfaces plot was created in MATLAB by using the "surf" function. The surfaces plot in this study used normalised matrices composed of vectors extracted from time series that had the seasonal oscillations removed by subtracting from the raw time series the smoothed variant of the selected time series. The subtraction acted as a high-pass filter. The smoothing used a moving average filter with a span of 49 values (in order to include 2 consecutive days). The vectors/columns of the matrices represent average hourly values and the normalisation process computed the z-score values of each vector (centre 0, standard deviation 1). The normalisation was necessary for representing on the same scale (for comparison purposes) the diurnal thermal amplitudes from both semesters of 2019 and from both environments, even if the real amplitude is higher in the warm semester and in the air.

Scalograms represent the graphic result of continuous wavelet analysis or wavelet coherence analysis. The wavelet transform involved the Morlet mother wavelet, and the continuous wavelet transform analysis used a rectified power spectrum [29]. We have chosen Morlet as the mother wavelet because it is frequently used in the analysis of time series in geosciences due to the good time–frequency localisation of events. The wavelet coherence analysis used the methodology of Grinsted et al. [30], which continued to implement the Morlet mother wavelet in the study of hydrological time series. For a detailed description of the wavelet analysis and equations, see our complementary studies [31,32].

XLSTAT (Statistical Software for Excel) was used for ANCOVA (analysis of covariance), which involved one quantitative dependent variable (parameter) and multiple quantitative explanatory variables (other parameters) for calculating the goodness-of-fit coefficients of the linear regression model, obtaining the Fisher's F test result (estimates the risk in assuming that the null hypothesis is wrong/that there is no effect of the explanatory variables) and producing the Type III SS Table (Sum of Squares analysis). This table estimates the contribution of each explanatory variable to the model by removing one variable at a time and evaluating its effect on the quality of the computed model (this removes the undesirable effect of the order in which the variables are selected in the process of model calculation). ANCOVA assumptions are: the dependent variable and covariate variables are measured on a continuous scale, there are no significant outliers, residuals are approximately normally distributed for the independent variable, there is a homogeneity of variances and there is a homogeneity of regression slopes.

An energy balance of the water system in this study was calculated according to the following formula:

$$temperature_{SVT} = (mass_{SVM}*temperature_{SVM} + mass_{TR}*temperature_{TR} + mass_{WW}*temperature_{WW}) / (mass_{SVM} + mass_{TR} + mass_{WW}) \quad (1)$$

where mass is the mass of the water flows calculated from discharge, SVT is Suceava River downstream (Tişăuţi), SVM is Suceava River upstream (Mihoveni), TR is the sum of Suceava River tributaries between the upstream and downstream monitoring stations on Suceava River and WW is the wastewater from the wastewater treatment plant of Suceava city.

3. Results and Discussion

In the monitored period of 2018 and 2019, water temperature of Suceava River ranged from 0 to 28 °C. It became supercooled during the 2018/2019 winter (minimum: −0.19 °C at Mihoveni (upstream) and −0.13 °C at Tişăuţi (downstream)), when frazil ice was forming, and warm during the summers (maximum: 28.57 °C at Mihoveni and 28.65 °C at Tişăuţi), due to heat waves in the atmosphere (Supplementary Figure S2). As the extreme values already revealed, Suceava River was generally warmer downstream of Suceava city (average: 11.97 °C upstream and 12.4 °C downstream). One can observe especially in the November–February months, that when water temperature upstream is near the freezing point, the same parameter downstream still exhibits diurnal cycles. These cycles in streamflows are created by similar cycles in air temperature and other mechanisms [33] and are still present downstream of Suceava city during cold days due to the heat input that the river received inside the city.

In order to find out if this increase in temperature is caused by the warmer air above Suceava city, we compared water and air temperature during an entire calendar year, 2019 (Figure 3). We selected two atmospheric datasets for this comparison. The first dataset is represented by the air temperature time series from Suceava Weather Station. Data from this monitoring station are less influenced by Suceava city due to its location outside the built-up area and are representative for the semi-natural environment that surrounds Suceava city and which represents the Suceava River catchment in Suceava Plateau.

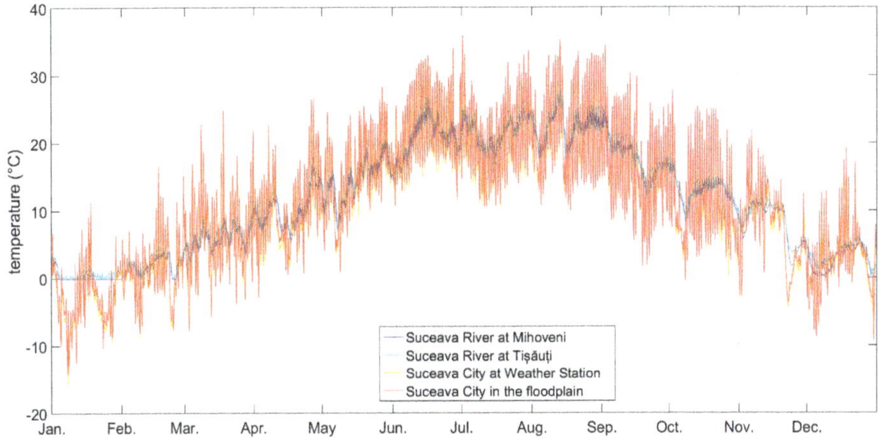

Figure 3. Comparison of water temperature at Mihoveni (upstream) and Tişăuţi (downstream) and air temperature in Suceava city area in 2019 (hourly data).

The second dataset is represented by time series obtained by averaging hourly values from the selected monitoring stations located in the urban floodplain of Suceava River (SV2, ITC, TRN, AMB, ACT—Table 1). This dataset is representative of the urban air of Suceava, as it was recorded in a complex residential, industrial and commercial area.

As expected, the thermal amplitude is higher in the air and the water temperature follows the air temperature (exceptions are the days with persistent negative air temperature, when water temperature stays around 0 °C due to the high energy loss necessary for freezing). In 2019, the average air temperature at Suceava Weather Station was 10.35 °C, while in the floodplain it was 11.05 °C. From the thermal difference of these stations (0.7 °C), only 0.5 °C might be explained by the vertical temperature gradient in the air due to the fact that the weather station is placed on a plateau at approximately 83 m above the mean elevation of the monitoring stations in the floodplain. Therefore, a surplus of 0.2 °C exists in the middle of the city as a result of the urban surfaces and activities. Water temperature was 11.54 °C upstream of Suceava city (Mihoveni) and 11.97 °C downstream (Tişăuţi). Studies on urban river temperature increase often find increases above 1 °C downstream of urban areas (e.g., 4–5 °C increase at some stations in North Carolina [34]). The amplitude of the temperature increase of our studied river is more similar to those found by Zeiger and Hubbart [35] in Missouri (0.2–0.7 °C). The difference between the mean values of the monitoring stations upstream and downstream in 2019 was 0.43 °C. The accuracy of the temperature sensors of the AT500 instruments used for measuring

Suceava River water temperature is ±0.1 °C. The standard error of the mean is often used to assess the error of continuous measurements [36,37], and the standard error of the mean of Suceava River water temperature measurements is 0.085 °C for the upstream time series and 0.084 °C for the downstream time series. Therefore, we conclude that there is a relevant water temperature increase downstream.

We can observe that Suceava River is warmer than the air above the city at both moments, of entering and of exiting its metropolitan area. A calculus done for the warm semester (April–September, average value from hourly data), in order to exclude winter months with water persisting around 0 °C when the air temperature is frequently negative, has revealed a similar context: air at weather station = 16.67 °C, air in the floodplain = 17.7 °C, water upstream = 17.84 °C and water downstream = 18.19 °C. Suceava River is heated above the air temperature as a result of its warmer banks, which absorbed the solar radiation, and this reason is available for its tributaries too, whose shallower water is more prone to overheating. This heating is stronger in constructed areas (Suceava city in the floodplain is 0.7 °C warmer than at Suceava Weather Station) due to the numerous man-made structures (buildings, roads, channels) that transfer their excess heat to the water nearby. Small urban streams have a low inertia and are more susceptible to urban influence [38].

During the communist era, a network of pipelines was built in order to transport hot water from the thermal power station towards the city inhabitants. Those pipelines are still used today not only for heating (during winter), but also for hot showers (in some homes). The pipelines transport warm water from the thermal power station (located in the floodplain, near the wastewater treatment plant) to various urban users. These pipelines are on land surface or underground and transfer some of their heat to the nearby soil and groundwater (hot water leaves the plant with 70–80 °C and arrives at the consumers, sometimes after more than 10 km of pipe, with 48 °C or less). The warm water is then discharged into the sewerage system in order to be collected at the wastewater treatment plant. The sewerage network often contains wastewater that is warmer than the groundwater. Warmer groundwaters diffusely recharge Suceava River and its urban tributaries in the floodplain (this recharge direction is proven by the small springs that appear along Suceava Riverbanks). No studies exist about the temperature of the urban groundwaters in our study area. Groundwater tends to be cooler than air in the summertime and warmer in wintertime, but even during summer, some groundwaters may become warmer because hot urban surfaces may transfer their heat through soils into phreatic waters.

It appears that the average heat transfer between Suceava River and Suceava city is from water to air after the water has been heated by the urban soil and constructions, but the analysis of the average diurnal profiles of water and air in our study area indicates that the water and air relationship is bidirectional. One can observe in Supplementary Figure S3a,b, that the diurnal cycle of the air temperature in the floodplain has steeper slopes than at Suceava Weather Station and the moments of the maximum and minimum values occur earlier. During the late evening and the night, air is colder in the floodplain. Air temperature in the urban floodplain was warmest, compared to Suceava Weather Station air temperature, at 2 p.m. (2.46 °C difference), and air was colder by a maximum of 0.63 °C at 9 p.m. All these differences indicate that the air in the floodplain is strongly influenced by the urban land use (constructed surfaces heat up faster and cool more easily that a vegetated area). Similarly, after Suceava River flows through the city, the diurnal profile of Suceava River water temperature has steeper slopes and earlier moments of the maximum and minimum temperatures than upstream. Water downstream is warmer than upstream, especially in the afternoon when land and air become hotter. Water downstream was warmer than upstream with a maximum of 0.99 °C (4 p.m.) and a minimum of 0.05 °C (3 a.m.). Minima of the air temperature, at Suceava Weather Station and in the floodplain, and of water temperature, upstream and downstream, occur at 6 a.m., 5 a.m., 8 a.m. and 7 a.m., respectively. Moments of the average daily maxima for the same sites are 3 p.m., 2 p.m., 7 p.m. and 4 p.m., respectively. These values show that the peak position in the diurnal cycle of water has been strongly shifted towards the moment of the maximum air temperature because the heat gain during the day is an active process, in opposition to the passive process of heat loss during the night.

As previously observed, the heat transfer between air and water is bidirectional, both during an average day and during a year. This led to the question of whether this relationship alters the warming of Suceava River downstream. A simple statistical analysis revealed that 26.4% of the hourly values of Suceava River at the downstream station are actually lower than at the upstream station, and 11.2% of the days in 2019 had an average value that was lower downstream than upstream. We analysed the hourly differences in the diurnal regime of temperature during opposite seasons (summer and winter) and found that a constantly higher water temperature downstream is representative for winter, while, during summer night-time, Suceava River is colder downstream of the city (Figure 4a,b).

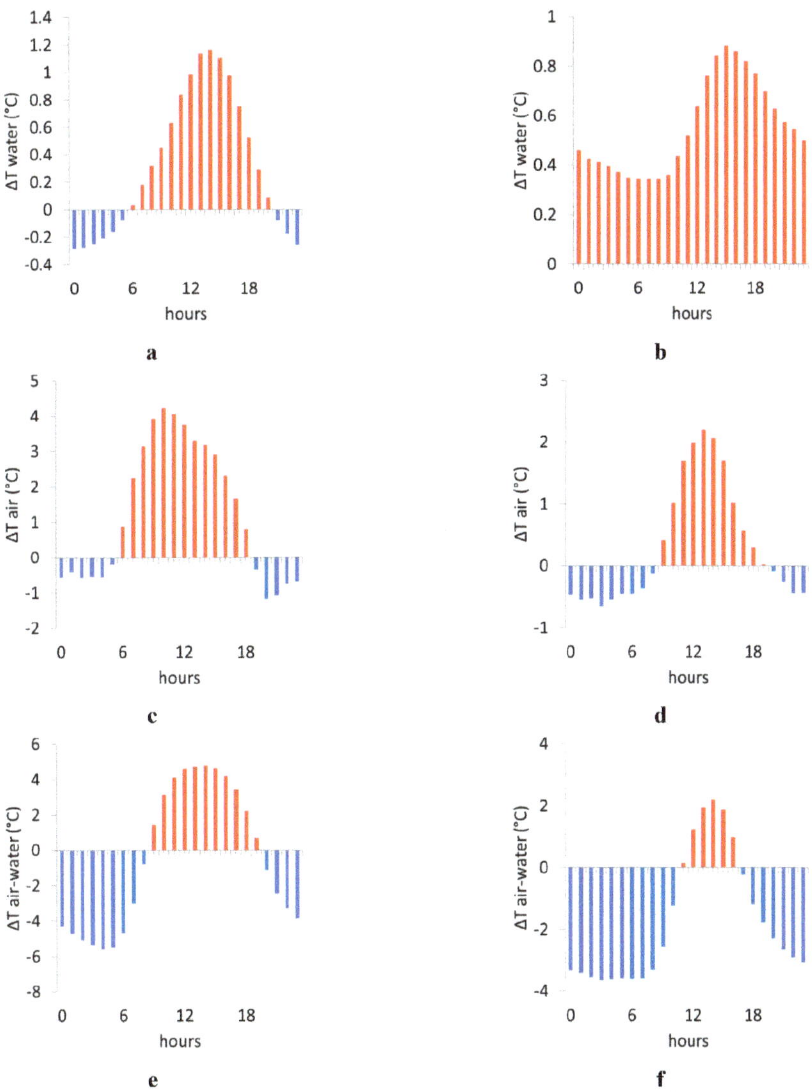

Figure 4. Hourly differences in the diurnal regime of temperature in 2019 during summer (left column) and winter (right column) for: Suceava River downstream—Suceava River upstream (**a**,**b**), floodplain air—Suceava Weather Station (**c**,**d**), floodplain air—Suceava River downstream (**e**,**f**).

During winter, the downstream–upstream thermal differences are lower than in summer. During the daytime, the heating of the active surface and of the air in the floodplain is much stronger than at Suceava Weather Station because of the built environment, but, during night-time, the floodplain becomes a space with frequent air temperature inversions of medium–low intensity (Figure 4c,d). These inversions are able, during most days in summer, to decrease the river water temperature downstream. During both seasons, the water of Suceava River is a heating source for the atmosphere of Suceava city, especially in winter (during 3/4 of an average day—Figure 4e,f).

Monthly average values in Figure 5a shows that, during March–June, air temperature in the urban floodplain of Suceava River is warmer than the stream water in both stations (or similar to it, in April). This might be attributed to the melting snows in the mountain area of Suceava River catchment and to the important rainfalls specific to this period of the year—all these create the colder water of Suceava River in the first half of the year. It is to be taken into consideration that 2019 was a dry year, with an annual sum of precipitation of 535.03 mm at Suceava station and 486.5 mm at Iţcani station (in the floodplain). We can assume that, during years with average or excess precipitation, the impact of the urban heat island on Suceava River temperature is easier to detect.

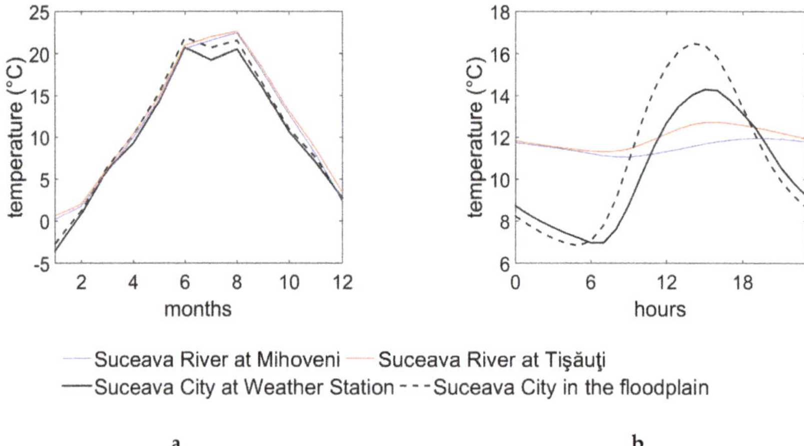

Figure 5. Comparisons of water and air temperatures in the Suceava city area in 2019: (**a**) annual regime of temperature, (**b**) average diurnal profiles.

Studies on other city–river pairs indicate that a river can be warmer than the urban air even during winters with frozen surface waters [17]. Even small, linear water bodies such as a river may have a thermal effect on the surrounding environment, as proven by the 22 m wide UK river in the study of Hathway and Sharples [39]. Suceava River is usually 50–60 m wide inside the city, during baseflow. A warm river enhances the urban heat island [17].

On average, Suceava River receives heat from the urban air from ~9 a.m. until ~7 p.m. (Figure 5b). During the daylight, Suceava River water temperature inside the city increases quickly, but, during the night, it decreases slowly due to the increasingly higher heat loss necessary to lose 1 °C at lower and lower temperatures (in this case, temperatures under 30 °C; see isobaric specific heat [38]). In this way, during mid to late summer and during the autumn, the stream water gains energy from one day to another because the heat loss during the night-time is smaller than the energy gain during the daytime.

The most intense heat absorption from the atmosphere to the aquatic environment is done during the long summer daytimes by active evaporation processes from the water surface and the floodplain in an urban atmosphere drier than the surroundings. The atmosphere loses heat at the level of its basal layer—the heat is transferred to water through evaporation (Supplementary Figure S1). The most intense heat transfer from the aquatic environment and floodplain to the urban atmosphere takes place during the long winter nights, foggy mornings and evenings, when the condensation or sublimation of excess vapor in the urban atmosphere of Suceava creates a caloric surplus given to the atmosphere. This contributes to mitigating night-time and morning radiative losses (Supplementary Figure S4).

In both situations (both in the summer—through the remarkable intensity of the evaporation process—and in winter—through the long duration of the foggy/humid air interval in Suceava valley), the presence of water vapor in the atmosphere plays a very important and, at the same time, an ambivalent role in the thermo–caloric relationship between the urban atmosphere of Suceava city and the water of Suceava River. Similar strong thermal connections between water and air were also noted in previous studies [40].

Boxplots of hourly water and air time series show the higher, urban variability of the air temperature in the constructed floodplain, especially in August (Figure 6). In January, the interquartile interval of the water time series downstream of the city is bigger due to the less extreme negative temperatures in the floodplain air and the contribution of various heat sources. The observed distribution of air temperatures in 2019 is one of a warm year—the average temperature was 10.35 °C at Suceava Weather Station (in comparison, at the same station, the average temperature during 1950–2012 was 7.86 °C [27]).

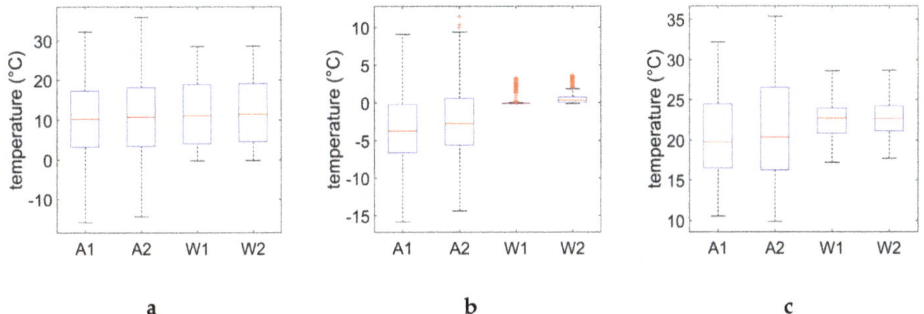

Figure 6. Boxplots of water and air temperature time series (hourly data) for the following time intervals (the red line represents the median): (**a**) year 2019, (**b**) January 2019, (**c**) August 2019 (A1—Suceava Weather Station, A2—Suceava city in the floodplain, W1—Suceava River upstream, W2—Suceava River downstream).

The average diurnal profiles of water and air vary in their shape from month to month and from one monitoring station to another (Figure 7). The annual variation of the monthly average diurnal profiles is similar for the air time series (Figure 7a,b), but the variation in water time series is different (Figure 7c,d). Suceava River downstream recorded a high variability of the moment when the diurnal maximum value occurs. That moment varied from 3 p.m. to 8 p.m., and the urban air recorded only a fluctuation of ±1 h of the moment of the diurnal maximum. One can observe that the annual variation of Suceava River water has shifted from an original variation towards a variation that is specific to that of the local air temperature. In water time series upstream and downstream, the biggest differences in the position of maxima values occur during the warm semester. During winter, probably due to the atmospheric forcing that moves water temperatures to 0 °C for a long time, the differences in the hourly position of maxima values is minimal (the same is true for minima values).

A shape change of the stream water average diurnal profile from upstream to downstream of Suceava city similar to that observed in our study was also revealed by a previous study [27] that analysed data from a shorter time interval (October 2012–January 2013): the hour of the maximum temperature shifted from the evening (upstream) to the afternoon inside the city and also downstream of Suceava city.

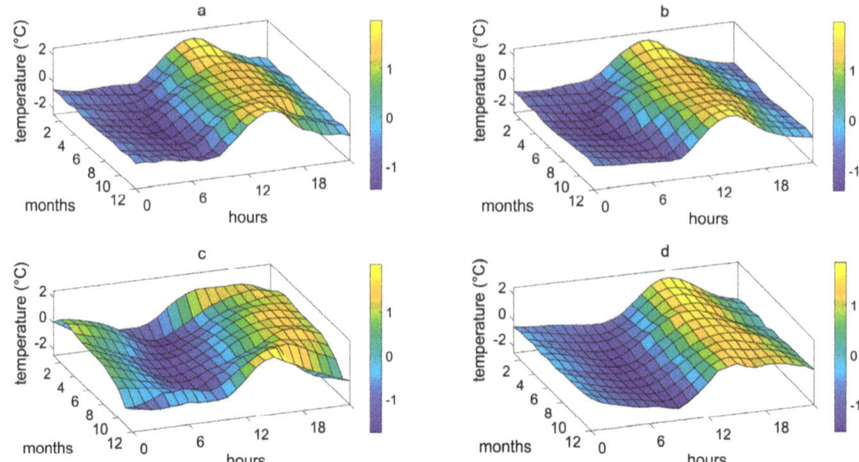

Figure 7. Surface plots of seasonally adjusted and normalised average hourly values of air and water temperature in Suceava city area in 2019: (**a**) air at Suceava Weather Station, (**b**) air in the floodplain, (**c**) Suceava River upstream and (**d**) Suceava River downstream.

The scalograms of the continuous wavelet transform analysis indicate that the diurnal cycle occurs in more consecutive days in the monitoring station from Tişăuţi—downstream (Figure 8). This persistence is sustained by heat transfer from the city into the river—this supplementary heat appears to attenuate the impact of external factors that tend to disrupt the repetition of a uniform diurnal cycle. The time interval with frequent significant (0.95 confidence level) diurnal cycles is longer downstream. During the cold semester, the diurnal cycles are less significant, especially in winter when, upstream, a diurnal periodicity may lack for many consecutive days. Also, the higher power of the diurnal periodicity downstream indicates that the diurnal cycle is easily shaped by the high thermal amplitude (day–night difference).

The wavelet coherence analysis of Suceava River water temperature at Mihoveni and Tişăuţi (Figure 9a) reveals very frequent similar diurnal oscillations in both monitoring stations (that co-vary in a non-linear way). The diurnal periodicities are generally in phase, as indicated by the phase arrows pointing right. The scalograms of the wavelet coherence analysis of one station in air and the other station in water (Figure 9b,c) show that there is a better covariance of the diurnal periodicities in the urban floodplain air and Suceava River water downstream, at Tişăuţi, than between the air relevant to the upstream catchment and Suceava River upstream, at Mihoveni. It is interesting to observe that the coherence between air and water downstream is stronger at the diurnal band of frequencies than the coherence between water upstream and water downstream. This indicates that the shape and timing of the diurnal water profile at Tişăuţi is imposed by the diurnal distribution of the air temperature, rather than the characteristics of water at Mihoveni. However, at lower frequencies, the relationship between water time series in both monitoring stations is stronger, as indicated by the power spectrum and the fact that the continuous significant coherent area (dark red area topped by the thin black line of the statistical significance area limit) starts from ~5 days (128 h), instead of ~21 days (512 h). The wavelet analysis is useful for the analysis of the non-linear covariation of the diurnal cycle. The higher variability of the diurnal cycle does not have an impact only on a scale of a few consecutive days, but also, often, at a monthly scale. Thus, for example, the hourly position of minima values in the average diurnal profiles of the air temperature at Suceava Weather Station and in the floodplain and of the water temperature at Mihoveni (upstream) and Tişăuţi (downstream) in July/August was 6/7 a.m., 5/6 a.m., 8/8 a.m. and 6/7 a.m., respectively.

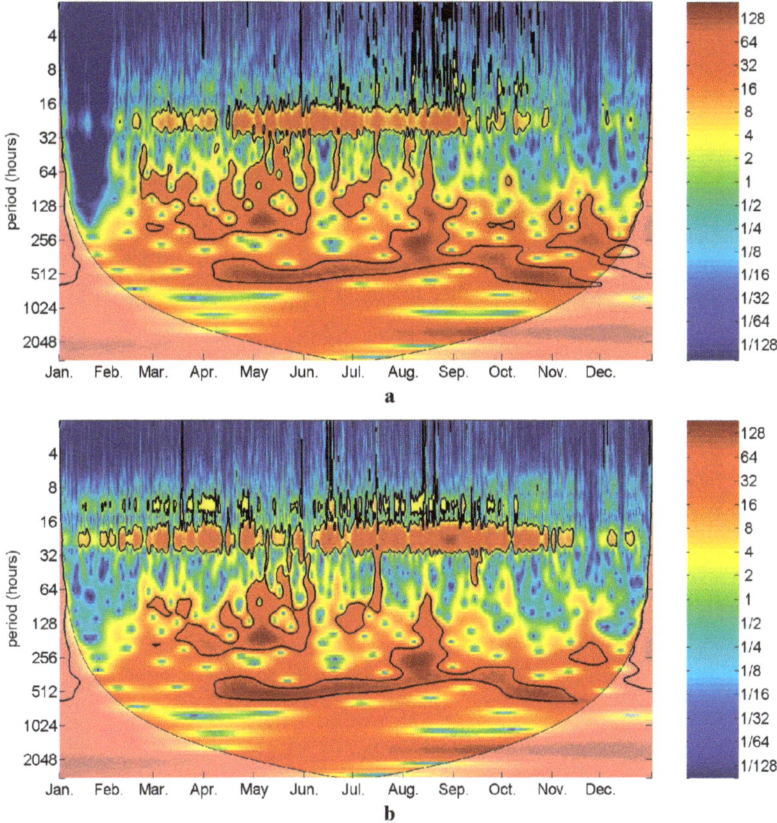

Figure 8. Continuous wavelet transform analysis of hourly water temperature time series of Suceava River in 2019: (**a**) Mihoveni, (**b**) Tişăuţi.

In order to find punctual sources of heat input for Suceava River, we analysed the temperature of some of its urban tributaries and the discharge characteristics of the wastewater treatment plant of Suceava city (the wastewater treatment plants are well-known sources of thermal pollution in rivers, worldwide [41] and in Suceava city [27]). With the exception of Şcheia River, the urban tributaries of Suceava River recorded water temperatures significantly greater than that of Suceava River (Figure 10) in the case study time interval.

During September 20–December 31, 2019, the average water temperature of (in upstream–downstream order; from hourly measurements) Suceava upstream, Şcheia, Cetăţii Creek, Collector and Suceava downstream was 8.54, 8.36, 11.64, 11.45 and 9.19 °C, respectively. Unlike the other urban tributaries, Şcheia flows mostly through a less urbanized landscape, before entering the city. This is why it has a temperature similar to that of Suceava River before entering the city. Şcheia lowers the temperature of Suceava River, as shown by measurements done ~0.4 km downstream, at Iţcani station, where the annual water temperature is 12.4 °C and the average of the case study time interval is 7.9 °C (calculated from daily data, used with precautions because the daily values are based on measurements done only twice per day).

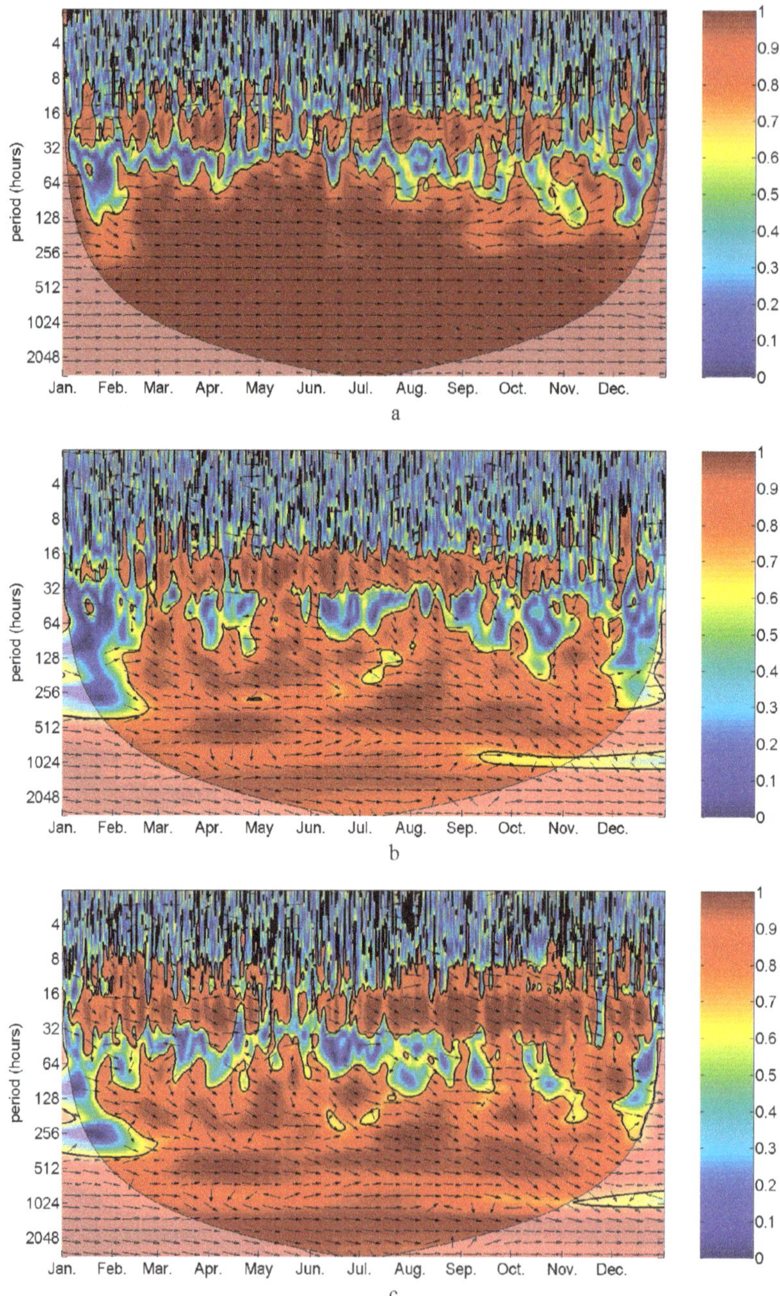

Figure 9. Wavelet coherence analysis of hourly water and air temperature time series of Suceava city in 2019: (**a**) Suceava River water temperature upstream and downstream, (**b**) air temperature at Suceava Weather Station and Suceava River water temperature upstream, (**c**) air temperature in the floodplain and Suceava River water temperature downstream.

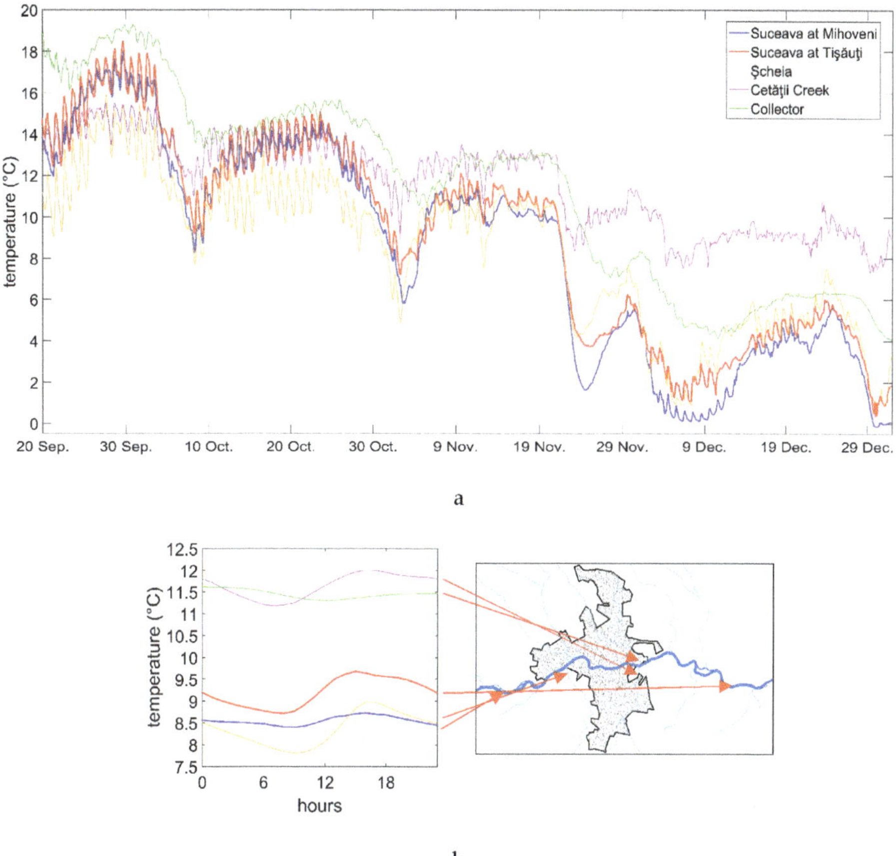

Figure 10. Comparisons of water temperature of Suceava River and its urban tributaries during September 20–December 31, 2019: (**a**) temporal evolution of hourly data, (**b**) average diurnal profiles of water temperature.

The temperature of the other minor streamflows that discharge into Suceava River indicate how sensitive these waters are to the urban alteration of their streambed and catchment, to the urban heat island and to the disposal of wastewaters. Concrete embankments are frequent inside the city—these are an aquatic grey infrastructure that removes the natural processes from the rivers, including the interaction with the subsurface flow through the hyporheic exchange [6].

In 2019, the wastewater treatment plant of Suceava city discharged approximately 0.34 m^3/s of wastewaters with an average of 16.31 °C. During September 20–December 31, 2019, it discharged 0.32 m^3/s with an average of 15.6 °C. The effluent of the wastewater treatment plant is colder than Suceava River only during summer months, when the tap water that becomes wastewater is colder than Suceava River (the main water supply for Suceava city is from Berchişeşti groundwater, located at the plateau border with the mountain area). The average temperature of the discharged wastewater was 20.71 °C during summer (with ~1.16 °C lower than Suceava River at Tişăuţi in the same time interval) and 12.24 °C during winter (with ~10.2 °C higher than Suceava River).

During September 20–December 31, 2019, we estimated the flow rate of Suceava River at Iţcani as 3.94 m^3/s. The average temperature of the tributaries that Suceava River receives between Mihoveni and Tişăuţi is computed from the average of the three measured minor streamflows (10.48 °C). This average is combined in a weighted average with the wastewater temperature (cumulative discharge is 0.57 m^3/s from 0.25 m^3/s streamflow + 0.32 m^3/s wastewater), resulting in a combined flow with a temperature of 13.35 °C (the discharge of tributaries is estimated as half of the 0.5 m^3/s estimate by using the discharge of Şcheia during September 20–December 31, 2019, which is approximately half of the average specific to this time interval of the year). The combined flow temperature is added in a weighted average (discharge used for weighting) to Suceava River temperature upstream (8.54 °C) and results in a new temperature of Suceava River downstream of 9.14 °C.

The measured temperature in the field was 9.19 °C, resulting in this calculus/the surface water input into Suceava River explaining 0.6 °C (92.3%) from the total 0.65 °C difference between Tişăuţi and Mihoveni. The remaining heat gain has to be attributed to the urban heat island (directly) and the groundwater input. It is to be remembered that the final heat gain is the result of complex processes of gaining or losing heat (e.g., through water input or exchange with atmosphere) and the estimated values are specific to the studied period of the end of 2019 only. Elements not taken into consideration (e.g., temperature of the raindrops) may have some role in the variability of Suceava River temperature. The indirect impact of the urban heat island on Suceava River temperature is exerted through urban tributaries. The thermal impact of the tributaries represents 31.75% (part of the above-mentioned 92.3%). It remains to be discovered how much of the temperature increase of tributaries over Suceava River temperature is caused only by the urban heat island, but a rough estimate is at least 50%. We can also estimate that the 0.05 °C left unexplained in the calculus of water temperature increase from upstream to downstream is also caused at least 50% by the urban heat island (direct influence). The measured impact of the wastewaters on the water temperature of Suceava River is not a surprise because the municipal treatment plants are considered as being probably the most important heat source for urban rivers [41]. Moreover, there was a constant increase in the temperature of wastewaters in cities reported, e.g., Tokyo [42].

The discharge of the wastewater treatment plant varies considerably, and the highest values are created by stormwaters collected during important rainfalls (Figure 11a). This is due to the fact that an exclusively pluvial drainage exists only in some small areas of the city. Most stormwaters are collected in the main (combined) sewerage and flow through the wastewater treatment plant. Therefore, the calculated impact of the wastewater is also representative of the impact of the stormwater runoff on the water quality of Suceava River. Temperature surges caused by stormwaters that washed warm city surfaces are difficult to detect, as most pluvial waters enter the combined sewers. Any detectable temperature surges caused by urban stormwater runoff must alter the average diurnal profile of Suceava River water temperature difference between the downstream station (Tişăuţi) and the upstream station (Mihoveni) (Figure 11b). An analysis of the hourly temperature difference between these stations shows that the standard deviation is 0.76 °C and the maximum occurs at 2 p.m. In order to identify possible water temperature surges, we analysed the days with a temperature difference greater than 1 °C and with atmospheric precipitation greater than 10 mm (this amount generates consistent runoff, especially during high-intensity rains). From the 15 times when the rainfalls exceeded 10 mm/day, only 2 times generated a runoff that strongly impacted the stream water temperature downstream of Suceava city (Figure 11c,d). During the surge event in August 2019, the temperature difference was 1.64 °C and the temperature increase induced by urban stormwater merged the diurnal peaks of 2 days into a larger peak with a duration of 24 h. The surge in July 2019 lasted 13 h and recorded the maximum temperature difference of 2019 (4.14 °C, which is 1.1 °C higher than the second highest value). The temperature surges occurred during warm summer days, similar to many cases reported in other studies [3,35]. Our surge values are similar to those of Rice et al. [34] (1.9–3.27 °C), Zeigler and Hubbart [35] (4 °C) or Nelson and Palmer [3] (2–7 °C). The surge durations in our study are greater than those reported previously (up to 10–10.5 h) [34,35].

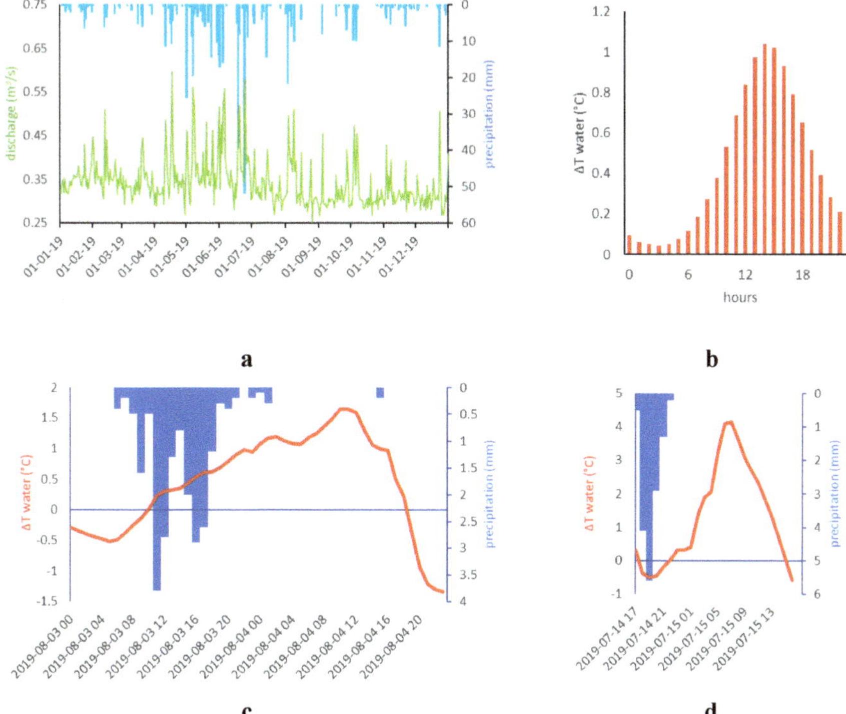

Figure 11. Links between rainfalls, wastewaters and the water temperature of Suceava River: (**a**) correlation between the rainfall amount and the wastewater treatment plant discharge, (**b**) average diurnal profile of Suceava River water temperature difference (downstream–upstream), (**c**) Suceava River water temperature difference (downstream–upstream) versus atmospheric precipitation at Suceava Weather Station during 3–4 August 2019 (48 h), (**d**) same as in (**c**), but during 14–15 July 2019 (24 h).

A correlation matrix (with Pearson coefficients) that took variables as daily averages during the entire year of 2019 into consideration (Supplementary Table S1) revealed very good correlations (>0.95) between Suceava River water temperature and air temperature or wastewater temperature. A good correlation in this linear estimate is between water temperature and solar radiation (>0.6). The good correlation of the solar radiation indicates the potentially high impact of other, unknown/unmeasured parameters (such as water turbidity) on the variability of water temperature and highlights the limits of any mathematical model.

ANCOVA tests, applied to some relevant parameters in order to find out which natural and anthropogenic factors influence the water temperature of Suceava River at Mihoveni (upstream) and Tişăuţi (downstream), revealed the results displayed in Table 2. The goodness-of-fit (R^2) of the linear model in explaining the upstream water temperature is 0.917, and it is 0.998 for the downstream case (in both cases, the risk in assuming that the selected explanatory variables explain the stream water temperature is <0.0001).

For the upstream case, the parameters that contribute more to the water temperature are air temperature, relative humidity and water level, but the contribution of the other parameters is not negligible. For the downstream case, the most important factors are water temperature from the upstream station and wastewater temperature, seconded by solar radiation and water level at the downstream monitoring station.

Table 2. Analysis of covariance (ANCOVA) of Suceava River water temperature upstream or downstream (dependent variable) versus other parameters (explanatory variables) - type III sum of squares analysis.

UPSTREAM	DF	Sum of Squares	Mean Squares	F	Pr > F
temperature WS	1	10,074.593	10,074.593	1885.322	<0.0001
precipitation WS	1	46.713	46.713	8.742	0.003
relative humidity WS	1	152.214	152.214	28.485	<0.0001
solar radiation fl.	1	75.037	75.037	14.042	0.000
water level M	1	109.142	109.142	20.424	<0.0001
DOWNSTREAM	**DF**	**Sum of Squares**	**Mean Squares**	**F**	**Pr > F**
temperature WS	1	0.000	0.000	0.002	0.964
precipitation WS	1	0.056	0.056	0.404	0.526
relative humidity WS	1	0.083	0.083	0.598	0.440
temperature fl.	1	0.009	0.009	0.062	0.803
solar radiation fl.	1	0.949	0.949	6.817	0.009
relative humidity fl.	1	0.226	0.226	1.622	0.204
precipitation I	1	0.003	0.003	0.023	0.879
water temperature M	1	611.040	611.040	4391.208	<0.0001
water level M	1	0.225	0.225	1.618	0.204
water level T	1	0.757	0.757	5.441	0.020
discharge WTP	1	0.000	0.000	0.002	0.963
temperature WTP	1	5.895	5.895	42.363	<0.0001

Abbreviations: WS—Suceava Weather Station, M—Mihoveni, I—Ițcani, T—Tişăuţi, fl.—floodplain, WTP—wastewater treatment plant.

In order to detect non-linear dependencies between multiple parameters, we applied a multiple wavelet coherence analysis. This uses the same characteristics of the previously discussed (simple) wavelet coherence and is a reliable method for the analysis of the non-linearly dependent processes [30]. The scalograms indicate the wavelet power of the first parameter explained by the other parameters (Figure 12). As in the case of ANCOVA, at the upstream station, the strongest links were detected between water temperature upstream and air temperature at Suceava Weather Station + water level upstream. However, at the downstream station, the most important factors linked to the evolution of the water temperature are air temperature in the floodplain and water temperature upstream (the latter has a stronger role than the temperature of the wastewaters, which was detected as relevant by ANCOVA). One can observe that the explanatory variables better explain the explained variable at the downstream station (the dark red areas are dominant and indicate stronger coherence, especially at the weekly and monthly scales).

An analysis of the scaled correlation between various time series was also conducted (Figure 13). The Pearson correlation coefficient was computed for every day (24 h/values generate 1 correlation coefficient and the result is a list of 365 correlation coefficients, 1 for each calendar day). We observed that, during 2019, the scaled correlation between water temperature at Tişăuţi (downstream) and air temperature in the floodplain is better (the average of the 365 correlation coefficients is 0.6188) than the scaled correlation between water temperature at Mihoveni (upstream) and air temperature at Suceava Weather Station (0.3407). The classical correlation between the entire time series gave an irrelevant difference (0.8819 versus 0.8898); therefore, the scaled correlation is more useful in the process of finding differences between upstream and downstream monitoring stations.

Figure 13b shows that the scaled correlation between the downstream/floodplain pair and air temperature in the floodplain is better when the air temperature is higher, for almost every month. Figure 13c indicates that the same scaled correlation is lower when water level is high, especially during the warm season, when high waters occur frequently.

All data indicate that Suceava River temperature increases downstream of Suceava city because of numerous factors whose importance vary in time and in space. This increase is acquired effectively in the lower 2/3 of the distance between Mihoveni and Tişăuţi (11.6 km straight line) and is caused by Suceava city, including its urban heat island.

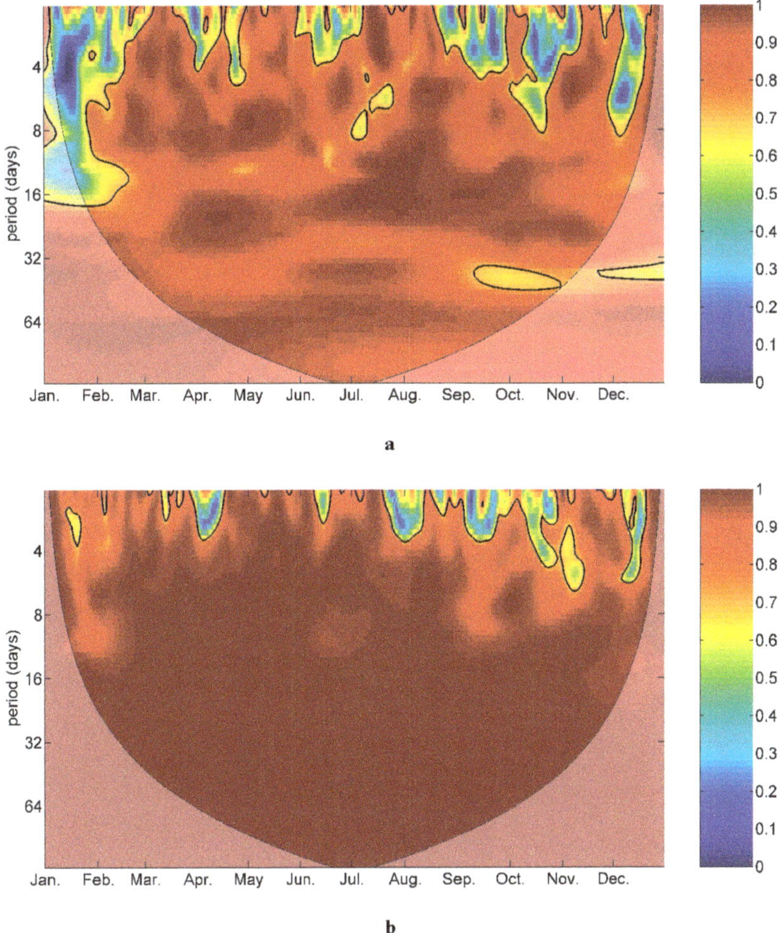

Figure 12. Multiple wavelet coherence analysis scalograms between Suceava river water temperature and air parameters: (**a**) water temperature upstream versus air temperature at Suceava Weather Station and water level upstream, (**b**) water temperature downstream versus air temperature in the floodplain and water temperature upstream.

The increase in water temperature has an impact on other parameters of the river. For example, lower dissolved oxygen concentrations were reported downstream of Suceava city, at Tişăuţi (compared to Mihoveni), partly caused by the increase in water temperature [31]. The lowered concentrations of dissolved oxygen are certainly diminishing the self-purification capacity of Suceava River that was attributed partly to this element [43,44].

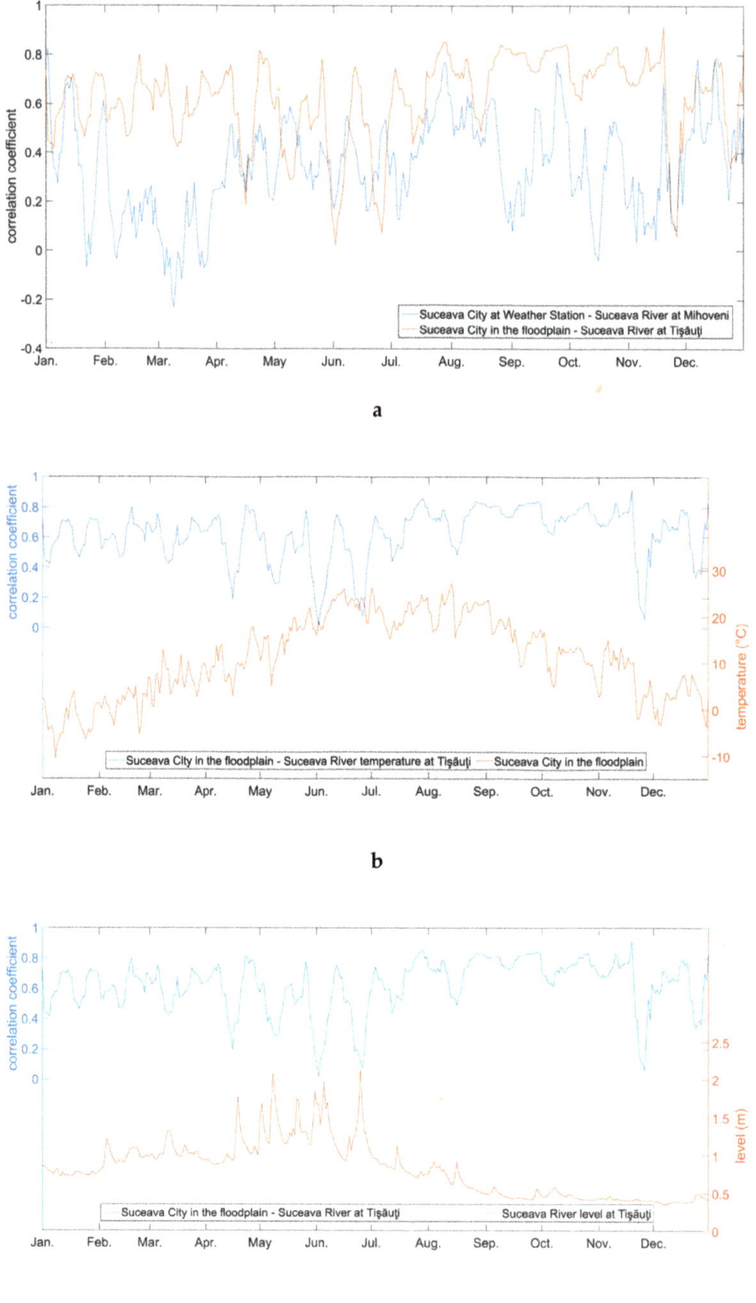

Figure 13. Comparative temporal evolutions of various scaled correlations' time series (smoothed with a span of 1 week/7 values) and water/air parameters of Suceava city area in 2019: (**a**) scaled correlations of upstream versus downstream time series, (**b**) downstream scaled correlations versus air temperature in the floodplain, (**c**) downstream scaled correlation versus water level at Tişăuţi.

4. Conclusions

In 2019, the average air temperature was 10.35 °C at Suceava Weather Station and 11.05 °C in Suceava River floodplain/centre of the city. Suceava River water temperature was 11.54 °C upstream of Suceava city (Mihoveni) and 11.97 °C downstream (Tişăuţi). The analysis of the average diurnal profiles of water and air in our study area indicates that the heat transfer between water and air is bidirectional. The peak position in the diurnal cycle of water has been strongly shifted towards the moment of the maximum air temperature of city. Monthly average values show that, in the first half of the year, air temperature in the urban floodplain of Suceava River was higher than the stream water temperature in both stations.

During a detailed spatial analysis (last quarter of 2019), the water temperature of some urban tributaries was ~2 °C higher than the water temperature of Suceava River. These tributaries are strongly impacted by the urban heat island and, together with the effluent of the wastewater treatment plant, increase the temperature of the main river.

The wavelet coherence analysis shows that there is a better covariance of the diurnal periodicities in the temperatures of urban floodplain air and Suceava River water downstream, at Tişăuţi, than between the air temperature nearby the city and Suceava River upstream, at Mihoveni. Simple and multiple wavelet coherence analyses and ANCOVA indicate that the evolution of Suceava River water temperature downstream of Suceava city is controlled mainly by urban air temperature, water temperature upstream, wastewater temperature and solar radiation.

Last, but not least, 2019 was a dry and warm year (atmospheric environment) and Suceava River had warm waters. Further analyses using high temporal resolution data are needed during average or wetter/colder years in order to better understand the relationship between the urban heat island and Suceava River. For a better understanding of the distinct influence of the urban heat island and wastewaters on the changes of Suceava River, a monitoring station on Suceava River immediately upstream of the wastewater treatment plant is necessary for future studies. Data provided from this new station will help in assessing the gradual changes in the diurnal thermal profile of Suceava River from the uppermost monitoring station towards downstream of Suceava city. Data collected in this and future studies could be integrated into computer models that might help local authorities mitigate the impact of the urban heat island on urban stream waters. Best management practices should be promoted for Suceava floodplain. Restoration treatments may lower a river temperature and increase its dissolved oxygen concentration, as proven in other urban areas [6]. Because riparian and floodplain forests keep the river temperature low [45], we recommend a forestation with willows (*Salix* sp.) of Suceava River floodplain inside Suceava city in order to mitigate the urban stream syndrome.

Supplementary Materials: The following are available online at http://www.mdpi.com/2073-4441/12/5/1343/s1, Figure S1: Urban heat island above Suceava city as revealed by hourly air relative humidity measurements in 2019 during winter - a - and summer - b - (values used in maps are averages per season) – UHI represents the area most impacted by the urban heat island. Figure S2: Comparative evolution of Suceava River water temperature upstream (Mihoveni) and downstream (Tişăuţi) of Suceava city (hourly data, June 28th, 2018 – December 31st, 2019). Figure S3: Comparisons of water temperature at Mihoveni (upstream) and Tişăuţi (downstream) and air temperature in Suceava city area in 2019: a. average diurnal profiles of water temperature, b. average diurnal profiles of air temperature. Figure S4: Relative humidity diurnal regime of the air at Suceava Weather Station and in the floodplain in 2019 during summer (a) and winter (b) (ΔRH is the difference between the floodplain values and those of Suceava Weather Station). Table S1: Correlation matrix of selected water and air parameters in Suceava city in 2019.

Author Contributions: Conceptualization, A.-E.B.; Data curation, A.-E.B. and D.M.; Investigation, A.-E.B. and D.M.; Methodology, A.-E.B. and D.M.; Resources, A.-E.B., A.P., D.I.O. and P.I.B.; Supervision, A.G.; Visualization, D.I.O., A.P. and P.I.B.; Writing—original draft, A.-E.B. and D.M.; Writing—review and editing, A.G. All authors have read and agreed to the published version of the manuscript.

Funding: This research was funded by CNCS–UEFISCDI, project number PN-III-P1-1.1PD-2016-2106.

Acknowledgments: Due to important field and laboratory work, Dumitru Mihăilă is to be considered, together with Andrei-Emil Briciu, as one of the principal authors. Wavelet coherence software was provided by A. Grinsted. Thanks for the data provided is addressed to three Romanian institutions, ANAR (Siret Water Basin Administration), ANM, ANPM, and to ACET Suceava. This study was supported by data obtained within the research project SQRTDA (Streamwater Quality Real-Time Data Analysis). This work was supported by a grant of Ministry of Research and Innovation, CNCS–UEFISCDI, project number PN-III-P1-1.1PD-2016-2106, within PNCDI III.

Conflicts of Interest: The authors declare no conflict of interest. The funders had no role in the design of the study; in the collection, analyses, or interpretation of data; in the writing of the manuscript, or in the decision to publish the results.

References

1. Hannah, D.M.; Garner, G. River water temperature in the United Kingdom: Changes over the 20th century and possible changes over the 21st century. *Prog. Phys. Geogr.* **2015**, *39*, 68–92. [CrossRef]
2. Wagner, T.; Midway, S.R.; Whittier, J.B.; DeWeber, J.T.; Paukert, C.P. Annual Changes in Seasonal River Water Temperatures in the Eastern and Western United States. *Water* **2017**, *9*, 90. [CrossRef]
3. Nelson, K.C.; Palmer, M.A. Stream temperature surges under urbanization and climate change: Data, models, and responses. *J. Am. Water Resour. Assoc.* **2007**, *43*, 2. [CrossRef]
4. Van Sickle, J.; Baker, J.; Herlihy, A.; Bayley, P.; Gregory, S.; Haggerty, P.; Ashkenas, L.; Li, J. Projecting the Biological Condition of Streams under Alternative Scenarios of Human Land Use. *Ecol. Appl.* **2004**, *14*, 368–380. [CrossRef]
5. Paul, M.J.; Meyer, J.L. Streams in the Urban Landscape. *Annu. Rev. Ecol. Syst.* **2001**, *32*, 333–365. [CrossRef]
6. Abdi, R.; Endreny, R.; Nowak, D.A. A model to integrate analysis of urban river thermal cooling in river restoration. *J. Environ. Manag.* **2020**, *258*, 110023. [CrossRef]
7. Galli, J.; Dubose, R. *Water Temperature and Freshwater Stream Biota: An Overview*; U.S. Environmental Protection Agency: Fort Meade, MD, USA, 1990.
8. Pluhowski, E.J. *Urbanization and Its Effect on the Temperature of the Streams on Long Island*; Professional Paper 627-D; U.S. Government Printing Office: Washington, DC, USA, 1970.
9. Brabec, E.; Schulte, S.; Richards, P.L. Impervious Surfaces and Water Quality: A Review of Current Literature and Its Implications for Watershed Planning. *J. Plan. Lit.* **2002**, *16*, 499–514. [CrossRef]
10. Zaltman, R.; Reid, G.W. Theoretical study on thermal pollution. *Geofís. Int.* **1972**, *12*, 1.
11. Daufresne, M.; Boët, P. Climate change impacts on structure and diversity of fish communities in rivers. *Glob. Chang. Biol.* **2007**, *13*, 2467–2478. [CrossRef]
12. Demars, B.O.; Russell Manson, J.; Ólafsson, J.S.; Gíslason, G.M.; Gudmundsdóttir, R.; Woodward, G.; Reiss, J.; Pichler, D.E.; Rasmussen, J.J.; Friberg, N. Temperature and the metabolic balance of streams. *Freshw. Biol.* **2011**, *56*, 1106–1121. [CrossRef]
13. Van Vliet, M.T.H.; Franssen, W.H.P.; Yearsley, J.R.; Ludwi, F.; Haddeland, I. Global river discharge and water temperature under climate change. *Glob. Environ. Chang.* **2013**, *23*, 450–464. [CrossRef]
14. Sugawara, H.; Narita, K.; Kim, M.S. Cooling effect by urban river. In Proceedings of the seventh International Conference on Urban Climate, Yokohama, Japan, 29 June–3 July 2009.
15. Qi, J.; Liu, J.; Song, X.; Guo, L. Field measurement of the influence of large urban river on urban thermal climate. *J. Harbin Inst. Tech.* **2011**, *43*, 56–59.
16. Cheng, L.; Guan, D.; Zhou, L.; Zhao, Z.; Zhou, J. Urban cooling island effect of main river on a landscape scale in Chongqing, China. *Sustain. Cities Soc.* **2019**, *47*, 101501. [CrossRef]
17. Moyer, A.N.; Hawkins, T.W. River effects on the heat island of a small urban area. *Urban Clim.* **2017**, *21*, 262–277. [CrossRef]
18. Kannel, P.R.; Lee, S.; Lee, Y.S.; Kanel, S.R.; Khan, S.P. Application of water quality indices and dissolved oxygen as indicators for river water classification and urban impact assessment. *Environ. Monit. Assess.* **2007**, *132*, 93–110. [CrossRef]
19. Abdi, R.; Endreny, T. A River Temperature Model to Assist Managers in Identifying Thermal Pollution Causes and Solutions. *Water* **2019**, *11*, 1060. [CrossRef]
20. Abdi, R.; Endreny, T.; Nowakc, D. i-Tree Cool River: An open source, freeware tool to simulate river water temperature coupled with HEC-RAS. *MethodsX* **2020**, *7*, 100808. [CrossRef]

21. Somers, K.A.; Bernhardt, E.S.; Grace, J.B.; Hassett, B.A.; Sudduth, E.B.; Wang, S.; Urban, D.L. Streams in the urban heat island: Spatial and temporal variability in temperature. *Freshw. Sci.* **2013**, *32*, 309–326. [CrossRef]
22. Mihăilă, D.; Briciu, A.-E. Actual climate evolution in the NE Romania. Manifestations and consequences. In Proceedings of the SGEM2012 Conference Proceedings, Albena, Bulgaria, 1 Augest 2012; Volume 4, pp. 241–252.
23. Ajeagah, G.; Cioroi, M.; Praisler, M.; Constantin, O.; Palela, M.; Bahrim, G. Bacteriological and environmental characterisation of the water quality in the Danube River Basin in the Galati area of Romania. *Afr. J. Microbiol. Res.* **2012**, *6*, 292–301.
24. Cojoc, G.M.; Romanescu, G.; Tirnovan, A. The importance of water temperature fluctuations in relation to the hydrological factor. case study—Bistrita river basin (Romania). *PESD* **2014**, *8*, 2. [CrossRef]
25. Briciu, A.-E.; Oprea-Gancevici, D.I. Diurnal thermal profiles of selected rivers in Romania. In Proceedings of the SGEM2015 Conference Proceedings, Albena, Bulgaria, 18 June 2015; Volume 1, pp. 221–228.
26. Briciu, A.-E.; Mihaila, D.; Mihaila, D. Short, medium and long term stochastic analysis of the Suceava River pollution evolution in the homonymous city. In Proceedings of the SGEM 2012 Conference Proceedings, Albena, Bulgaria, 1 Augest 2012; Volume 3, pp. 809–816.
27. Briciu, A.-E. *Studiu de hidrologie urbană în arealul municipiului Suceava;* (Urban hydrology study in Suceava municipality area); Ştefan cel Mare University Publishing House: Suceava, Romania, 2017.
28. Lu, G.; Wong, D. An adaptive inverse-distance weighting spatial interpolation technique. *Comput. Geosci.* **2008**, *34*, 1044–1055. [CrossRef]
29. Ng, E.K.W.; Chan, J.C.L. Geophysical applications of partial wavelet coherence and multiple wavelet coherence. *J. Atmos. Oceanic Technol.* **2012**, *29*, 1845–1853. [CrossRef]
30. Grinsted, A.; Moore, J.C.; Jevrejeva, S. Application of the cross wavelet transform and wavelet coherence to geophysical time series. *Nonlinear Process. Geophys.* **2004**, *11*, 5–6. [CrossRef]
31. Briciu, A.-E.; Graur, A.; Oprea, D.I.; Filote, C. A Methodology for the Fast Comparison of Streamwater Diurnal Cycles at Two Monitoring Stations. *Water* **2019**, *11*, 2524. [CrossRef]
32. Briciu, A.-E. Changes in Physical Properties of Inland Streamwaters Induced by Earth and Atmospheric Tides. *Water* **2019**, *11*, 2533. [CrossRef]
33. Gribovszki, Z.; Szilágyi, J.; Kalicz, P. Diurnal fluctuations in shallow groundwater levels and streamflow rates and their interpretation—A review. *J. Hydrol.* **2010**, *385*, 371–383. [CrossRef]
34. Rice, J.S.; Anderson, W.P., Jr.; Thaxton, C.S. Urbanization influences on stream temperature behavior within low-discharge, headwaters streams. *Hydrol. Res. Lett.* **2011**, *5*, 27–31. [CrossRef]
35. Zeiger, S.J.; Hubbart, J.A. Urban Stormwater Temperature Surges: A Central US Watershed Study. *Hydrology* **2015**, *2*, 193–209. [CrossRef]
36. Orr, H.G.; Simpson, G.L.; des Clers, S.; Watts, G.; Hughes, M.; Hannaford, J.; Dunbar, M.J.; Laizé, C.L.R.; Wilby, R.L.; Battarbee, R.W.; et al. Detecting changing river temperatures in England and Wales. *Hydrol. Process.* **2015**, *29*, 752–766. [CrossRef]
37. Jonkers, A.R.; Sharkey, K.J. The Differential Warming Response of Britain's Rivers (1982–2011). *PLoS ONE* **2016**, *11*, e0166247. [CrossRef]
38. Cox, R.A.; Smith, N.D. The specific heat of sea water. *Proc. Roy. Soc. Lond.* **1959**, *252*, 51–62.
39. Hathway, E.A.; Sharples, S. The interaction of rivers and urban fromin mitigating the urban heat island effect: A UK case study. *Build. Environ.* **2012**, *58*, 14–22. [CrossRef]
40. Lolli, S.; D'Adderio, L.P.; Campbell, J.R.; Sicard, M.; Welton, E.J.; Binci, A.; Rea, A.; Tokay, A.; Comerón, A.; Barragan, R.; et al. Vertically Resolved Precipitation Intensity Retrieved through a Synergy between the Ground-Based NASA MPLNET Lidar Network Measurements, Surface Disdrometer Datasets and an Analytical Model Solution. *Remote Sens.* **2018**, *10*, 1102. [CrossRef]
41. Kinouchi, T.; Yagi, H.; Miyamoto, M. Increase in stream temperature related to anthropogenic heat input from urban wastewater. *J. Hydrol* **2007**, *335*, 78–88. [CrossRef]
42. Kinouchi, T. Impact of long-term water and energy consumption in Tokyo on wastewater effluent: Implications for the thermal degradation of urban streams. *Hydrol. Process.* **2007**, *21*, 1207–1216. [CrossRef]
43. Briciu, A.-E.; Toader, E.; Romanescu, G.; Sandu, I. Urban Streamwater Contamination and Self-purification in a Central-Eastern European City. Part I. *Rev. Chim.* **2016**, *67*, 1294–1300.

44. Briciu, A.-E.; Toader, E.; Romanescu, G.; Sandu, I. Urban Streamwater Contamination and Self-purification in a Central-Eastern European City. Part, B. *Rev. Chim.* **2016**, *67*, 1583–1586.
45. Sun, N.; Yearsley, J.; Voisin, N.; Lettenmaier, D.P. A spatially distributed model for the assessment of land use impacts on stream temperature in small urban watersheds. *Hydrol. Process.* **2015**, *29*, 2331–2345. [CrossRef]

Article

Tracking Lake and Reservoir Changes in the Nenjiang Watershed, Northeast China: Patterns, Trends, and Drivers

Baojia Du [1,2], Zongming Wang [1,3], Dehua Mao [1,*], Huiying Li [4] and Hengxing Xiang [1,2]

1 Key Laboratory of Wetland Ecology and Environment, Northeast Institute of Geography and Agroecology, Chinese Academy of Sciences, Changchun 130102, China; dubaojia@neigae.ac.cn (B.D.); zongmingwang@neigae.ac.cn (Z.W.); xianghengxing@neigae.ac.cn (H.X.)
2 University of Chinese Academy of Sciences, Beijing 100049, China
3 National Earth System Science Data Center, Beijing 100101, China
4 School of Management Engineering, Qingdao University of Technology, Qingdao 266033, China; lihy@qut.edu.cn
* Correspondence: maodehua@neigae.ac.cn; Tel.: +86-431-8554-2254; Fax: +86-431-8854-2298

Received: 17 March 2020; Accepted: 31 March 2020; Published: 13 April 2020

Abstract: In terms of evident climate change and human activities, investigating changes in lakes and reservoirs is critical for sustainable protection of water resources and ecosystem management over the Nenjiang watershed (NJW), an eco-sensitive semi-arid region and the third-largest inland waterbody cluster in China. In this study, we established a multi-temporal dataset documenting lake and reservoir (area $\geq$ 1 km^2) changes in this region using an object-oriented image classification method and Landsat series images from 1980 to 2015. Using the structural equation model (SEM), we analyzed the diverse impacts of climatic and anthropogenic variables on lake changes. Results indicated that lakes experienced significant changes with fluctuations over the past 35 years including obvious declines in the total area (by 42%) and number (by 51%) from 1980 to 2010 and a slight increase in the total lake area and number from 2010 to 2015. More than 235 lakes in the size class of 1–10 km^2 decreased to small lakes (area < 1 km^2), while 59 lakes covering 243.75 km^2 disappeared. Total reservoir area and number had continuous increases during the investigated 35 years, with an areal expansion of 54.9% from 919 km^2 to 1422 km^2, and a number increase by 65.3% from 78 to 129. The SEM revealed that the lake area in the NJW had a significant correlation with the mean annual precipitation (MAP), suggesting that the MAP decline clarified most of the lake shrinkage in the NJW. Furthermore, agricultural consumption of water had potential impacts on lake changes, suggested by the significant relationship between cropland area and lake area.

Keywords: lakes; reservoirs; Nenjiang watershed; climate change; Landsat

---

1. Introduction

Inland waterbodies (lakes and reservoirs) are closely related to human life due to their diverse ecosystem services [1,2]. Lakes and reservoirs in the drylands, although covering only a small proportion of the landscape, play irreplaceable roles in fragile environments and for local residents [3]. Meanwhile, they are sensitive to climate change and human disturbances [4]. Monitoring changes in lakes and reservoirs and investigating their driving forces are thus of great significance to sustainable water resource management and regional economic development.

Due to diverse driving forces, lakes around the world have experienced changes in both area and number during the past decades. Thaw and "breaching" of permafrost have caused a widespread decline in the Arctic lake number and area [5]. Because of warmer air temperatures, which allow higher evaporation rates, as the world's largest surface freshwater system, the total area of the Great Lakes of North America experienced a continuous decrease from the 1980s to 2005 [6]. Under the background of global warming, glaciers in China's Tibetan Plateau (TP) showed a general retreat, while the glacial lakes expanded notably [7]. However, a drier climate and degraded permafrost have led to lake shrinkage in some basins of the TP [8]. Lakes over Mongolia experienced areal decline mostly related to a drier climate, while lakes in Inner Mongolia had notable shrinkage which was attributed mainly to human consumption of water resources, particularly coal mining [9]. Moreover, the lake area in the Jianghan Plain of China decreased dramatically from the 1950s to 1998, which was mainly caused by agricultural cultivation.

During the past 60 years, the number of reservoirs has increased notably, and there are about 50,000 large reservoirs nowadays around the world [10]. For example, China constructed nearly 45,000 reservoirs in the Yangtze River Basin to meet the large demand for water resources. By 2013, 98,000 reservoirs had been built in China [11]. Previous studies [12,13] indicated that most of large river systems and lakes in China were affected by reservoirs. Due to the construction of large reservoirs in upstream areas, water was impounded upstream, which seriously affected the water supply of lakes in the middle and lower reaches [14]. Therefore, it is necessary to track and analyze the changes in lakes and reservoirs at different scales and regions to form a scientific management response [15,16].

Optical images from different satellite sensors were widely used to monitor spatiotemporal variations of inland waterbodies at multiple geographic scales [17–22]. The Moderate Resolution Imaging Spectroradiometer (MODIS) data have been widely used to assess the water extent at daily to 16-day timescales despite a spatial resolution at 250 or 500 m [23]. Yet, areal changes of small lakes or reservoirs with irregular shapes were not accurately delineated due to the coarse resolution of source data. Other optical sensors, such as Quickbird (DigitalGlobe, Longmont, Colorado, USA) and IKONOS (DigitalGlobe, Longmont, Colorado, USA), provide finer images comparable to aerial photography for the extraction of lake or reservoir boundaries [24]. However, those data are limited in application at broad scale due to the high costs, narrow swath size, and so on [25,26]. Landsat series images have a fine spatial resolution (30 or 80 m) compared with previously mentioned data and have provided the longest temporal and spatial records for surface observations since their first launch in 1972 [27]. Consequently, Landsat imagery has been widely-used remote sensing data in examining changes in lakes or reservoirs.

As one of the third-largest waterbody clusters, an eco-environmentally fragile area, and an important base for grain production, the Nenjiang watershed (NJW) plays an important role in ecological conservation and national grain security in China [28]. However, the changes in lakes and reservoirs across the NJW during recent decades have rarely been examined and their correlations with climate change and human disturbances have rarely been quantitatively investigated. Doing so is critical to understanding the regional water cycle and sustainable water resource management. Therefore, this study aims to (1) employ Landsat 8 images to investigate the current status of lakes and reservoirs in the NJW, (2) evaluate changes in area and number of lakes and reservoirs using consistent Landsat images (i.e., 1980, 1990, 2000, 2010, and 2015), and (3) quantify the roles of climatic factors and artificial variables in driving lake changes.

2. Materials and Methods

2.1. Study Area

The NJW, covering an area of 297 × 103 km^2, is located in the core region of Northeast China (Figure 1), with latitudes from 44°1′48″ N to 51°42′1″ N and longitudes from 119°12′1″ E to 127°54′2″ E [28]. Obvious terrain variances can be found in this area with elevations ranging from 120 to 1740 m above sea level. The highest elevation is in the northwest at the Greater and Lesser Khingan Mountains, and the lowest elevation is in the southeast at the Songnen Plain. This is why the lakes and reservoirs are mainly observed in the southeast. The NJW is dominated by a temperate semi-arid continental climate with an mean annual precipitation of around 470 mm and a mean annual air temperature of 4 °C [29]. Most precipitation (82%) occurs in months from June to September which is a critical growth period for vegetation in this region. As an important grain base in China, the NJW plays an critical role in promoting regional economic development and ensuring national food security [30,31]. Moreover, this region provides important habitats for threatened species and migratory waterfowls.

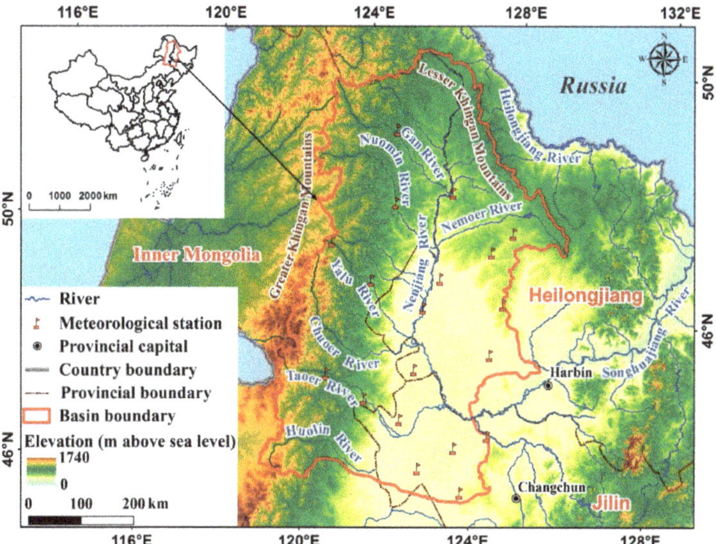

Figure 1. Geographic location and general situation of the Nenjiang watershed.

2.2. Data Source and Processing

2.2.1. Satellite Data

In this study, multi-temporal Landsat images from multispectral scanner (MSS), thematic mapper (TM), enhanced thematic mapper plus (ETM+), and operational land imager (OLI) sensors were acquired from the United States Geological Survey (USGS) to investigate the changes of lakes and reservoirs from 1980 to 2015. Specifically, we used 28, 27, 27, 29, and 26 scenes of images to extract the lakes and reservoirs for five dates of 1980, 1990, 2000, 2010, and 2015, respectively, (Figure 2). A total of 137 images acquired for months from June to September were used for the classification because the vegetation in this period is easier to be identified and the lakes and reservoirs have the largest amount. All of the images have a little cloud cover (less than 5%) and lakes/reservoirs on these images are clearly visible. In addition, in order to analyze the impact of floods on the area and the number of lakes and reservoirs in 1998 and 2013, we used 30 scenes of TM and 30 scenes of OLI images to extract the lakes and reservoirs in 1998 and 2013, respectively. Prior to image classification, data preprocessing including geometric, topographic, and radiometric corrections was performed for all the images using the ENVI 5.3 (Exelis Visual Information Solutions, Inc. Boulder, USA) software package. To ensure the data consistency, all images were re-projected to the 1984 WGS UTM zone 51N projection.

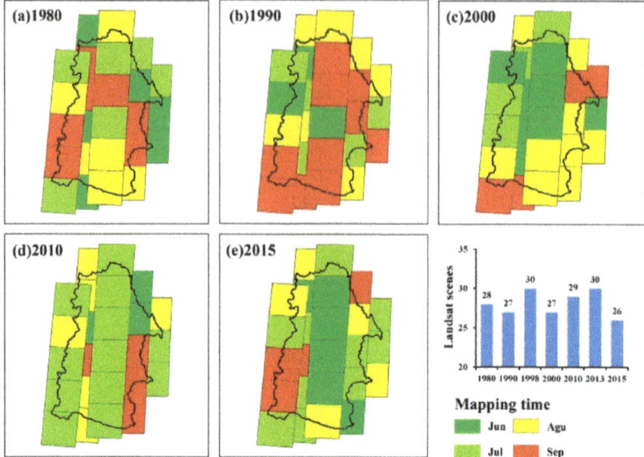

Figure 2. Landsat images selection in optimum months for different dates.

2.2.2. Meteorological Data

In order to analyze the relationships of lake/reservoir area with climatic factors, the daily climatic data, including extreme air temperature and precipitation, mean wind speed, and sunshine hours during 1980–2015, were collected from the meteorological records of the China Meteorological Data Service Center. Figure 1 shows the locations of the meteorological stations in the NJW. Spatial patterns of the mean annual air temperature (MAAT) and mean annual precipitation (MAP) were interpolated from those meteorological records using an Anusplin software considering the elevation differences [32]. We used the daily meteorological data and the Food and Agriculture Organization of the United Nations (FAO) Penman–Monteith model [8,33] to calculate the actual evapotranspiration (ET). ET is calculated as follows:

$$ET_0 = \frac{0.408\Delta(R_n - G) + \gamma \frac{900}{T+273} U_2 (e_s - e_a)}{\Delta + \gamma(1 + 0.34U_2)} \quad (1)$$

$$ET = 9.78 + 0.0072 \times ET_0 \times PPT + 0.051 \times PPT \times LAI \quad (2)$$

where ET_0 is the potential evapotranspiration, R_n is the net radiation, G is the soil heat flux, e_s is the saturation vapor pressure, e_a is the actual vapor pressure, $e_s - e_a$ is the saturation vapor pressure deficit of the air, T is the air temperature at 2 m height, U_2 is the wind speed at 2 m height, Δ is the slope of the saturation vapor pressure temperature relationship, γ is the psychrometric constant, PPT is precipitation, and the calculation of LAI is referenced from the method of Lu et al. [34]. These parameters are directly calculated or derived from the average daily maximum and minimum temperature, the daily average temperature, the daily actual vapor pressure, the daily wind speed data, the actual duration of sunshine hours, the relative humidity data, and other empirical metrics.

In order to analyze the potential impact of climate change in the next 35 years (2015–2050) on regional lake area and number changes, we downloaded the Representative Concentration Pathways 5 (RCP 5) datasets from the China Agrosys Platform (http://stdown.agrivy.com). The future climate change data were generated based on the Fifth Generation Coupled Global Climate Model (CGCM 5) from the Canadian Centre for Climate Modeling and Analysis. The datasets constructed using original meteorological observations for each station included daily mean temperatures, daily precipitation, and daily potential evaporation. We converted daily mean temperature, daily precipitation, and daily potential evapotranspiration into annual mean temperature, annual mean precipitation, and annual potential evapotranspiration for our analysis.

2.2.3. Other Data

The areal data of the cropland area and gross domestic product (GDP) index collected from local statistical yearbook (http://www.stats.gov.cn).

2.3. Data Analysis

2.3.1. Extracting Lakes and Reservoirs

We focused on lakes and reservoirs greater than 1 km^2 in terms of the 30 and 80 m resolutions of the satellite images [35]. An object-based image analysis (OBIA) was used rather than the traditional pixel-based classification method, because OBIA classifies objects instead of individual pixels [36–38]. In the process of OBIA classification, the spectrum, spatial information, texture, and geometric features characterized by remote sensing images were fully utilized. The lake and reservoir extents of the NJW in 1980, 1990, 1998, 2000, 2010, 2013, and 2015 were extracted in eCognition Developer 8.6 [14,39]. The normalized difference water index (*NDWI*) is the most popular used index for automated inland waterbodies delineation [35,40,41]. In addition, we tested different versions of *NDWI* (*mNDWI*, *NDRW*, and *NDWI*) [21,42] and found that the selected *NDWI* is more effective in our study area. Therefore, *NDWI* was applied to extract lakes and reservoirs, which was defined as:

$$NDWI = (Green - NIR)/(Green + NIR) \tag{3}$$

where *Green* and *NIR* represent the reflectance of the green and Near Infrared (*NIR*) bands, respectively [25]. The data processing steps for the lake and reservoir inventory are shown in Figure 3. The extraction of lakes and reservoirs consisted of three major steps: image multi-resolution segmentation, *NDWI* threshold testing, and classification rule designing. By defining the length/width index and the rectangular fit index built into the software, rivers and artificial ponds were removed. Then, we used visual interpretation to extract the reservoirs.

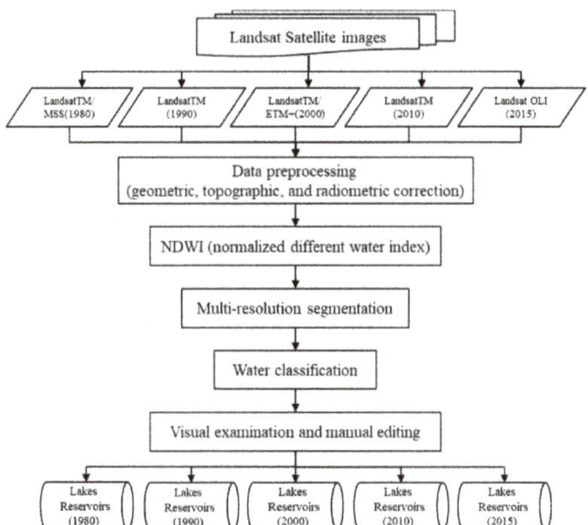

Figure 3. Flowchart for the lake and reservoir mapping process.

To validate the accuracy of extracted lakes and reservoirs, we calculated error matrices based on 1492 validation samples selected from Google Earth and Landsat sensors. The detailed number of validation samples for the five dates is given in Table 1. The overall accuracies of the lake and reservoir classification for the five dates were respectively evaluated and the standard errors bars (uncertainties) were estimated [43]. The overall accuracies of lakes and reservoirs were larger than 90%, and the kappa coefficients were larger than 0.87.

Table 1. Collected validation samples for the five dates.

Date	1980	1990	2000	2010	2015
Sources	Google Earth Image/MSS	TM	TM/ETM+	TM	OLI
Lakes	126	158	170	244	256
Reservoirs	38	47	50	62	65
Non-water bodies	35	40	79	65	57
Total Samples	199	245	299	371	378

MSS—multispectral scanner, TM—thematic mapper, ETM+—enhanced thematic mapper plus, OLI—operational land imager.

2.3.2. Temporal Analysis of Lakes and Reservoirs

To fully understand the changes of lakes and reservoirs of different sizes, the lakes and reservoirs were categorized into four classes: 1–10 km^2, 10–50 km^2, 50–100 km^2, and >100 km^2. In order to examine the areal change rate of the lake or reservoir in different periods, an indicator of the lake or reservoir area dynamic degree, as shown Equation (4), was used to analyze their changes [44,45].

$$K = \frac{U_b - U_a}{U_a} \times \frac{1}{T} \times 100\% \quad (4)$$

where K is the dynamic indicator for the lake or reservoir area; U_a and U_b are the area of the lake and reservoir at the start date and end date, respectively; and T is the time scale under consideration.

2.3.3. Assessing the Roles of Climatic Factors and Anthropogenic Causes in Lake Changes

The roles of climatic factors and artificial variables in changes of lakes during the study period were quantified using structural equation modeling (SEM) by the AMOS 22 software (IBM, Armonk, NY, USA) [39]. This paper analyzed the changes in MAAT, MAP, and ET to examine the influences of climate change on lake changes. The statistical data of cropland area and GDP were selected to investigate the impacts of artificial variables on lake changes [46,47]. Table 2 illustrates the optimum values for these indicators necessary for the SEM.

Table 2. Measures used to test the goodness of model.

Measure	Optimum Values	Reference
RMSEA (root mean square error of approximation)	Less than 0.08	Li et al. (2019) [39]
λ^2/df (chi-square/degree of freedom)	Less than 3	James (2007) [48]
GFI (goodness of fit index)	0.90 and above	Melucci et al. (2019) [49]
CFI (comparative fit index)	0.90 and above	David et al. (2000) [50]

3. Results

3.1. Spatial Pattern of Lakes and Reservoirs in 2015

Figure 4 shows the lake and reservoir distribution in 2015 across the NJW. Lakes and reservoirs were identified dominantly in the southeastern part of the NJW. A total of 233 lakes (area $\geq$ 1 km^2) covering an area of 2110 ± 53 km^2 in 2015 were extracted from satellite images. Most of the lakes (89.4%) had an area of smaller than 10 km^2. There were three lakes with area greater than 100 km^2, and their accumulated area accounted for approximately 33.8% of the total lake area in the NJW. The Chagan Lake was the largest lake with an estimated area of 292 km^2.

There were 129 reservoirs (area $\geq$ 1 km^2) in the NJW with a total area of 1422 ± 56 km^2 in 2015. Of these, 108 reservoirs had an area smaller than 10 km^2. There were two reservoirs with an area larger than 100 km^2, with the accumulated area accounting for approximately 41.5% of the total reservoir area in the NJW. The Nierji Reservoir was the largest reservoir with an estimated area of 360 km^2.

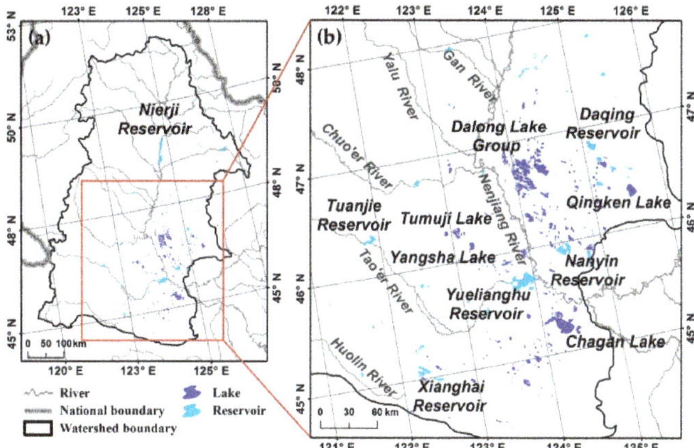

Figure 4. Spatial distribution of lakes and reservoirs in the Nenjiang watershed (NJW) in 2015.

3.2. Temporal Changes of Lakes and Reservoirs from 1980 to 2015

Figure 5 suggests that dramatic changes in the total area and number of lakes and reservoirs over the NJW occurred from 1980 to 2015. During the observed 35 years, the total lake area decreased by 38.7% from 3440 ± 72 km^2 to 2110 ± 53 km^2, whilst the total number of lakes decreased by 233 from the original 468 in 1980. Specifically, lake changes had significant fluctuations over the past 35 years, including obvious declines in total area (42%) and number (51%) from 1980 to 2010 and slight increases in the total lake area and number from 2010 to 2015. Reservoirs in the NJW experienced continuous expansion during 1980–2015. The total number of reservoirs increased from 78 to 129 with the total area expansion being 55%, from 919 ± 53 km^2 to 1422 ± 56 km^2.

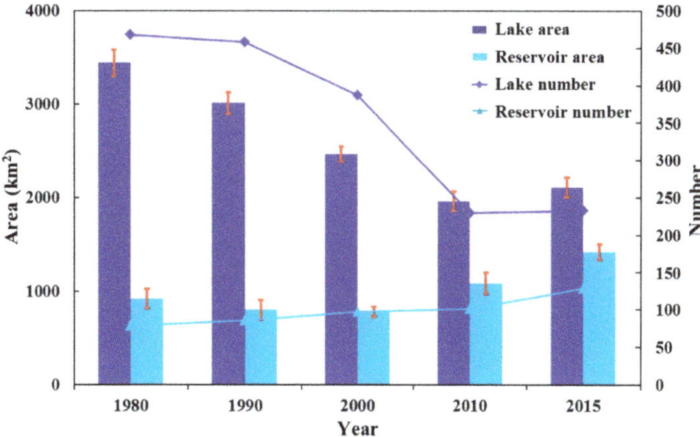

Figure 5. Decadal variations in the total area and number of lakes and reservoirs (area ≥ 1 km^2) in the NJW from 1980 to 2015. The error bars represent the 95% confidence level.

3.2.1. Lake Changes

For a further understanding of the temporal changes of lakes, the spatial heterogeneity of lake area changes across the NJW was investigated (Figure 6). Changes in the lake area presented clear variations. During the first period, 1980–1990, most of the expanded lakes were mainly distributed in the southeastern part of the NJW, while the lakes that shrunk were mainly distributed in the central part (Figure 6a). A rapid shrinkage of the total lake area and a decline in lake number occurred during 1990–2000. The larger lake shrinkage occurred in the southern NJW (Figure 6b), while lakes in the Chagan Lake, Yangsha Lake, and Tumuji Lake zones exhibited expanding trends during this period. During 2000–2010, lakes in the NJW with large areal loss were distributed mainly in the Dalong Lake zones, while expanded lakes were identified mainly at the intersection point between the Nenjiang River and Taoerhe River (Figure 6c). During 2010–2015, it is noteworthy that the characteristic of lake expansion was relatively obvious in the NJW compared to the other periods. Lakes with shrinkage were mainly distributed in the southeastern part, whereas lake expansions mostly occurred in the eastern part (Figure 6d).

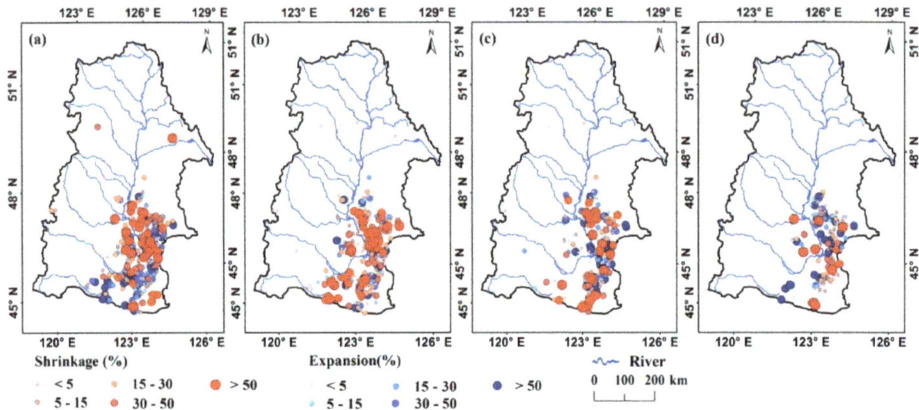

Figure 6. Spatial variations of lake changes in the Nenjiang watershed during different periods. Red and blue dots present lake shrinkage and expansion, respectively, while the dot extent denotes change proportion in the corresponding period. (**a–d**) represent the periods 1980–1990, 1990–2000, 2000–2010, and 2010–2015, respectively.

The detailed lake changes in each class are shown in Table 3. From 1980 to 1990, the largest areal decline of lakes was observed for the 50–100 km^2 size class. Interestingly, the Dalong Lake, which was in the size class of 50–100 km^2, separated into five lakes in the size class of 10–50 km^2 during this period. Specifically, 10 lakes with a size of 1–10 km^2 covering a total area of 34 km^2 in the central region disappeared in this period. During 1990–2000, lake shrinkage occurred with the total lake area declining by 18%, and 53 lakes (area $\leq$ 50 km^2) vanished during this decade. During 2000–2010, both the total number and area of lakes in the size class of 50–100 km^2 increased, while both the total number and area of lakes in other size classes decreased. From 2010 to 2015, both the total area and number of lakes in the size classes 1–10, 10–50, and 50–100 km^2, increased.

Table 3. Lake number, area, and changes in different classes between 1980 and 2015.

	Year or Period	Area Classes (km^2)				Total
		1–10	10–50	50–100	>100	
Number of lakes	1980	419	43	3	3	468
	1990	414	40	1	3	458
	2000	343	39	0	4	386
	2010	198	27	2	3	230
	2015	199	29	2	3	233
Change in number (%)	1980–1990	−1	−7	−67	0	−2
	1990–2000	−17	−3	0	33	−15
	2000–2010	−42	−31	100	−25	−41
	2010–2015	0	7	0	0	1
	1980–2015	−53	−33	−33	0	−50
Lake area (km^2)	1980	1369 ± 11	916 ± 17	197 ± 20	958 ± 24	3440 ± 72
	1990	1145 ± 10	884 ± 15	81 ± 17	900 ± 17	3010 ± 59
	2000	934 ± 8	729 ± 19	0.0	804 ± 13	2467 ± 40
	2010	548 ± 7	535 ± 16	111 ± 17	767 ± 13	1961 ± 53
	2015	595 ± 7	656 ± 14	145 ± 15	714 ± 17	2110 ± 53
Change in area (%)	1980–1990	−16 **	−4 **	−59 **	−6.0 **	−13 **
	1990–2000	−18 **	−18 **	−100 **	−10.7 **	−18 **
	2000–2010	−41 **	−27 **	0	−4.5 **	−20 **
	2010–2015	9 **	23 **	30 **	−7.0 **	8 **
	1980–2015	−57 **	−28 **	−26 **	−25.5 **	−39 **

Note: * denotes change at a significant level of 0.05, ** denotes significant at 0.01.

3.2.2. Reservoir Changes

Figure 7 shows the spatiotemporal patterns of reservoir changes in the NJW during different periods. During 1980–1990, the reservoirs that shrunk were distributed mainly in the central and northeast parts, whereas the expanded reservoirs were mainly distributed in the southern and eastern parts of the NJW (Figure 7a). During 1990–2000, the expanded reservoirs were mainly distributed in the northeast and southern parts, whereas the reservoirs that shrunk were mainly distributed in the central and southern parts of the NJW (Figure 7b). During 2000–2010, significant reservoir expansions were identified across the NJW, while the larger reservoirs that shrunk were mainly distributed in the southern part of the NJW (Figure 7c). After 2010, the reservoirs that shrunk were mainly distributed in the southern and eastern parts. Larger expanded reservoirs were mainly identified in the northern part and the Taoer River basin (Figure 7d).

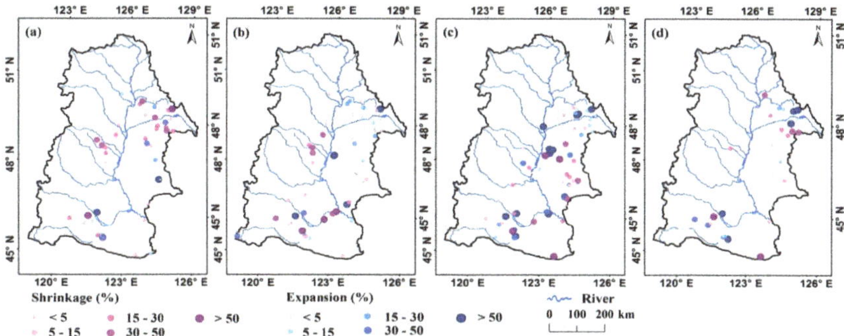

Figure 7. Spatial distribution of reservoir changes in the Nenjiang watershed during different periods. Purple and blue dots present reservoir shrinkage and expansion, respectively, while the dot extent denotes change proportion in the corresponding period. (**a–d**) represent the periods 1980–1990, 1990–2000, 2000–2010, and 2010–2015, respectively.

During the period of 1980–1990, the total area of reservoirs in size classes of 10–50 km^2 and 50–100 km^2 increased by 121%, with an areal increase from 149 ± 10 to 328 ± 16 km^2, and by 133% from 52 ± 14 to 121 ± 17 km^2, respectively. The total areas of reservoirs in the size classes of 1–10 km^2 and >100 km^2 decreased from 205 ± 9 to 200 ± 6 km^2 (−2%) and from 513 ± 20 to 151 ± 14 km^2 (−71%), respectively (Table 4). During 1990–2000, the total areas of reservoirs in the size classes of 1–10 km^2 and 50–100 km^2 increased by 23%, with an areal increase from 200 ± 6 to 247 ± 5 km^2, and by 123% from 121 ± 17 to 270 ± 15 km^2, respectively. The total area of reservoirs in the size classes of 10–50 km^2 decreased from 328 ± 16 to 263 ± 8 km^2 (−20%). During the period of 2000–2010, the total area of reservoirs increased from 780 ± 28 to 1086 ± 58 km^2. The total areas of reservoirs in the size classes of 1–10 km^2 and 10–50 km^2 increased by 30% and 24%, respectively. The total area of reservoirs in the size class of 50–100 km^2 decreased by 50%. During the period of 2010–2015, the total reservoir area increased by 31% with an increase in area from 1086 ± 58 to 1422 ± 56 km^2. Detailed reservoir changes for these different classes are shown in Table 4.

Table 4. Reservoir number, area, and changes in different classes between 1980 and 2015.

	Year or Period	Area Classes (km^2)				Total
		1–10	10–50	50–100	>100	
Number of reservoirs	1980	69	6	1	2	78
	1990	68	15	2	1	86
	2000	73	19	4	0	96
	2010	84	13	2	2	101
	2015	108	17	2	2	129
Change in number (%)	1980–1990	−1	150	100	−50	10
	1990–2000	7	27	100	0	13
	2000–2010	15	−32	−50	100	4
	2010–2015	29	31	0	0	28
	1980–2015	57	183	100	0	65
Reservoir area (km^2)	1980	205 ± 9	149 ± 10	52 ± 14	513 ± 20	919 ± 53
	1990	200 ± 6	328 ± 16	121 ± 17	151 ± 14	800 ± 53
	2000	247 ± 5	263 ± 8	270 ± 15	0	780 ± 28
	2010	320 ± 6	324 ± 16	134 ± 17	308 ± 19	1086 ± 58
	2015	259 ± 4	397 ± 16	175 ± 18	591 ± 18	1422 ± 56
Change in area (%)	1980–1990	−2	121 **	133 **	−71 **	−13 **
	1990–2000	23 **	−20 **	123 **	−100 **	−3
	2000–2010	30 **	24 **	−50 **	0	39 **
	2010–2015	−19 **	23 **	31 **	92 **	31 **
	1980–2015	27 **	167 **	237 **	15 **	55 **

Note: * denotes change at a significant level of 0.05, ** denotes significant at 0.01.

3.3. Roles of Climatic Factors and Artificial Variables in Driving Lake Changes

As shown in Figure 8, the model passed the reliability test, convergent validity test, and discriminant validity test. The dominant climatic factor that influenced lake changes was the MAP ($\beta = 0.66, p < 0.001$), followed by the ET ($\beta = 0.39, p < 0.05$) and MAAT ($\beta = 0.15, p < 0.1$). Agricultural consumption of water had a significant effect on lake changes, suggested by the significant relationship between cropland area and lake area ($\beta = 0.17, p < 0.1$).

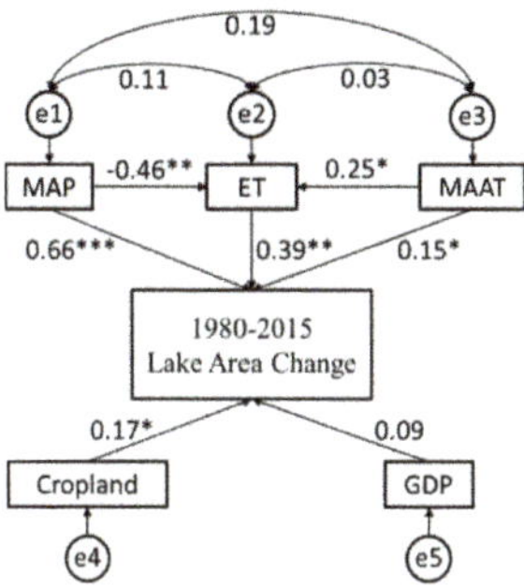

Figure 8. Structural model and path coefficient. Arrows show the effect path and direction. Numbers adjacent to arrows are path coefficients, (β). The path coefficients (β) characterize the extent of the effect of the examined variable on the lake changes, while the larger value indicates a strong positive or negative effect. P represents a significant level. * represents p-value < 0.1, ** represents p-value < 0.05, and *** represents p-value < 0.01. MAP denotes the mean annual precipitation, MAAT represents the mean annual air temperature, ET represents the evapotranspiration, and GDP represents the gross domestic product. The "e" represents the error (residual term) of path analysis of observed variables in the structural model. The numbers on the arrows are the values of the standardized regression weights of the model.

4. Discussion

Lake and reservoir mapping is affected by the resolution of the used satellite data [26]. We focused only on lakes and reservoirs with area greater than 1 km^2 to reduce as much as possible the errors induced by a coarse resolution of 30 and 80 m. Although some images out of the optimal season were used, these data mainly covered the Greater and Lesser Khingan Mountains where few lakes and reservoirs were identified. This did not yield large uncertainties in our analysis. This study integrated OBIA and visual interpretation instead of automatic classification to extract lakes and reservoirs, which ensured data accuracy and effective analysis.

Both climate change and human activities contributed to the changes of lakes in the NJW. On the one hand, the climate over the study area is a semi-arid continental climate. Water supply and output for these lakes dominated with precipitation and evapotranspiration, respectively. SEM analysis revealed that the lake shrinkage in the NJW had a significant correlation with the MAP ($\beta = 0.66$, $p < 0.001$) (Figure 8), followed by the ET ($\beta = 0.39$, $p < 0.05$), and the MAAT ($\beta = 0.15$, $p < 0.1$). This suggests that precipitation had a significant statistical relationship with lake area and potentially had the largest impacts on the lake area. During the investigated 35 years, a warmer climate was identified for the NJW with a significant increase of MAAT ($p < 0.05$) (Figure 9a). A reduced water supply from precipitation and output by increased ET characterized most of the lake shrinkages in the NJW. Specifically, lakes in the NJW showed areal changes with a decrease from 1980 to 2010 and then an increase from 2010 to 2015 (Figure 5), which is consistent with the changed trend of MAP during the 35 years (Figure 9a). The MAP could be regarded as the main climatic factor to explain the lake shrinkage in the NJW from 1980 to 2015 in terms of its decline.

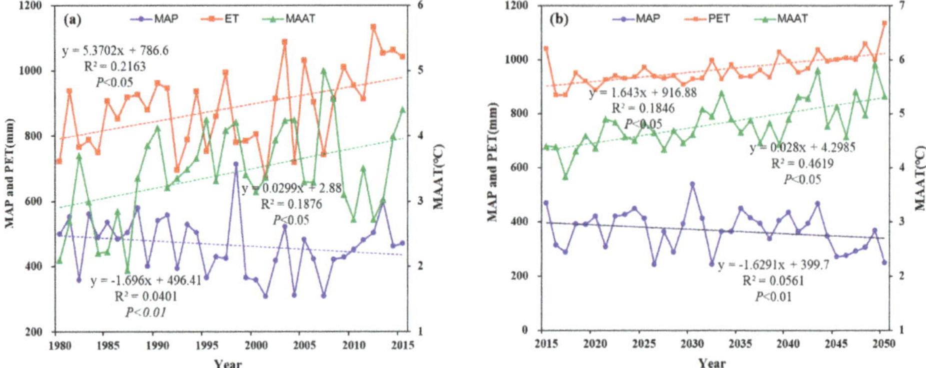

Figure 9. Past changes and future projections of climatic factors in the NJW. (**a**) Changes of mean annual air temperature (MAAT), mean annual precipitation (MAP), and evapotranspiration (ET) from 1980 to 2015 and (**b**) changes of future MAAT, MAP, and PET from 2015 to 2050.

Future projections of climate change using the Intergovernmental Panel on Climate Change (IPCC) models (RCP 5) indicate that the warming and drying trend will continue in the NJW (Figure 9b). Both the MAAT and PET show increasing trends with a rate of 0.03 °C yr^{-1} and 1.64 mm yr^{-1} ($p < 0.05$), while the MAP exhibits a significant decline with a rate of −1.63 mm yr^{-1} ($p < 0.01$). If this trend continues, the total area of lakes in the NJW may continue to decrease, and some small lakes will disappear.

This study found that flood events can markedly affect the lake area and number, especially for the small lakes (1–10 km^2). According to the hydrological records, the most serious floods during the recent century occurred in 1998 and 2013 [51]. Compared with the total lake area in 2010 (normal flow year), the total lake area in the NJW respectively increased by 2946 km^2 in 1998 and 895 km^2 in 2013 during the flooding events. It is clear that extreme precipitation events (floods) have accelerated the expansion of the total lake area (Figure 10). In arid and semi-arid regions, a multidimensional view on the prevention and exploitation of floods is required. Lakes and reservoirs have huge water storage capacity, and thus they could serve as hydrological buffers to prevent floods and provide irrigation water for sustainable agricultural development [10].

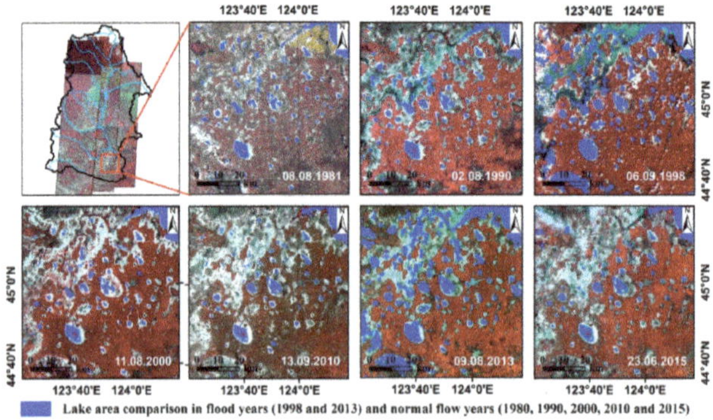

Figure 10. Increased lake area due to floods in 1998 and 2013.

On the other hand, the study area is an important grain production base in China. Agricultural demand for water had potential impacts on lakes. Besides climate change, human activities have imposed marked impacts on lakes and reservoirs [52]. While our study revealed that climatic factors drive striking lake and reservoir changes, human-induced changes should be responded to as quickly as possible. In this study, a clear correlation ($\beta = 0.17$, $p < 0.1$) was observed between lake area and crop area (Figure 8). Lake shrinkage caused by artificial variables is mainly attributed to the agricultural water consumption in the NJW [50,53]. Due to the construction of large reservoirs upstream water was therefore impounded upstream, which seriously affected the water supply of lakes in the middle and lower reaches. In particular, the largest reservoir, Nierji Reservoir, constructed over the upstream area of the Nenjiang River exerted a marked influence on downstream lakes. Due to reclamation from lakes and agricultural development during the past 35 years, some lakes in the size class of 1–10 km^2 decreased to small lakes with areas smaller than 1 km^2. For example, a large area of shallow waters was reclaimed for planting rice (Figure 11). Since 1990, the area of paddy fields in the NJW has increased significantly by 90.26 km^2 from 3.46 to 93.73 km^2. The total cropland area increased from 459.42 to 948.18 km^2. With the population increase and agricultural land expansion in this region, more and more open water resources have been applied to agricultural irrigation. It is evident that the human impacts on lake and reservoir changes could not be ignored.

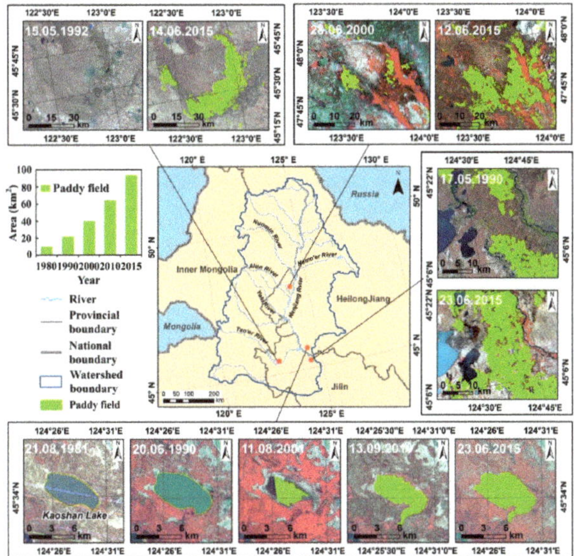

Figure 11. Image examples for lakes and reservoirs affected by the expanded paddy fields.

Lake shrinkage and disappearance lead to ecological and environmental degradation, such as aggravating the degree of sandstorms and desertification and reducing the number of wild animals [4,54]. To optimize water distribution programs through scientific water conservancy projects is very important. Therefore, appropriate measures need to be implemented by managers to further reduce the decline of lake areas [55], such as enhancing the drainage capacity of reservoirs in drought years but the storage capacity of reservoirs in flood years. In addition, we should control the expansion of paddy fields to relieve water stress and guarantee regional sustainable development in the NJW [56].

5. Conclusions

In this paper, we established a multi-temporal dataset of lakes and reservoirs in the NJW using long time-series Landsat images from 1980 to 2015 to document their changes on a decadal scale and quantified the contribution degree of MAAT, MAP, ET, cropland, and GDP to change in lake area. A notable decline in the total lake area by 1330 km^2 in the period of 1980 to 2015 was identified, while the lake number decreased contemporaneously. In contrast to lake shrinkage, the total area and number of reservoirs in the NJW experienced continuous increases. We identified 51 newborn reservoirs with total area of 504 km^2 in 2015 compared to 1980. SEM analysis revealed that decrease of the MAP is the dominant factor driving the changes of lakes, followed by ET and MAAT, especially for those of small size. Furthermore, the human impacts on lake and reservoir changes could not be ignored. Timely and appropriate policies and measures are required to reduce lake shrinkage and respond to the degraded environment. The results and analysis in this study are expected to provide guidance for sustainable management of water resource in the NJW.

Author Contributions: B.D. and D.M. designed the analytical framework of this study. B.D. performed the data analysis and drafted the manuscript. H.L. and H.X. provided methodological advice. D.M. and Z.W. made major revisions of the manuscript. All authors have read and agreed to the published version of the manuscript.

Funding: This study was jointly funded by the National Key Research and Development Program of China (2019YFA0607100, 2016YFC0500408, 2016YFC0500201, and 2016YFA0602301), the National Natural Science Foundation of China (41771383), and the funding from Youth Innovation Promotion Association, Chinese Academy of Sciences (2017277, 2012178) and National Earth System Science Data Center of China (http://northeast.geodata.cn).

Acknowledgments: We appreciate the facility that make Landsat 8 OLI images accessible through the USGS. We thank the three anonymous reviewers for the constructive comments and suggestions, which help improve the quality of this manuscript.

Conflicts of Interest: The authors declare no conflict of interest.

References

1. Bai, J.; Chen, X.; Li, J.; Yang, L.; Fang, H. Changes in the area of inland lakes in arid regions of central Asia during the past 30 years. *Environ. Monit. Assess.* **2011**, *178*, 247–256. [CrossRef] [PubMed]
2. Li, B.; Yang, G.; Wan, R.; Zhang, L.; Zhang, Y.; Dai, X. Using fuzzy theory and variable weights for water quality evaluation in Poyang Lake, China. *Chin. Geogr. Sci.* **2017**, *27*, 39–51. [CrossRef]
3. Duan, Z.; Bastiaanssen, W.G.M. Estimating water volume variations in lakes and reservoirs from four operational satellite altimetry databases and satellite imagery data. *Remote Sens. Environ.* **2013**, *134*, 403–416. [CrossRef]
4. Williamson, C.E.; Saros, J.E.; Vincent, W.F.; Smol, J.P. Lakes and reservoirs as sentinels, integrators, and regulators of climate change. *Limnol. Oceanogr.* **2009**, *54*, 2273–2282. [CrossRef]
5. Smith, L.C.; Sheng, Y.; MacDonald, G.; Hinzman, L. Disappearing arctic lakes. *Science* **2005**, *308*, 1429. [CrossRef]
6. McBean, E.; Motiee, H. Assessment of impact of climate change on water resources: A long term analysis of the Great Lakes of North America. *Hydrol. Earth Syst. Sci.* **2008**, *12*, 239–255. [CrossRef]
7. Zhiguo, L. Glacier and lake changes across the Tibetan Plateau during the past 50 years of climate change. *J. Resour. Ecol.* **2014**, *5*, 123–132. [CrossRef]
8. Mao, D.; Wang, Z.; Yang, H.; Li, H.; Thompson, R.J.; Li, L.; Song, K.; Chen, B.; Gao, H.; Wu, J. Impacts of Climate Change on Tibetan Lakes: Patterns and Processes. *Remote Sens.* **2018**, *10*, 358. [CrossRef]
9. Tao, S.; Fang, J.; Zhao, X.; Zhao, S.; Shen, H.; Hu, H.; Tang, Z.; Wang, Z.; Guo, Q. Rapid loss of lakes on the Mongolian Plateau. *Proc. Natl. Acad. Sci. USA* **2015**, *112*, 2281–2286. [CrossRef]
10. Lehner, B.; Liermann, C.R.; Revenga, C.; Vörösmarty, C.; Fekete, B.; Crouzet, P.; Döll, P.; Endejan, M.; Frenken, K.; Magome, J.; et al. High-resolution mapping of the world's reservoirs and dams for sustainable river-flow management. *Front. Ecol. Environ.* **2011**, *9*, 494–502. [CrossRef]
11. Ministry of Water Resources. *Bulletin of First National Census for Water*; China Water and Power Publisher: Beijing, China, 2013.
12. Yang, X.; Lu, X. Drastic change in China's lakes and reservoirs over the past decades. *Sci. Rep.* **2014**, *4*, 6041. [CrossRef] [PubMed]
13. Mao, D.; Wang, Z.; Wu, J.; Wu, B.; Zeng, Y.; Song, K.; Yi, K.; Luo, L. China's wetlands loss to urban expansion. *Land Degrad. Dev.* **2018**, *29*, 2644–2657. [CrossRef]
14. Chang, B.; Li, R.; Zhu, C.; Liu, K. Quantitative Impacts of Climate Change and Human Activities on Water-Surface Area Variations from the 1990s to 2013 in Honghu Lake, China. *Water* **2015**, *7*, 2881–2899. [CrossRef]

15. Zheng, Z.; Li, Y.; Guo, Y.; Xu, Y.; Liu, G.; Du, C. Landsat-Based Long-Term Monitoring of Total Suspended Matter Concentration Pattern Change in the Wet Season for Dongting Lake, China. *Remote Sens.* **2015**, *7*, 13975–13999. [CrossRef]
16. Wu, G.; Liu, Y. Mapping Dynamics of Inundation Patterns of Two Largest River-Connected Lakes in China: A Comparative Study. *Remote Sens.* **2016**, *8*, 560. [CrossRef]
17. Wang, M.; Du, L.; Ke, Y.; Huang, M.; Zhang, J.; Zhao, Y.; Li, X.; Gong, H. Impact of Climate Variabilities and Human Activities on Surface Water Extents in Reservoirs of Yongding River Basin, China, from 1985 to 2016 Based on Landsat Observations and Time Series Analysis. *Remote Sens.* **2019**, *11*, 560. [CrossRef]
18. Feyisa, G.L.; Meilby, H.; Fensholt, R.; Proud, S.R. Automated Water Extraction Index: A new technique for surface water mapping using Landsat imagery. *Remote Sens. Environ.* **2014**, *140*, 23–35. [CrossRef]
19. Tulbure, M.G.; Broich, M.; Stehman, S.V.; Kommareddy, A. Surface water extent dynamics from three decades of seasonally continuous Landsat time series at subcontinental scale in a semi-arid region. *Remote Sens. Environ.* **2016**, *178*, 142–157. [CrossRef]
20. Beaton, A.; Whaley, R.; Corston, K.; Kenny, F. Identifying historic river ice breakup timing using MODIS and Google Earth Engine in support of operational flood monitoring in Northern Ontario. *Remote Sens. Environ.* **2019**, *224*, 352–364. [CrossRef]
21. Carroll, M.; Wooten, M.; DiMiceli, C.; Sohlberg, R.; Kelly, M. Quantifying Surface Water Dynamics at 30 Meter Spatial Resolution in the North American High Northern Latitudes 1991–2011. *Remote Sens.* **2016**, *8*, 622. [CrossRef]
22. Yamazaki, D.; Trigg, M.A.; Ikeshima, D. Development of a global ~90m water body map using multi-temporal Landsat images. *Remote Sens. Environ.* **2015**, *171*, 337–351. [CrossRef]
23. Li, Q.; Lu, L.; Wang, C.; Li, Y.; Sui, Y.; Guo, H. MODIS-Derived Spatiotemporal Changes of Major Lake Surface Areas in Arid Xinjiang, China, 2000–2014. *Water* **2015**, *7*, 5731–5751. [CrossRef]
24. Karlsson, J.M.; Jaramillo, F.; Destouni, G. Hydro-climatic and lake change patterns in Arctic permafrost and non-permafrost areas. *J. Hydrol.* **2015**, *529*, 134–145. [CrossRef]
25. Wendleder, A.; Friedl, P.; Mayer, C. Impacts of Climate and Supraglacial Lakes on the Surface Velocity of Baltoro Glacier from 1992 to 2017. *Remote Sens.* **2018**, *10*, 1681. [CrossRef]
26. Khadka, N.; Zhang, G.; Thakuri, S. Glacial Lakes in the Nepal Himalaya: Inventory and Decadal Dynamics (1977–2017). *Remote Sens.* **2018**, *10*, 1913. [CrossRef]
27. Mishra, N.; Haque, M.O.; Leigh, L.; Aaron, D.; Helder, D.; Markham, B. Radiometric Cross Calibration of Landsat 8 Operational Land Imager (OLI) and Landsat 7 Enhanced Thematic Mapper Plus (ETM+). *Remote Sens.* **2014**, *6*, 12619–12638. [CrossRef]
28. Wang, Z.; Wang, Z.; Zhang, B.; Lu, C.; Ren, C. Impact of land use/land cover changes on ecosystem services in the Nenjiang River Basin, Northeast China. *Ecol. Process.* **2015**, *4*, 11. [CrossRef]
29. Feng, X.; Zhang, G.; Yin, X. Hydrological responses to climate change in Nenjiang river basin, northeastern China. *Water Resour. Manag.* **2011**, *25*, 677–689. [CrossRef]
30. Qiu, B.; Lu, D.; Tang, Z.; Chen, C.; Zou, F. Automatic and adaptive paddy rice mapping using Landsat images: Case study in Songnen Plain in Northeast China. *Sci. Total Environ.* **2017**, *598*, 581–592. [CrossRef]
31. Zhang, Y.; Zang, S.; Sun, L.; Yan, B.; Yang, T.; Yan, W.; Meadows, M.E.; Wang, C.; Qi, J. Characterizing the changing environment of cropland in the Songnen Plain, Northeast China, from 1990 to 2015. *J. Geogr. Sci.* **2019**, *29*, 658–674. [CrossRef]
32. Liu, Z.-H.; McVicar, T.R.; LingTao, L.; Van Niel, T.G.; Yang, Q.-K.; Li, R.; Mu, X.-M. Interpolation for time series of meteorological variables using ANUSPLIN. *J. Northwest A F Univ.* **2008**, *36*, 227–234.
33. Yoo, S.-H.; Choi, J.-Y.; Jang, M.-W. Estimation of design water requirement using FAO Penman–Monteith and optimal probability distribution function in South Korea. *Agric. Water Manag.* **2008**, *95*, 845–853. [CrossRef]
34. Lü, Y.; Fu, B.; Feng, X.; Zeng, Y.; Liu, Y.; Chang, R.; Sun, G.; Wu, B. A policy-driven large scale ecological restoration: Quantifying ecosystem services changes in the Loess Plateau of China. *PLoS ONE* **2012**, *7*, e31782. [CrossRef] [PubMed]
35. Korzeniowska, K.; Korup, O. Object-Based Detection of Lakes Prone to Seasonal Ice Cover on the Tibetan Plateau. *Remote Sens.* **2017**, *9*, 339–361. [CrossRef]
36. Walter, V. Object-based classification of remote sensing data for change detection. *ISPRS J. Photogramm.* **2004**, *58*, 225–238. [CrossRef]

37. Blaschke, T.; Lang, S.; Hay, G. *Object-Based Image Analysis: Spatial Concepts for Knowledge-Driven Remote Sensing Applications*; Springer: Berlin, Germany, 2008.
38. Jia, M.; Wang, Z.; Wang, C.; Mao, D.; Zhang, Y. A new vegetation index to detect periodically submerged Mangrove forest using single-tide sentinel-2 imagery. *Remote Sens.* **2019**, *11*, 2043. [CrossRef]
39. Li, H.; Mao, D.; Li, X.; Wang, Z.; Wang, C. Monitoring 40-Year Lake Area Changes of the Qaidam Basin, Tibetan Plateau, Using Landsat Time Series. *Remote Sens.* **2019**, *11*, 343. [CrossRef]
40. Nie, Y.; Sheng, Y.; Liu, Q.; Liu, L.; Liu, S.; Zhang, Y.; Song, C. A regional-scale assessment of Himalayan glacial lake changes using satellite observations from 1990 to 2015. *Remote Sens. Environ.* **2017**, *189*, 1–13. [CrossRef]
41. Li, H.; Gao, Y.; Li, Y.; Yan, S.; Xu, Y. Dynamic of Dalinor Lakes in the Inner Mongolian Plateau and Its Driving Factors during 1976–2015. *Water* **2017**, *9*, 749. [CrossRef]
42. Mueller, N.; Lewis, A.; Roberts, D.; Ring, S.; Melrose, R.; Sixsmith, J.; Lymburner, L.; McIntyre, A.; Tan, P.; Curnow, S.; et al. Water observations from space: Mapping surface water from 25 years of Landsat imagery across Australia. *Remote Sens. Environ.* **2016**, *174*, 341–352. [CrossRef]
43. Olofsson, P.; Foody, G.M.; Herold, M.; Stehman, S.V.; Woodcock, C.E.; Wulder, M.A. Good practices for estimating area and assessing accuracy of land change. *Remote Sens. Environ.* **2014**, *148*, 42–57. [CrossRef]
44. Yuan, F.; Sawaya, K.E.; Loeffelholz, B.C.; Bauer, M.E. Land cover classification and change analysis of the Twin Cities (Minnesota) Metropolitan Area by multitemporal Landsat remote sensing. *Remote Sens. Environ.* **2005**, *98*, 317–328. [CrossRef]
45. Zhou, W.; Troy, A.; Grove, M. Object-based land cover classification and change analysis in the Baltimore metropolitan area using multitemporal high resolution remote sensing data. *Sensors* **2008**, *8*, 1613–1636. [CrossRef] [PubMed]
46. Enders, C.K.; Mansolf, M. Assessing the fit of structural equation models with multiply imputed data. *Psychol. Methods* **2018**, *23*, 76. [CrossRef]
47. Meydan, C.; Sesen, H. *Structural Equation Modeling AMOS Applications*; Detay Yayıncılık: Ankara, Turkey, 2011.
48. Schreiber, J.B. Core reporting practices in structural equation modeling. *Res. Soc. Adm. Pharm.* **2008**, *4*, 83–97. [CrossRef]
49. Melucci, M.; Paggiaro, A. Evaluation of information retrieval systems using structural equation modeling. *Comput. Sci. Rev.* **2019**, *31*, 1–18. [CrossRef]
50. Gefen, D.; Straub, D.; Boudreau, M.-C. Structural equation modeling and regression: Guidelines for research practice. *Commun. Assoc. Inf. Systems* **2000**, *4*, 7. [CrossRef]
51. Li, S.; Gang, A. A diagnostic study of northeast cold vortex heavy rain over the Songhuajiang-Nenjiang River Basin in the Summer of 1998. *Chin. J. Atmos. Sci.* **2001**, 342–353.
52. Zhang, F.; Tiyip, T.; Johnson, V.C.; Kung, H.-t.; Ding, J.-l.; Sun, Q.; Zhou, M.; Kelimu, A.; Nurmuhammat, I.; Chan, N.W. The influence of natural and human factors in the shrinking of the Ebinur Lake, Xinjiang, China, during the 1972–2013 period. *Environ. Monit. Assess.* **2014**, *187*, 4128. [CrossRef]
53. Zhang, G.; Yao, T.; Shum, C.; Yi, S.; Yang, K.; Xie, H.; Feng, W.; Bolch, T.; Wang, L.; Behrangi, A. Lake volume and groundwater storage variations in Tibetan Plateau's endorheic basin. *Geophys. Res. Lett.* **2017**, *44*, 5550–5560. [CrossRef]
54. Zhong, L.; Liu, L.; Liu, Y. Natural Disaster Risk Assessment of Grain Production in Dongting Lake Area, China. *Agric. Sci. Procedia* **2010**, *1*, 24–32. [CrossRef]
55. Mao, D.; He, X.; Wang, Z.; Tian, Y.; Xiang, H.; Yu, H.; Man, W.; Jia, M.; Ren, C.; Zheng, H. Diverse policies leading to contrasting impacts on land cover and ecosystem services in Northeast China. *J. Clean. Prod.* **2019**, *240*, 117961. [CrossRef]
56. Mao, D.; Luo, L.; Wang, Z.; Wilson, M.C.; Zeng, Y.; Wu, B.; Wu, J. Conversions between natural wetlands and farmland in China: A multiscale geospatial analysis. *Sci. Total Environ.* **2018**, *634*, 550–560. [CrossRef] [PubMed]

Article

The Long-Term Effects of Land Use and Climate Changes on the Hydro-Morphology of the Reno River Catchment (Northern Italy)

Donatella Pavanelli [1,*], Claudio Cavazza [2], Stevo Lavrnić [1] and Attilio Toscano [1]

1 Department of Agricultural and Food Sciences, Alma Mater Studiorum—University of Bologna, Viale Giuseppe Fanin 50, 40127 Bologna, Italy
2 Reno Catchment Technical Service, Emilia-Romagna Region, 40100 Bologna, Italy
* Correspondence: donatella.pavanelli@unibo.it

Received: 3 July 2019; Accepted: 28 August 2019; Published: 3 September 2019

Abstract: Anthropogenic activities, and in particular land use/land cover (LULC) changes, have a considerable effect on rivers' flow rates and their morphologies. A representative example of those changes and resulting impacts on the fluvial environment is the Reno Mountain Basin (RMB), located in Northern Italy. Characterized by forest exploitation and agricultural production until World War II, today the RMB consists predominantly of meadows, forests and uncultivated land, as a result of agricultural land abandonment. This study focuses on the changes of the Reno river's morphology since the 1950s, with an objective of analyzing the factors that caused and influenced those changes. The factors considered were LULC changes, the Reno river flow rate and suspended sediment yield, and local climate data (precipitation and temperature). It was concluded that LUCL changes caused some important modifications in the riparian corridor, riverbed size, and river flow rate. A 40–80% reduction in the river bed area was observed, vegetation developed in the riparian buffer strips, and the river channel changed from braided to a single channel. The main causes identified are reductions in the river flow rate and suspended sediment yield (−36% and −38%, respectively), while climate change did not have a significant effect.

Keywords: farmland abandonment; LULC changes; climate change; runoff/suspended sediment changes; river morphology dynamics; Italian Apennines

1. Introduction

The development and specific characteristics of rivers and streams are influenced by surrounding landscapes [1,2]. Our current understanding of rivers' dynamics incorporates a conceptual framework of spatial nested controlling factors in which climate, geology, and topography at large scales influence geomorphic processes that shape channels at intermediate scales [3]. However, direct human impact on the environment cannot be neglected at a local scale, especially in the last century. In particular, land use/land cover (LULC) changes have a significant impact on basin water cycles and soil erosion dynamics. Human factors also include water abstraction for irrigation, flow regulation, the construction of reservoirs, and mining. An extensive bibliography analyzing the effects of dams and reservoirs on the geomorphic responses of rivers has been produced [4–7]. However, the influence of other drivers, such as climate and LULC changes has been much less documented over time, although recent studies highlight their importance in inducing river changes [8–11].

It can be said that agriculture and land abandonment are two complementary aspects of human impacts on the landscape. Land abandonment can affect net soil losses [12], while recolonization of natural vegetation can lead to a reduction in soil loss and a progressive improvement of soil characteristics [13]. Moreover, land abandonment and agriculture can also lead to changes of river stream morphologies, in particular narrowing and incision [13]. Liebault and Piegay [14] observed on the Roubion River (France), that colonization of unstable gravel bars tends to channel minor floods which, in turn, form a new and narrower channel in the existing river bed. A number of studies [3,15] have documented statistical links between LULC and stream conditions, using multisite comparisons and empirical models.

Hydrological alteration is one of the principal environmental factors by which LUCL influences stream ecosystems [3]. It alters the runoff-evapotranspiration balance, can cause an increase or decrease in flow rate's magnitude and frequency, and often lowers river's base flow. In addition, hydrological alteration contributes to a change of channel dynamics, including increased erosion of the channel and its surroundings, and a less frequent overbank flooding [3]. Zhang and Schilling [16] noted that increasing streamflow in the Mississippi River was mainly due to an increase in base flow, which in turn was a consequence of LULC (conversion of perennial vegetation to seasonal row crops). Many researchers have studied the effects of LULC changes on river flow and most of them have indicated that intensified afforestation will reduce both runoff peak and total runoff volume [17]. On the other hand, even a modest riparian deforestation in highly forested catchments can result in the degradation of a stream habitat, owing to sedimentary input. A comparison of different catchments showed that an increased forest area results in lower concentrations of suspended sediments, inferior turbidity at base flow, lower bed-load transport, and less embeddedness [3].

Numerous studies have demonstrated that LULC, the abandonment of rural activities, and consequently, a decrease of human pressure on mountain areas, has contributed to increase of vegetation cover [18–20]. In the case of Reno River mountain basin, Pavanelli et al. [21] documented that recolonization of natural vegetation and a consequent increase of actual evapotranspiration, was the key hydrological variable that caused the decrease of the river flow rate. However, not many studies have addressed the effect of the redevelopment of natural vegetation at the river-basin scale on river flow, sediment yield, and riverbed morphology. Picco et al. [22] noted a consistent increase of riparian vegetation within the corridor of the Piave River (Northern Italy) during the last five decades, concluding that it depended on human activities, both in the main channel and at basin scale. LULC and local climate change (e.g., precipitation, temperature and evapotranspiration), may induce notable alterations of watershed hydrology [13,23]. Several studies showed that precipitation increase alone is insufficient to explain increasing flow rate trends in agricultural watersheds [24,25], since changes in agricultural land use can also result in increased flow rate.

Collectively, these studies provide strong evidence for the importance of the surrounding landscape, and human activities for the hydrological and morphological characteristics of rivers [3,15]. However, it is often difficult to separate human from naturally driven activities [26]. In addition, most of these articles were mainly conducted on small spatial scales, within a few hundred meters of a stream, without considering larger spatial units. Finally, only few studies have addressed the effect of redevelopment of natural vegetation at the river basin scale on river discharge, sediment yield, and bed river morphology.

Therefore, the aim of this paper was to explore relationships between local climate change that occurred in the last century and agricultural land abandonment (and consequent LULC changes), which culminated in the 1950s, on the one hand; and modifications in morphology and hydrology of the Reno River (Northern Italy), on the other hand.

2. Materials and Methods

2.1. Study Area

The Reno River, located in the Northern Italy, flows into the Adriatic Sea. It is the sixth biggest Italian river, with a catchment area of 5965 km^2 and a length of 211.8 km. The mountain hydrographic network of the Reno is rather ramified and dense, and it is composed of eight major rivers, 12 secondary rivers and 600 torrents. This study concentrates on one part of the Reno River; namely, the Reno River

Mountain Basin (RMB), located in the Northern Italian Apennines (Emilia Romagna and Tuscany Regions), with a catchment area of 1061 km^2 and length of 80 km. The RMB's average altitude is of 639 m a.s.l.; it ranges from a maximum elevation of 1945 to a minimum one of 60.35 m a.s.l. at the dam of Chiusa of Casalecchio (44°47′ N, 11°28′ E), which is the RMB outlet (Figure 1).

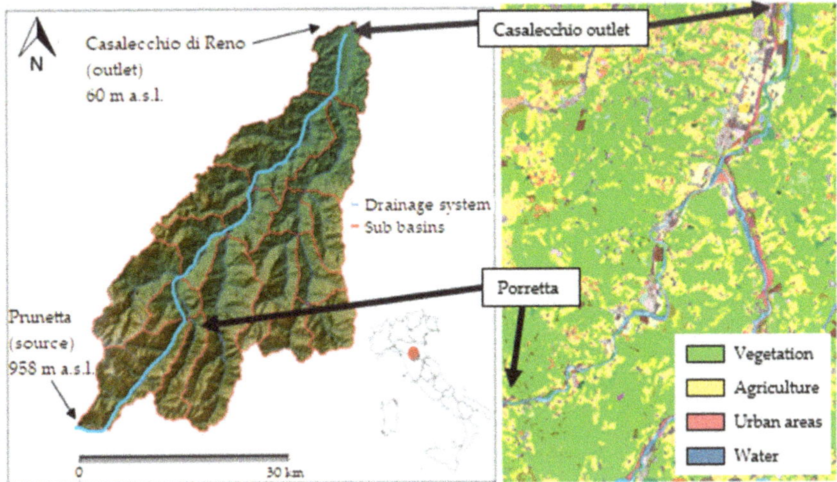

Figure 1. Orography map of the Reno River Mountain Basin with the main river channel in blue (**left**) and the simplified land use/land cover (LULC) of the area in 2003 (**right**).

The RMB can be considered representative of the environmental and anthropogenic changes that have occurred in the Italian central and northern Apennines in the last century. The Apennine agricultural area of the Emilia Romagna Region (RER) decreased by almost 50% from 1960 to 2000 [27]. After 1950, the population moved to the cities and valley, a phenomenon that affected the whole Italian Apennines. Until World War II, this area had an average population density of 85 inhabitants km^{-2}, with agro-forestry and pastoral farming as the main activities. Currently the density is reduced to less than 70% of the previous one [27]. After World War II, due to industrialization and the development of agricultural mechanization, the landscape rapidly transformed. Agriculture remained where it was cost-effective, while the rest gave way to permanent meadows, scrub, and woodland [28]. Currently, land cover is characterized by oak woods, beeches, shrubs, and pastures at higher altitudes. Chestnut woods are present at medium altitudes and on coppices, pastures, and crops on hillsides. Crops, vineyards, orchards and urban areas cover the catchment valley (Figure 1).

The RMB consists mainly of erodible sedimentary rocks. Land cover, runoff, soil erosion, and suspended sediment in the river are closely related to each other. The bedrock consists of resistant limestone, sandstone, and meta-sandstone in the upper part of the watershed and of weakly cemented marl, mudstone, sandstone, and conglomerate in the middle and lower part of the watershed [29]. The fluvial terraces are preserved from the outlet (Casalecchio) to about 20 km upstream (~150 m a.s.l.). Upstream of that point, landslides and earthflows preclude significant preservation of terraces.

The RMB average precipitation is 1305 mm year^{-1} (Table 1), with the following distribution: Winter 337 mm, spring 307 mm, summer 182 mm, and autumn 411 mm. The average temperature is 10.7 °C; July is the hottest month, with peaks up to 29.5 °C (1998). Winters are generally very cold, with the average minimum monthly temperature dropping to −8.9 °C (January 1942). The fluvial regime of the Reno River is linked to rainfall, with floods occurring in autumn and spring. Seasonal floods are characterized by short times of concentration (time it takes to reach the basin outlet), owing to the long and narrow shape of the catchment. The evaluations of the 30 and 200-year recurrence interval floods at Casalecchio (Chiusa) are 1541 and 2280 m^3 s^{-1}, respectively.

Table 1. Available data set and its properties.

Parameter	Data Record	Number of Years Available	Area	Lenght	Data Source
River flow rate ($m^3\ s^{-1}$)	1921–2013	87	1061 km^2	80 km	SIMI, ARPA
Suspended sediment yields ($Mg\ km^{-2}$)	1942–1978	31	1061 km^2	80 km	SIMI
Land Cover RMB	1954&2003	2	1061 km^2	na	GAI flight, Quickbird
Land Cover Reno riparian buffer strips (R1–R3)	1954&2003	2	27.3 km^2	54.5 km	GAI flight, Quickbird & field survey
Reno riverbed morphology (R1–R3)	1954&2003	2	27.3 km^2	54.5 km	GAI flight, Quickbird & field survey
RMB Precipitation (mm)	1921–2013	87	1061 km^2	na	SIMI, ARPAE
Maximum and minimum temperature (°C) *	1928–2007 1936–2007	7670	na	na	SIMI, ARPAE

* The two periods correspond to the two meteorological stations—one in the valley and another in the mountains, that were analyzed separately in order to highlight differences between them.

The dam "Chiusa of Casalecchio" is the RMB outlet. It is the oldest hydraulic building in Europe, and has allowed the social and economic development of the city of Bologna through hydraulic energy since the XII century. It is also included in the UNESCO program's list of Patrimony Messengers of a Culture of Peace. The Chiusa dam controls the downstream base level, making the reach from the source to the Chiusa geomorphologically independent. Census of the hydraulic works detected 51 weirs on that reach, and most of them were built in the first decades of the 20th century. Similarly, in Rio Maggiore, a tributary of Reno, there is a total of 162 dams, which equals a 10 dam km^{-2} density [28]. Moreover, in the early 1900s, five hydroelectric dams were built in the tributaries of the Reno. Even though dams and reservoirs do not influence river water budget on the longer time scale, they do act as river sediment traps and can affect the river flood regulation [21]. During the last few decades, part of water has been diverted for domestic and irrigation purposes; however, these withdrawals represent a very limited percentage (3%) of the Reno river's flow rate [21].

2.2. *Methodology*

The parameters considered in this study were LULC (since the 1950s), river morphology (in 1954 and 2003), and hydro-climate changes (since the 1920s) of the RMB. They were used to assess different environmental changes that occurred in the RMB and whether if they were due to agricultural land abandonment and a consequent renaturalization after the 1950s. The main hypothesis of this research is that the year 1960 is the date from which effects of these changes can actually be seen.

2.2.1. Land Use Changes

As already said, the year 1960 was taken as a point when effects of big LULC changes started to be visible, and the year 1954 was taken as representative of the period before agricultural land abandonment. The two LULC maps are both to the scale 1:25000 (Soil Use Maps RER 1954 and 2003) were derived from the black and white aerial photographs flight G.A.I. 1954 (Military Geographic Institute-Italy) and the satellite images from 2003 (Quickbird). They were analyzed and compared with Geographic Information System software ARCVIEW 3.2 (Environmental Systems Research Institute, Redlands, CA, USA) to evaluate the RMB LULC changes. The relatively heterogeneous classes of the two different maps were reclassified according to the CORINE land cover classes [30] in order to be compared. Then, 20 and 51 land cover classes of the 1954 and 2003 Soil Use Maps respectively, were grouped into urban areas, water bodies, fallows, forests, and crops. This approach reduced the error due to the different sources of images of the maps. In Figure 1 fallows and forests are grouped together to highlight natural vegetation and renaturalization effects.

2.2.2. Morphological River Changes

To assess changes to the river belt and river morphology, aerial photos from the year 1954 (GAI-IGMI) and satellite images from Quickbird, 2003 (Figure 2), were used. The effects of LULC changes and human impact on the Reno river terraces and banks were analyzed on a reach from

Porretta Terme (349 m a.s.l.) to Casalecchio di Reno (60 m a.s.l.), with a length of 54.5 km and a mean bed river slope of 0.53%. The area was within 250 meters from each side of the river bed, done using GIS software. The transects (Figure 2 left) were drawn using the same reference points for both 1954 and 2003, to estimate variations in the river bank width, vegetation, and stream bed. The river's morphological changes, the width, the river's channel area, and the riparian LULC of each transect, were evaluated. The river channel was identified in the images as a non-vegetated part of river bed corresponding to the physical confine of the normal water flow. Banks, on the other hand, are subject to water flow only during high water stages and a riparian buffer strip is the vegetated area near a stream. All of them constitute the river corridor.

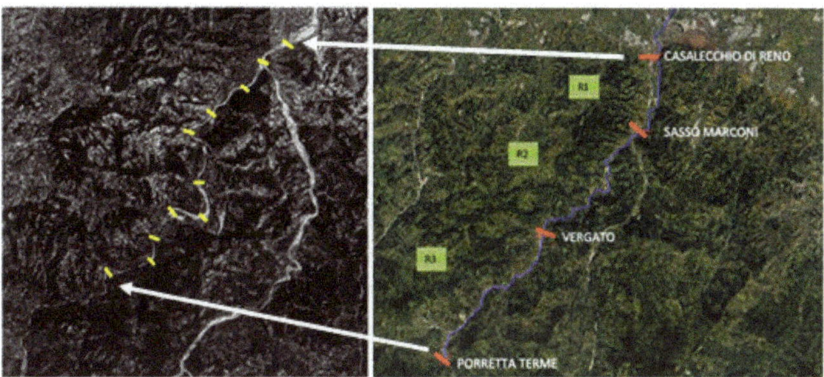

Figure 2. The Reno river transects (in yellow) between Casalecchio di Reno and Porretta Terme on an IGMI-GAI photo from 1954 (**left**), and the reaches R1, R2, and R3 on a satellite image from Quickbird, 2003 (**right**).

The studied 54.5 km portion of the Reno river was split into three reaches (Figure 2, right), on the basis of similar morphological characteristics (mean width and slope of the river and its banks): R1 from Porretta Terme to Vergato (21 km long), R2 from Vergato to Sasso Marconi (23 km long) and R3 from Sasso Marconi to Casalecchio Chiusa (10.5 km long). The properties of the three reaches were obtained from field survey of the transects and from the 1954 and 2003 images. The cartographic and photo-interpretation data were managed with GIS ARCVIEW 3.2 software used for map drawing and calculating the extent of the areas covered by vegetation. On the other hand, data on the river morphology (e.g., type of riverbed material) and vegetation type in the transects were obtained during field surveys.

2.2.3. Climate and Hydrological Data

The Reno River hydrological and RMB climate data (Table 1), were processed on a monthly, yearly and seasonal basis, and they were divided into two periods—before and after 1960, the year that was taken as a date from which effects of LULC changes and renaturalization can be seen. The monthly data of the river flow rate (Q) and the suspended sediment yields (SSY) are came from samples collected at the outlet of the RMB (Casalecchio di Reno gauge, 60 m a.s.l.) by the Italian Hydrographical Service (SIMI) and by the Regional Agency for Environmental Protection of the Emilia Romagna Region (ARPAE).

In addition, precipitation and temperature data collected by the SIMI and ARPAE were analyzed to evaluate impact of climate change on the RMB. For the minimum and maximum monthly temperatures of the two RMB stations: One in the mountains (Monteombraro, at 704 m. a.s.l.) and one in the valley (Anzola at 42 m. a.s.l.) were examined (Table 1).

2.2.4. Data Analysis

Statistical analyses were performed on the data records with the STATGRAPHICS®Centurion XVI software (StatPoint Technologies, Inc., The Plains, VA, USA). Statistical tests were used to verify whether the river flow rate data (Q_{mean} and Q_{max}) detected before 1960 (1921–1959) were statistically different from the data collected after 1960 (1960–2013). Two samples were compared using F test, discriminant analysis, box and whisker plots, *t*-tests, the Kolmogorov–Smirnov Test, and analysis of variance tests. Discriminant analysis was run on flow rate samples to verify if each value was correctly classified in the two periods. The *t*-tests were run to compare means of the two samples and to test the null hypothesis that the two means were equal. The results were verified with a Kolmogorov–Smirnov test and the distributions of the two samples were compared, for both the Q_{mean} and the Q_{max}. Finally, a box and whisker plot was run to demonstrate a significant difference ($p < 0.05$) between the flow rate data coming from the two periods considered.

Moreover, the river flow rate, SSY and climate data were also analyzed with linear trend analysis in order to show linear regression in the examined periods with 95% confidence limits and the prediction limits of the least squares fit model. Trend-line slopes (b) were calculated by the least-square linear fitting method.

3. Results and Discussion

3.1. LULC Changes

At present, more than 60% of the entire mountain Reno catchment is covered with forest [31], while in the past, forest was scarce due to exploitation (Figure 3). Cultivated land in the RMB catchment decreased from about 37% in 1954 to 5% in 2003, partially losing space to forest, which is, in fact, mainly 50–60 years old (Figure 4). In detail, forests and fallows increased from 39.5% to 57% and from 19% to 28%, respectively. Urban areas increased from 0.45% (1954) to 6.5% (2003), with the majority of population concentrated in the Reno valley. Lastly, water surfaces were found to be 40% smaller—they reduced from 9.3 km^2 in 1954 to 5.7 km^2 in 2003 (Figure 4a). The disappearance of rocky outcrops that were predominantly clayey badlands, the so called "Calanchi," is also interesting: They decreased from 1.67% to 0.46% of the RMB area, a phenomenon that was repeated throughout the Apennines.

Figure 3. Photos of Tresca Mountain (1473 m. a.s.l.) near Porretta Terme (RMB): In the past, say 1914 (**left**), forests were sparse due to exploitation, while currently the area is forested (**right**).

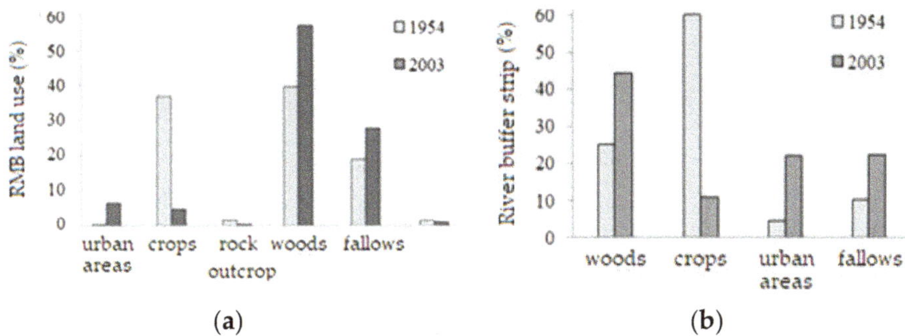

Figure 4. LULC changes in the RMB between 2003 and 1954 (**a**), and the LULC of the riparian buffer strips (**b**).

LULC changes noted in the riparian buffer strips between 1954 and 2003 were an increase of forests, urban areas, and uncultivated land, at the expense of cultivated land (that decreased from 60.25% to 11%) (Figure 4b). That decrease was the most prominent near the city of Bologna (reach R3, Figure 2), where cultivated land decreased from 85% (1954) to 20% (2003), while urban areas and woods showed an increase, from 7% to 42%, and from 8% to 38%, respectively.

The most widespread species (40–50%) in the riparian buffer strips are *Populus nigra* and *Salix alba*. Besides them, other species present are *Quercus pubescens* (about 20%) and *Salix alba* (10–15%), that form mixed populations, and different arboreal and shrub forms. *Alnus glutinosa* (about 10%) is present in the upper river banks where the anthropic impact is less. The shrub layer is composed of *Sambucus nigra*, *Corylus avellana*, *Cornus sanguinea*, and *Prunus spinosa*. Exotic species that have found, in this fluvial habitat, the ideal conditions of development, are *Robinia pseudoacacia* and *Acer negundo*, while among the shrubby, an infesting species is *Amorpha fruticose*.

3.2. Hydrology Changes

The monthly mean flow rate of the Reno River is 23.4 $m^3\ s^{-1}$, while the maximum monthly value recorded was 143 $m^3\ s^{-1}$ in December 1959 (Table 2). Between 1921 and 2013, the mean yearly flow rate was reduced by 11 $m^3\ s^{-1}$ (b = −0.12) or 36% (Figure 5), while the Q_{max} reduction was about 30.3%. The correlation coefficient (R^2) values of linear regression for Q_{mean} and Q_{max} were 0.22 and 0.08, respectively. Since the *p*-values were both less than 0.05, there was a statistically significant relationship at 95% confidence level. The monthly flow rate values are given in the box and whisker plot (Figure 6a). Starting from the hypothesis that the renaturalization and consequent hydrological changes caused by agricultural land abandonment in the 1950s were detectable after 1960, river flow rate data was divided into two sub-periods: Before and after this date. The Q_{mean} values were 26.03 $m^3\ s^{-1}$ and 21.08 $m^3\ s^{-1}$, for the 1921–1959 and 1960–2013 periods, respectively (Figure 6b). The F-test and *p*-value were, respectively, equal to 9.79 and 0.0026, according to the ANOVA test. Since the *p*-value was less than 0.05, there was a statistically significant difference between the two flow rate means.

Table 2. Summary statistics of mean and maximum annual flow rate (Q) and suspended sediment yield (SSY)—average yearly data.

Parameter	Years Available	Average	Minimum	Maximum
Q_{mean} ($m^3\ s^{-1}$)	77	23.4 ± 7.0	13.3 (in 1938)	42.4 (1937)
Q_{max} ($m^3\ s^{-1}$)	77	68.7 ± 25.96	21.8 (in 2007)	143.1 (in 1959)
SSY (Mg km^{-2})	31	935 ± 440	217 (in 1962)	2250 (in 1951)

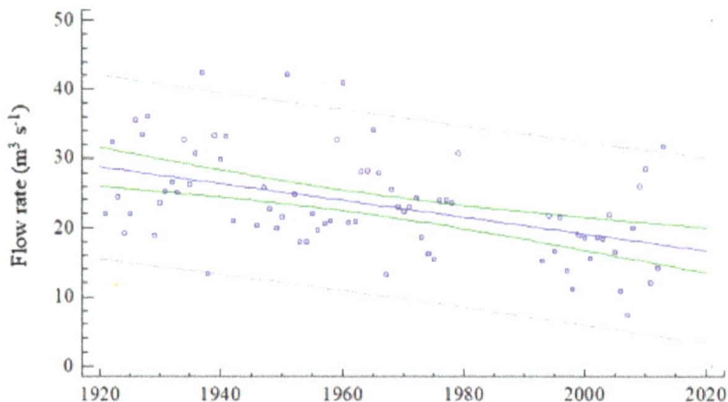

Figure 5. Time series of the Reno River mean yearly flow rate and linear trends with 95% confidence limits (green) and the prediction limits (grey) of the least squares fit model.

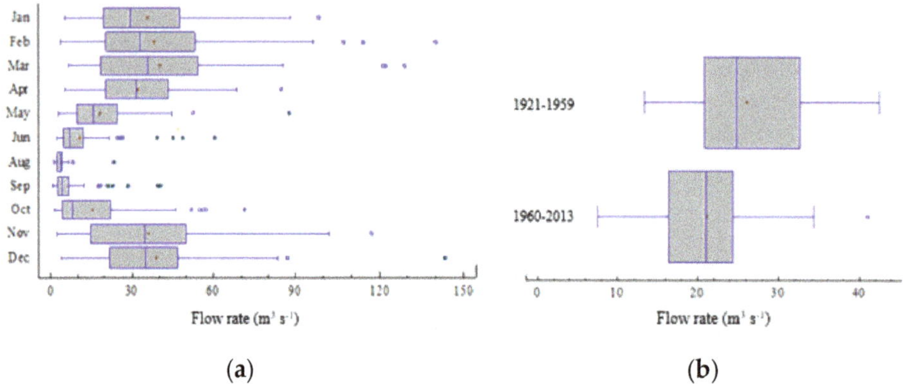

Figure 6. Box and whisker plots of monthly flow rates between 1921 and 2013 (**a**), and of the subpopulations: 1921–1959 and 1960–2013 (**b**).

Discriminant analysis was run on the two groups of Q_{mean}: Before and after 1960. From the 77 values used to fit the model, 72.7% were correctly classified in the groups, out of which 66.7% and 78% were for 1921–1959 and 1960–2013, respectively. A statistically significant difference ($p < 0.05$) was found between the two groups, for both Q_{mean} and Q_{max} (Figure 6b). It was evident that the lowest flow rate values were mostly concentrated during the 1960—2013 sub-period. In fact, there was a marked change in the trend of Q_{max} and Q_{mean} around 1960. Additionally, significant differences between the two periods were shown by discriminant analysis, as dispersion and as data trend. Dispersion of the mean flow rate data for the first period (1921–1959) was higher in respect to the second one (1960–2013), as evidenced by correlation coefficients R^2 (0.04 and 0.20 respectively), and therefore the two groups are statistically different. This difference, particularly the higher value of dispersion data of the first period, indicated presence of floods events, including disastrous ones [30].

3.3. Suspended Sediment Yield (SSY)

LULC change is an important factor that affects soil erosion-runoff and SSY. The relationships between climate and LULC changes on one side and SSY on the other, was investigated by PJ Ward et al. [32], with use of geo-referenced model WATEM/sedem. The authors found that the sediment increase in the Meuse, a northern European river, was almost entirely due to LULC change (conversion of forests to agricultural land). They concluded that increase of riparian buffer strip and development of riparian vegetation can result in the reduction in SSY, as they can be used as barrier to soil runoff [33].

Similarly, in the Apennines, SSY can be used to assess a real loss of the catchment soil due to runoff, rills erosion, gullies, and badlands, but also losses due to agricultural land abandonment and LULC change [33,34]. Currently, soil erosion prevails in the badlands and agricultural areas that are concentrated on the slopes, and that are easily accessible to mechanization. The average yearly SSY in the period 1942–1978 was 934.3 Mg km^{-2}. The maximum monthly value was 960 Mg km^{-2} in December 1976 and the yearly maximum value was 2225 Mg km^{-2} in 1951 (Table 2). Yearly and seasonal SSYs are given in the Figure 7a. It can be seen that the analysis of the SSY linear trend indicated a 17.5% reduction during the year 31 of data, or a 38% reduction in the period 1921–2013.

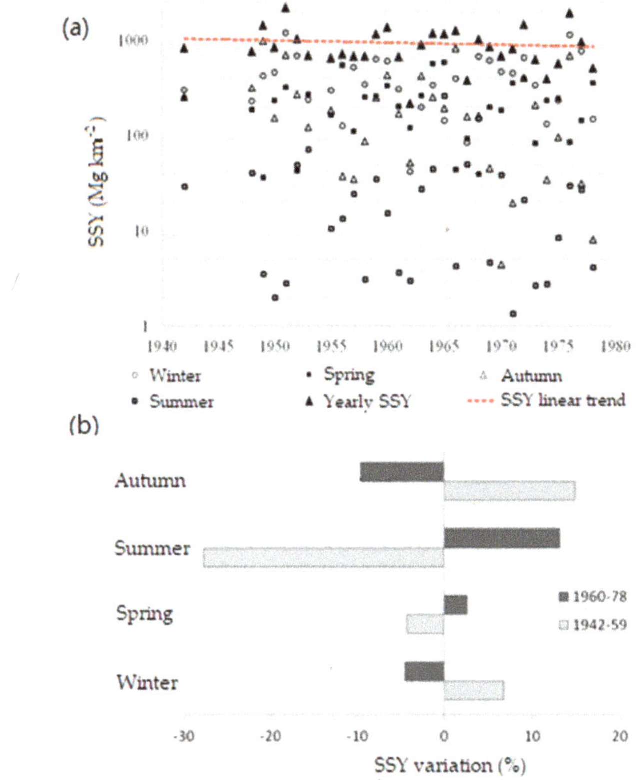

Figure 7. Seasonal SSY and the linear trend of yearly SSY (**a**). Seasonal SSY variations of 1942-59 and 1960-78 years respect to seasonal average of 1942 to 1978 years (**b**).

For the period 1942-1978, the seasonal SSY averages were: 426 Mg km^{-2} (winter), 231 Mg km^{-2} (spring), 32 Mg km^{-2} (summer) and 244 Mg km^{-2} (autumn) (Figure 7a). The average SSYs for the two sub-periods 1942–1959 and 1960–1978 (Figure 7b) were 981 and 902 Mg km^{-2} respectively. Moreover, some interesting seasonal variations occurred. For example, compared to the 1942–1978 seasonal average, SSY in the second period was lower in winter and autumn (−4.5% and −9.6%, respectively). On the other hand, it increased in spring and summer by 2.6% and 13.2% respectively (Figure 7b).

In general, the highest average monthly values of SSY were in February and November (213 and 188 Mg km^{-2}, respectively), but they decreased by 37% and 32% in the two following decades. If seasons of the two periods are compared, it can be noted that SSY reduced in summer and spring, while it increased in winter and autumn (Figure 7b). Finally, linear relations between the yearly SSY and Q_{mean} and Q_{max}, showed that SSY is better correlated to the maximum flow rate (Figure 8).

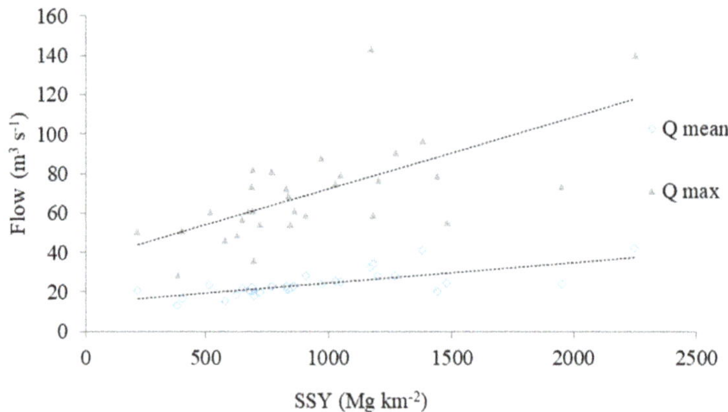

Figure 8. Linear relation between SSY and mean (Q_{mean}) and maximum (Q_{max}) flow rate.

The average annual erosion in the RMB, calculated on the basis of 31 year of SSY data, was about 9.3 Mg ha^{-1} year^{-1} or 0.6 mm year^{-1}. Various bibliography estimates available for the Region of Emilia Romagna give different results. The European Agency for the Environment, using the model Pesera [35], estimated a soil loss of 2.42 Mg ha^{-1} year^{-1}, slightly below the average Italian value (3.11 Mg ha^{-1} year^{-1}). In addition, the Emilia Romagna Region (including the RMB) was defined as a soil erosion risk area [36]. It was found that around 21% of the region has a medium to high risk of soil loss. The average loss in higher grounds of the region was estimated to be around 6 Mg ha^{-1} year^{-1} [37]. The difference between the erosion value reported in this study and other researchers' ones is mainly due to the heterogeneity of estimation models and basic data being estimated. Considering the current conditions of vegetation cover, they are all lower than those calculated on the basis of historical data from the 1942–1978 period.

3.4. Climate Change

An important aspect of a basin water budget is climate: Temperature and precipitation trends. In fact, temperature variations influence hydrology and river flow rate, and since they have an impact on evapotranspiration from vegetation, water surfaces and soil. Figure 9a shows a linear increase for both minimum and maximum mean yearly temperature in the RMB. That is visible for the mountain and valley gauge. The minimum temperature (T_{min}) showed a similar rising trend for the two stations: +4 °C/100 years (R^2 = 0.49) for the Mountain gauge and +5 °C/100 years (R^2 = 0.59) for the Valley gauge. On the other hand, T_{max} trends were more complex—decreasing in the mountain areas (−1.9 °C/100 years; R^2 = 0.15) and showing an increase in the valley (+2.8 °C/100 years; R^2 = 0.21). Since only the minimum, hence night, temperatures increased in the mountain areas where vegetation is also more developed than in the valley, and since night-time evapotranspiration is much smaller than in the day-time, the temperature increase most probably did not have a big influence on the observed flow rate reduction.

Precipitation in the RMB was lowered by 10.67% between 1921 and 2013 (Figure 9b), corresponding to 145 mm reduction in 92 years. However, that value is not statistically significant (R^2 = 0.034). A strong reduction (from about 0.6 to 0.4) in the catchment runoff coefficient (flow rate/precipitation), that was observed in the last 90 years (R^2 = 0.43) (Figure 10), and is mainly due to reduction of the flow rate.

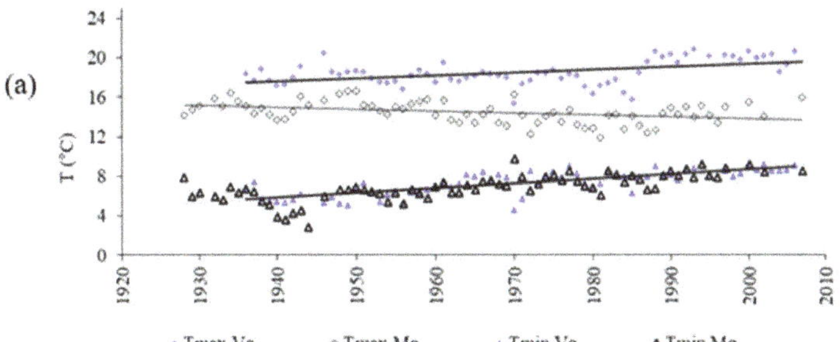

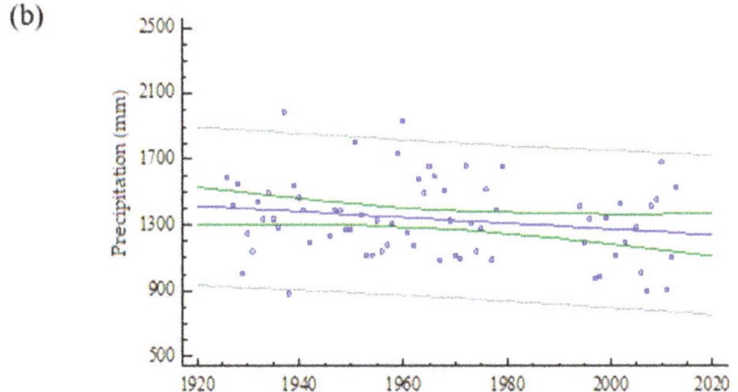

Figure 9. Local climate change: Linear trends of maximum and minimum temperature (T_{max} and T_{min}) for mountain (Mg) and valley (Vg) gauges (**a**); yearly RMB precipitation with linear trend with 95% confidence limits (green) and the prediction limits (grey) of the least squares fit model (**b**).

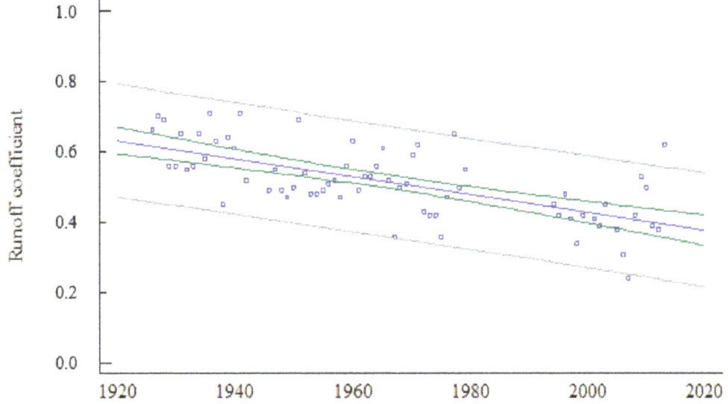

Figure 10. Runoff coefficient observed in the last 90 years (R^2 = 0.43) with linear trends, prediction and confidence limits.

3.5. Morphological Stream Changes: Riparian Buffer Strips

The main proprieties and changes of the three reaches considered, based on aerial and satellite images, and field surveys, are reported in Table 3. Figure 11 gives the relationship (exponential equation, R^2 = 0.75) between the average width and mean altitude of the Reno valley. The two values that are above the curve are of two villages (Porretta and Pioppe di Salvaro) that are in a larger area due to tributary torrents. The changes of the Reno River reaches (R1–R3) concern the banks and the river morphology (Figure 12), from upstream to downstream:

- R1 is a more torrential reach of the river; the valley is narrower and currently characterized by woods and meadows. The normal riverbed flow occupied 210 ha in 1954, but it reduced to 80 ha in 2003 (Table 3). It was and is predominantly covered by gravel pits, sand deposits, and rock outcrops. Riparian forests currently cover an overall surface of 53.8 ha (Figure 12), while in 1954 they were absent due to farming activities.
- R2 is the middle reach. Its normal flow riverbed in 1954 occupied 206 hectares, and it was predominantly made of gravel pits and sand deposits. Instead, in 2003, the area occupied by the riverbed reduced to 78 ha (Table 3). On stabilized alluvial deposits of the river stream, where occasional floods occur, there are typical igrophilous-forests, consisting of elms, poplars, and willows (Figure 13).
- R3 is a fluvial stretch in which the valley widens and then turns to the Po Valley. The riverbed area decreased from 299 ha to 61 ha between 1954 and 2003 (Table 3). In 1954 the riverbed consisted of gravel bars and sand deposits, while currently clay and silt prevail. In addition, the river channel changed from braided to a single one. Riparian forest showed a strong development: it was discontinuous in 1954 and inadequate as a buffer zone, while currently it is well developed and forms a continuous wooded area (Figure 12). The fluvial park and most of riparian forests are now protected by the EU Habitats Directive (Habitat Code 92A0).

Table 3. Main features of the Reno reaches.

Feature	R1		R2		R3	
	1954	2003	1954	2003	1954	2003
Length (km)	20.9		23.0		10.6	
Mean bed slope (%)	0.53		0.45		0.31	
Average valley width (m)	166		423		1476	
Average channel width (m)	90–100	30–40	90–300	30–70	300	90
Average channel area (ha)	210	80	206	78	299	61

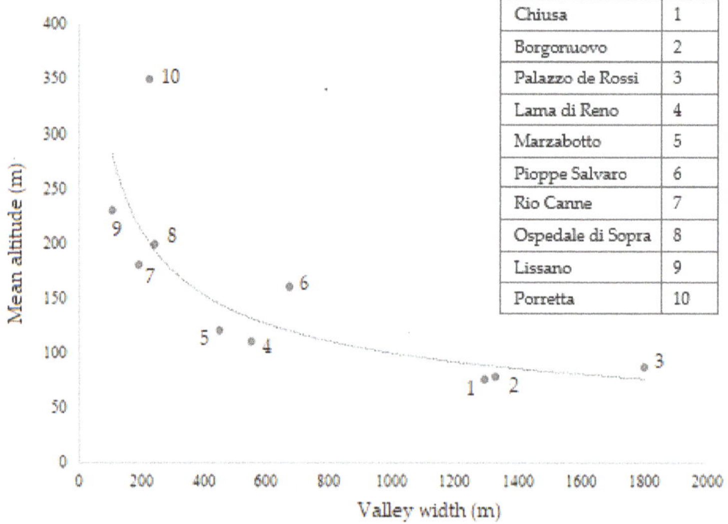

Figure 11. Reno River transect between Casalecchio Chiusa and Porretta Terme, with corresponding locations.

Figure 12. Reach R1 near Porretta that is more torrential (**left**) and R3 near Chiusa Casalecchio with fine sediments (**right**): Riparian forest shows a strong development.

Figure 13. Reach 2 near Marzabotto village. Wide stream and area covered with crops in 1954 (**left**), but urban areas and a narrow stream in 2003 (**right**).

Based on the RMB's soil use maps, it was estimated that the riverbed area decreased by about 40% from 1954 to 2003. However, based on the river transects, the reduction was as high as 80%. Fluvial banks with woods along the Reno were mostly absent in 1954, and the area was used for farming. In 2003 riparian forests appear to be well developed along the entire stream in the RMB. This is especially visible on the right-hand side (looking downstream) of the river, where fluvial terraces are narrower or absent, and are therefore under a lower human impact. In general, it is observed that the river bed gives way to riparian forests. For example, a reduction in the active riverbed corresponds to formation and/or expansion of the riparian buffer strips, and the stream reaches are colonized by riparian vegetation.

4. Conclusions

This study has examined relationships between two major geomorphological changes (channel narrowing and formation of wide vegetated banks) that took place in the Reno River mountain basin in the last century, and the hydrological, climatic, and basin re-naturalization factors that have contributed to these changes. The two phenomena are strongly and positively covariant, indicative of cause and effect, and in fact, wide vegetated banks are formed at the expense of the riverbed. While riparian buffer strips were mostly absent in 1954, currently they are well developed along the entire stream. In addition, the shape of the river channel changed from braided to a single one and the width of the river bed was reduced by around 80%. The riverbed in the past, mostly consisted of gravel bars and sand deposits, and currently clay and silt prevail at the basin outlet.

Based on this study, the steering factors for those changes and significant aspects were:

- LUCL changes between 1954 and 2003: A reduction in the agricultural land use (from 37 to 5%), an increase of forest cover (from 40% to 57%), and development of riparian vegetation.
- Considerable reduction in SSY (−38%) and flow rate (−36%) during the last 90 years, and a consequent change of runoff coefficient (reduction from about 0.6 to 0.4), was an important parameter for hydraulic watershed management.

The effect of agricultural land abandonment that occurred in the 1950s can be recognized after 1960, confirming the initial hypothesis that this year can be taken as a starting point for the basin change. After that date a decrease in the Reno flow rate was observed and dispersion of data is significantly reduced. All statistical analyses confirm that the hydrological flow data measured after 1960 (period 1960 to 2013) are significantly different from those measured when the basin was still heavily agricultural (period 1921 to 1958). However, although this study has identified the human factor as one of the main causes of the above-mentioned changes, it can often be challenging to separate human from naturally driven activities, and future research is needed in order to do it. The geomorphological evolution of the Reno River shows how these changes are mainly related to the hydrological dynamics and catchment re-naturalization. Although climate changed in the period studied (precipitation reduction of 10.67% and 4–5 °C increase of T_{min}) it had little bearing on the observed environmental changes.

This study, applied to a typical North Apennine river, illustrates the effectiveness of combining historical data (hydrological and climate data, so as aerial and satellite images) on the one hand, and the use of modern technology (geographic information systems) and direct surveys on the other. Combining these techniques can certainly contribute to a sustainable management of river systems. Although further research is needed, this study gives an insight to the past and present factors that regulate water course, its hydrology, and morphology. An in-depth knowledge of this factors can certainly make it possible to predict the evolution and dynamics of the Reno river flow and its morphology.

Author Contributions: Conceptualization, D.P.; methodology, D.P. and C.C.; validation, D.P.; formal analysis, D.P.; data curation, D.P. and C.C.; writing—original draft preparation, D.P.; writing—reviewing and editing, D.P., S.L., and A.T.; visualization, D.P. and S.L.; supervision, D.P. and A.T.

Funding: This research received no external funding.

Acknowledgments: This research was undertaken as part of the collaboration agreement for teaching and study with the Emilia Romagna Region, Civil Protection, Basin Authority of the Reno River and ARPAE-RER, (Italy).

Conflicts of Interest: The authors declare no conflict of interest.

References

1. Hynes, H.B.N. The stream and its valley. *Verh. Internat. Verein. Limnol.* **1975**, *19*, 1–15. [CrossRef]
2. Vannote, R.L.; Minshall, W.G.; Cummins, K.W.; Sedell, J.R.; Cushing, C.E. The river continuum concept. *Can. J. Fish. Aquat. Sci.* **1980**, *37*, 130–137. [CrossRef]
3. Allan, J.D. Landscapes and riverscapes: The influence of land use on stream ecosystems. *Annu. Rev. Ecol. Evol. Syst.* **2004**, *35*, 257–284. [CrossRef]
4. Graf, W.L. Downstream hydrologic and geomorphic effects of large dams on American rivers. *Geomorphology* **2006**, *79*, 336–360. [CrossRef]
5. Schmidt, J.C.; Wilcock, P.R. Metrics for assessing the downstream effects of dams. *Water Resour. Res.* **2008**, *44*, W04404. [CrossRef]

6. Burke, M.; Jorde, K.; Buffington, J.M. Application of a hierarchical framework for assessing environmental impacts of dam operation: Changes in stream flow, bed mobility and recruitment of riparian trees in a western North American river. *J. Environ. Manag.* **2009**, *90*, S224–S236. [CrossRef]
7. Martínez-Fernández, V.; Maroto, J.; García de Jalón, D. Fluvial corridor changes over time in regulated and nonregulated rivers (Upper Esla River, NW Spain). *River Res. Appl.* **2017**, *33*, 214–223. [CrossRef]
8. Piégay, H.; Walling, D.E.; Landon, N.; He, Q.; Liébault, F.; Petiot, R. Contemporary changes in sediment yield in an alpine montane basin due to afforestation (the upper Drôme in France). *Catena* **2004**, *55*, 183–212. [CrossRef]
9. Keestra, S.D.; Van Huissteden, J.; Vandenberghe, J.; Ol, V.D.; De Gier, J.; Pleizier, I.D. Evolution of the morphology of the river Dragonja (SW Slovenia) due to land-use changes. *Geomorphology* **2005**, *69*, 191–207. [CrossRef]
10. Pont, D.; Piégay, H.; Farinetti, A.; Allain, S.; Landon, N.; Liébault, F.; Dumont, B.; Richard-Mazet, A. Conceptual framework and interdisciplinary approach for the sustainable management of gravel-bed rivers: The case of the Drôme River basin (S.E. France). *Aquat. Sci.* **2009**, *71*, 356–370. [CrossRef]
11. Piqué, G.; Batalla, R.J.; Sabater, S. Hydrological characterization of dammed rivers in the NW Mediterranean region. *Hydrol. Process.* **2015**, *30*, 1691–1707. [CrossRef]
12. Debolini, M.; Schoorl, J.M.; Temme, A.; Galli, M.; Bonari, E. Changes in agricultural land use affecting future soil redistribution patterns: a case study in Southern Tuscany (Italy). *Land Degrad. Dev.* **2013**, *26*, 574–586. [CrossRef]
13. García-Ruiz, J.M.; Lana-Renault, N. Hydrological and erosive consequences of farmland abandonment in Europe, with special reference to the Mediterranean region, a review. *Agric. Ecosyst. Environ.* **2011**, *140*, 317–338. [CrossRef]
14. Liébault, F.; Piégay, H. Causes of 20th century channel narrowing in mountain and piedmont rivers of southeastern France. *Earth Surf. Process. Landf.* **2002**, *227*, 425–444. [CrossRef]
15. Tomer, M.D.; Schilling, K.E. A simple approach to distinguish land-use and climate-change effects on watershed hydrology. *J. Hydrol.* **2009**, *376*, 24–33. [CrossRef]
16. Zhang, Y.K.; Schilling, K.E. Increasing streamflow and baseflow in Mississippi River since the 1940s: Effect of land use change. *J. Hydrol.* **2006**, *324*, 412–422. [CrossRef]
17. Zhang, T.; Zhang, X.; Xia, D.; Liu, Y. An analysis of land use change dynamics and its impacts on hydrological processes in the Jialing River basin. *Water* **2014**, *6*, 3758–3782. [CrossRef]
18. Poyatos, R.; Latron, J.; Llorens, P. Land use and land cover change after agricultural abandonment—The case of a Mediterranean mountain area (Catalan pre-Pyrenees). *Mt. Res. Dev.* **2003**, *23*, 362–368. [CrossRef]
19. Vicente-Serrano, S.M.; Lasanta, T.; Romo, A. Analysis of spatial and temporal evolution of vegetation cover in the Spanish Central Pyrenees: Role of human management. *Environ. Manag.* **2004**, *34*, 802–818. [CrossRef]
20. Lasanta-Martínez, T.; Vicente-Serrano, S.M.; Cuadrat-Prats, J.M. Mountain Mediterranean landscape evolution caused by the abandonment of traditional primary activities: A study of the Spanish Central Pyrenees. *Appl. Geogr.* **2005**, *25*, 47–65. [CrossRef]
21. Pavanelli, D.; Capra, A. Climate change and human impacts on hydroclimatic variability in the Reno River catchment, Northern Italy. *Clean Soil Air Water* **2014**, *42*, 535–545. [CrossRef]
22. Picco, L.; Comiti, F.; Mao, L.; Tonon, A.; Lenzi, M.A. Medium and short term riparian vegetation, island and channel evolution in response to human pressure in a regulated gravel bed river (Piave River, Italy). *Catena* **2017**, *149*, 760–769. [CrossRef]
23. Morán-Tejeda, E.; Zabalza, J.; Rahman, K.; Gago-Silva, A.; López-Moreno, J.I.; Vicente-Serrano, S.; Lehmann, A.; Tague, C.L.; Beniston, M. Hydrological impacts of climate and land-use changes in a mountain watershed: Uncertainty estimation based on model comparison. *Ecohydrology* **2014**, *8*, 1396–1416. [CrossRef]
24. Schilling, K.E.; Libra, R.D. Increased baseflow in Iowa over the second half of the 20th century. *J. Am. Water Res. Assoc.* **2003**, *39*, 851–860. [CrossRef]
25. Raymond, P.A.; Oh, N.H.; Turner, R.E.; Broussard, W. Anthropogenically enhanced fluxes of water and carbon from the Mississippi River. *Nature* **2008**, *451*, 449–452. [CrossRef] [PubMed]
26. Fuller, I.; Macklin, M.G.; Richardson, J.M. The geography of the Anthropocene in New Zealand: Differential river catchment response to human impact. *Geogr. Res.* **2015**, *53*, 255–269. [CrossRef]

27. RER (Regione Emilia Romagna) 5th General Census of Agriculture (in Italian) 2000. Available online: http://agricoltura.regione.emilia-romagna.it/entra-in-regione/agricoltura-in-cifre/censimenti-generali-dell-agricoltura/censimenti-generali-agricoltura) (accessed on 3 May 2019).
28. Pavanelli, D.; Cavazza, C.; Correggiari, S.; Rigotti, M. Overland flow control via surface management techniques over the last century in the tuscan-emilian Apennines range: The Rio Maggiore case study. In Proceedings of the COST Action 634 Erosion International Conference, Prague, Czech Republic, 1–3 October 2007; pp. 157–176.
29. Eppes, M.C.; Bierma, R.; Vinson, D.; Pazzaglia, F. A soil chronosequence study of the Reno valley, Italy: Insights into the relative role of climate versus anthropogenic forcing on hillslope processes during the mid-Holocene. *Geoderma* **2008**, *147*, 97–107. [CrossRef]
30. European Environment Agency Corine Land Cover (CLC) 2006. Available online: https://land.copernicus.eu/pan-european/corine-land-cover/clc-2006?tab=metadata (accessed on 28 March 2019).
31. RER (Regione Emilia Romagna) Summary of Land Use—Forest Surface (in Italian) 2003. Available online: http://ambiente.regione.emilia-romagna.it/it/parchi-natura2000/foreste/quadro-conoscitivo/inventari-e-carte-forestali/inventario-forestale/estratto_del_AssLeg_90-2006.pdf) (accessed on 4 May 2019).
32. Ward, P.J.; van Balen, R.T.; Verstraeten, G.; Renssen, H.; Vandenberghe, J. The impact of land use and climate change on late Holocene and future suspended sediment yield of the Meuse catchment. *J. Environ. Qual.* **2008**, *37*, 1894–1908. [CrossRef]
33. Pavanelli, D.; Cavazza, C. River suspended sediment control through riparian vegetation: A method to detect the functionality of riparian vegetation. *CLEAN Soil Air Water* **2010**, *38*, 1039–1046. [CrossRef]
34. Pavanelli, D.; Pagliarani, A. Monitoring water flow, Turbidity and Suspended Sediment Load, from an Apennine Catchment, Italy. *Biosyst. Eng.* **2002**, *83*, 463–468. [CrossRef]
35. Gobin, A.; Govers, G.; Kirkby, M.J.; Le Bissonnais, Y.; Kosmas, C.; Puigdefabregas, J.; Van Lynden, G.; Jones, R.J.A. *PESERA Pan European Soil Erosion Risk Assessment Project Technical Annex*; European Commission: Brussels, Belgium, 1999.
36. RER (Regione Emilia Romagna) Regional Programme of the Rural Development 2007–2013 of Emilia Romagna, Analysis of the Social-Economic, Agricultural and Environmental Context 2007. Available online: http://agricoltura.regione.emilia-romagna.it/psr/doc/organismi-e-strumenti/monitoraggio-e-valutazione/doc-ex-ante/rapporto-di-valutazione-ex-ante-testo-completo) (accessed on 18 April 2019). (In Italian).
37. Grimm, M.; Jones, R.J.A.; Rusco, E.; Montanarella, L. *Soil Erosion Risk in Italy: A Revised USLE Approach*; European Soil Bureau Research Report No.11, EUR 20677 EN; Office for Official Publications of the European Communities: Luxembourg, 2003.

Article

Hydrologic Impacts of Land Use Changes in the Sabor River Basin: A Historical View and Future Perspectives

Regina Maria Bessa Santos [1], Luís Filipe Sanches Fernandes [1], Rui Manuel Vitor Cortes [1] and Fernando António Leal Pacheco [2,*]

1 Centre for the Research and Technology of Agro-Environment and Biological Sciences, University of Trás-os-Montes and Alto Douro, 5001-801 Vila Real Ap. 1013, Portugal

2 Chemistry Research Centre, University of Trás-os-Montes and Alto Douro, 5001-801 Vila Real Ap. 1013, Portugal

* Correspondence: fpacheco@utad.pt

Received: 3 July 2019; Accepted: 10 July 2019; Published: 15 July 2019

Abstract: The study area used for this study was the Sabor river basin (located in the Northeast of Portugal), which is composed mostly for agroforestry. The objectives were to analyze the spatiotemporal dynamics of hydrological services that occurred due to land use changes between 1990 and 2008 and to consider two scenarios for the year 2045. The scenarios were, firstly, afforestation projection, proposed by the Regional Plan for Forest Management, and secondly, wildfires that will affect 32% of the basin area. In this work, SWAT (Soil and Water Assessment Tool) was used to simulate the provision of hydrological services, namely water quantity, being calibrated for daily discharge. The calibration and validation showed a good agreement for discharge with coefficients of determination of 0.63 and 0.8 respectively. The land use changes and the afforestation scenario showed decreases in water yield, surface flow, and groundwater flow and increases in evapotranspiration and lateral flow. The wildfire scenario, contrary to the afforestation scenario, showed an increase in surface flow and a decrease in lateral flow. The Land Use and Land Cover (LULC) changes in 2000 and 2006 showed average decreases in the water yield of 91 and 52 mm·year^{-1}, respectively. The decrease in water yield was greater in the afforestation scenario than in the wildfires scenario mainly in winter months. In the afforestation scenario, the large decrease varied between 28 hm^3·year^{-1} in October and 62 hm^3·year^{-1} in January, while in the wildfires scenario, the decrease was somewhat smaller, varying between 15 hm^3·year^{-1} in October and 49 hm^3·year^{-1} in January.

Keywords: SWAT; water balance components; Land Use and Land Cover changes; wildfires; afforestation

---

1. Introduction

Land Use and Land Cover (LULC) are considered the most critical factors affecting the intensity and frequency of surface flow, as well as soil erosion and the loss of nutrients [1,2]. Watershed-level studies have indicated that rapid LULC changes could have significant impacts on water resources [3] and were reported to be an essential factor for controlling water resources on local and global scales during the last century [4,5]. The factors responsible for land use changes are population density fluctuations, changes in the agricultural policy, and the conditions of national and international markets. The main changes occurred due to the abandonment of farmland in less productive mountain areas, the expansion of some subsidized crops to marginal lands, and intense soil erosion during extreme rainstorm events.

Several studies developed in the Mediterranean region showed that inadequate land use and soil cover accelerate water erosion processes and, consequently, lead to land degradation [1,2,6]. Soil erosion is one of the leading causes of the reduction of water quality due to the amount of sediment that arrives at the watercourses and reservoirs [1]. However, many authors have demonstrated that both runoff and sediment loss decrease exponentially as the percentage of vegetation cover increases [1,6]. Thus, forests are often used as a management strategy to improve the provision of ecosystem services in watersheds, because they have a substantial widespread positive influence on climate, hydrology, soils, and biodiversity [7–10].

The studies developed in the 19th century, based on the too-hasty generalization of single point observation, believed that natural and planted forests increased total flow and base flow [3]. Nowadays, studies prove that forests have an impact on the water balance at the basin scale, as forest water consumption is generally higher than that of other vegetation types [3,10,11]. The rise in shrub and forest cover produces declines in water yield, surface, and groundwater flows, and an increase in transpiration. The decrease in the groundwater flow in forests can be justified by the fact that trees with deep roots and high transpiration rates may act as pumps that remove water from the soil and return it to the atmosphere [1,11,12].

In the Mediterranean, erosion is the consequence of complex interactions between environmental and human-related factors. The erosion processes are products of the occurrence of intense rainstorms and long-lasting droughts, high evapotranspiration, the presence of steep slopes, topographic diversity, and recent tectonic activity as well as the recurrent use of fire, deforestation, overgrazing, farming, and construction activities [1,2]. In agreement with the CORINE program, Spain and Portugal are the Mediterranean countries in the European Union facing the highest risk of erosion [13,14]. CORINE means "coordination of information on the environment", and one of the priorities of the program is the evaluation of natural resources and environmental problems in the southern part of the European Community (e.g., soil erosion, water resources, land cover, and coastal problems) [14]. In Portugal, areas of high erosion risk cover almost one-third of the country [15]. For this reason, soil erosion is one of the most intensively studied issues in the Mediterranean region, Portugal included [1,2,6,15]. Erosion and land degradation became a problem in Portugal when arable farming expanded into marginal areas, namely cereal cultivation until the middle of the twentieth century [1]. The introduction of modern agriculture led to the abandonment of traditional or semi-traditional agriculture in mountainous areas as well as in areas with difficult access, resulting in fundamental transformations to the landscape, characterized by the spread of natural vegetation, including both shrubland and forestland [1]. The natural afforestation has resulted in a decline in water resources and surface flow and has decreased soil loss and sediment delivery, as well as caused a progressive improvement in soil characteristics [1]. It has been extensively documented that the rise in shrub and forest cover has promoted lower soil losses and sediment yields than in arable land [2,6].

However, wildfires have been responsible for sudden increases in erosion rates, and they can also result in land degradation and sometimes desertification over the long-term [16]. The occurrence of forest fires affecting thousands of hectares each year is a significant problem in the Mediterranean basin [2,16]. In Portugal, fires are essentially human-caused, and their extension and severity are dependent on extreme weather [17]. In central Portugal, at Pedrógão Grande-Góis, a tragic wildfire occurred on June 17, 2017, with an official death toll of 64 people, almost 500 buildings destroyed, and a continuous patch of more than 42 thousand hectares burned in one week [18]. The areas burned correlated well with socioeconomic and environmental characteristics (e.g., population density and land use, climate, weather, topography, and vegetation cover) [16,17]. However, in this tragic fire, the climate and meteorology played key roles in the initiation and spreading of the wildfires [18]. The warmer and drier than average spring made the landscape prone to the occurrence of large fires and extreme weather conditions favored the ignition and spread of wildfires [18].

It is generally accepted that fires increase runoff, soil erosion, and nutrient and pesticide transport to the river [16,19]. In the Mediterranean region, the most substantial erosion rates and nutrient losses tend to occur during the first rainstorms after wildfire occurrence during the dry season. The increases in runoff are due to the reduced infiltration capacity of the soil caused by the removal of vegetation and soil organic matter by fire [16]. Wildfires lead to the release of nutrients through the combustion of vegetation and exposure of the soil to erosive factors. Consequently, wildfires are responsible for hydrologic problems and the degradation of water quality caused by excessive input of nutrients, such as phosphorus and nitrogen [16,20]. The excess nutrients affect primary production and, consequently, the eutrophication of aquatic systems. The relationship between the risk of eutrophication and nutrient exports from burned areas has been widely documented [16,19,20].

Hydrological models are computational tools that perform a mathematical representation of hydrological processes, such as infiltration of water into the soil, recharge of aquifers, runoff and drainage network flow [21,22], as well as hydrochemical processes such as weathering or contaminant transport [23–31]. They are also the basis of decision support systems, which can help watershed managers in the control of extreme events such as floods [32,33], or in the assessment of water resource availability [34–40] and threats to water quality [41–48]. The SWAT (Soil and Water Assessment Tool) was the hydrological model used to model the physical processes that occurred in the Sabor river basin. The SWAT model is one of the most widely used water quality watershed and river basin scale models worldwide [49]. It has been applied from small hydrographic basins to the continental scale. Some examples of applications are in North America [50], Europe [51], and Australia [52,53]. These studies were done in four main categories: hydrologic modelling, sediment transport, nutrient and pesticide transport, and scenario analyses.

Within this framework, the following three demands were addressed: (i) establishment of the hydrologic baseline with calibration over a 39-year period (1960–1999), (ii) assessment of the land cover and land use changes between 1960 and 2008 and their effects on water balance components, and (iii) assessment of the forecast of the afforestation and wildfire scenarios for 2045, also on water balance components. A majority of articles discussing environmental effects of land use change and dealing with scenario creation only assess some water balance components, namely water yield or surface flow. This work intends to study the changes occurring not only in water yield or surface flow but also in evapotranspiration, lateral flow, and in groundwater flow.

2. Materials and Methods

2.1. Study Area

The Sabor river basin is located in the northeast of Portugal and drains into the International Douro basin (Figure 1). The Sabor river has an extension of 212.6 km. Its source is located in Spain at an altitude of about 1600 m, and its mouth is located in the Douro River at about 88 m [54]. The Sabor river basin covers an area of about 3834.5 km^2, of which 3170.7 km^2 is in Portugal. The average slope of the basin is 16.2% according to the digital elevation model [54]. The slope below 10% is located in the upper zones and plateaus of the basin, and the most rugged slope is located on the scarps and banks of the Sabor river and its tributaries. The main tributaries are the Maças river (drainage area: 720 km^2) and the Angueira river (drainage area: 540 km^2). In the Sabor river, the Sabor hydroelectric system was built. It comprises two retention dams, known as upstream and downstream walls, which are located at 12.6 and 3 km from the mouth of the Sabor river, respectively [55].

The climate is close to the Mediterranean type. It is characterized by warm-dry summers and precipitation concentrated in the winter and spring seasons. The average annual basin values of rainfall and evapotranspiration are approximately 730 and 540 mm·year^{-1}, respectively [56].

The soil characterization in the Sabor river basin is mainly lithosols (87% of the catchment area), but also with the presence of cambisols (7.4%), alisols (3%), anthrosols (1.4%), and fluvisols (0.7%) [57]. The 1990 CORINE Land Cover data published by the European Environmental Agency [58] identified the area as having 59% agricultural areas, 32% semi-natural areas, 9% forest, and less than 1% artificial areas and water bodies (Figure 1b). Agriculture and livestock farms are the main economic activities in the region, with olive and almond being the main crops [59]. According to the 2011 demographic census, in the Sabor river basin, the population density was 20 inhabitants per km^2 [60].

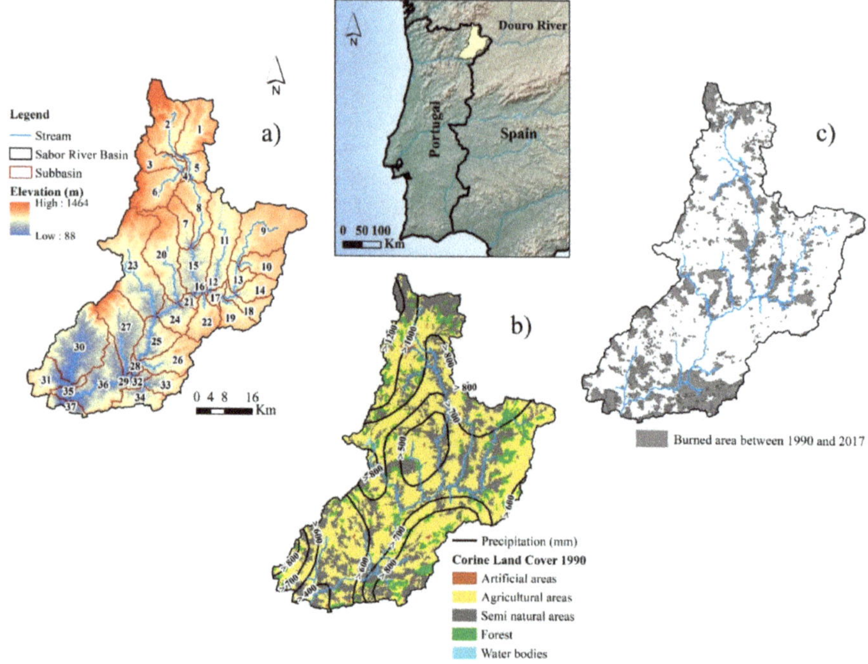

Figure 1. The spatial distribution of (**a**) the drainage network, the topography, and the sub-basins of the Sabor river basin; (**b**) the temperature and the CORINE Land Cover in 1990; and (**c**) the area burned between 1990 and 2017.

2.2. CORINE Land Cover Changes and Wildfires

In the study of LULC, changes were used to form the maps of the CORINE Land Cover for 1990, 2000, and 2006 [58]. These maps were reclassified to SWAT land cover classes, which is the format read from ArcSWAT. The reclassification is found in the Supplementary Material (worksheet 1). Figures 2 and 3 illustrate the changes between the SWAT land cover classes of 1990 and 2000 and between the SWAT land cover classes of 1990 and 2006. These changes represented 6% of the basin area for the SWAT land cover classes of 1990 and 2000 (Figure 2a), and 10% of the basin area between 1990 and 2006 (Figure 2b). For better visualization and analysis, in Figure 3, only areas greater than 500 hectares are shown. The major changes in the area of SWAT land cover classes from 1990 to 2000 comprise the replacement of range—brush by forest (7080 ha), essentially coniferous (FRSE, 4009 ha) and deciduous forests (FRSD, 2212 ha), as well as burned areas (BARR, 1643 ha). The major change in the area of SWAT land cover classes from 1990 to 2006 was the increase in range—brush (RNGB, 14,384 ha) mainly from range—grasses (RNGE, 3635 ha), agricultural land—generic (AGRL, 2584 ha), forest—mixed (FRST, 2395 ha), forest—deciduous (FRSD, 1915 ha), forest—evergreen (FRSE, 1725 ha), and burned area (BARR, 1421 ha). As well as the replacement of agricultural land—row crops (AGRR) in agricultural land—generic (AGRL, 4155 ha).

The cartography of the burned areas covering the period 1990–2017 is available at the Conservation of Nature and Forests [61] (Table 1). In this period in the Sabor river basin, 103,033 ha was burned, which corresponds to 32% of the basin area (Figure 1c). The largest burned area (approximately 13,000 ha) occurred in 2013, followed by burned areas above 6700 ha, which occurred in 1994, 1998, 2000, 2012, and 2017 (Figure 4).

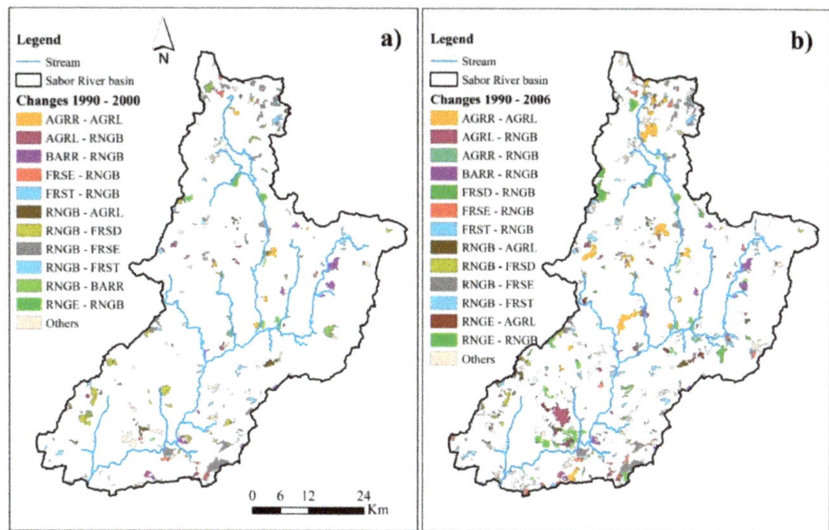

Figure 2. The spatial distribution of changes (**a**) between the CORINE Land Cover 1990 and 2000; and (**b**) between the CORINE Land Cover 1990 and 2006, in the Sabor river basin. Only the areas with changes greater than 500 hectares are presented. The SWAT codes are agricultural land—row crops (AGRR), agricultural land—generic (AGRL), range—brush (RNGB), range—grasses (RNGE), barren (BARR), forest—deciduous (FRSD), forest—evergreen (FRSE), and forest—mixed (FRST).

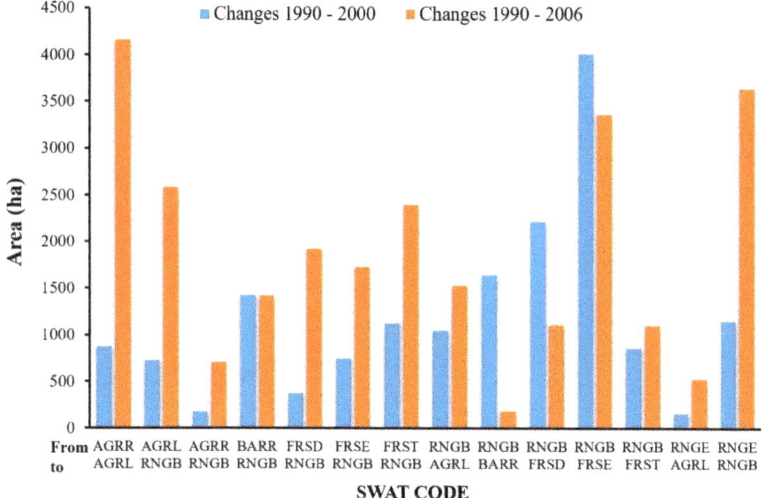

Figure 3. Graphic of changes between the CORINE Land Cover in 1990 and 2000, and between the CORINE Land Cover in 1990 and 2006 in the Sabor river Basin. Only the areas with changes greater than 500 hectares are presented. The SWAT codes are agricultural land—row crops (AGRR), agricultural land—generic (AGRL), range—brush (RNGB), range—grasses (RNGE), barren (BARR), forest—deciduous (FRSD), forest—evergreen (FRSE), and forest—mixed (FRST).

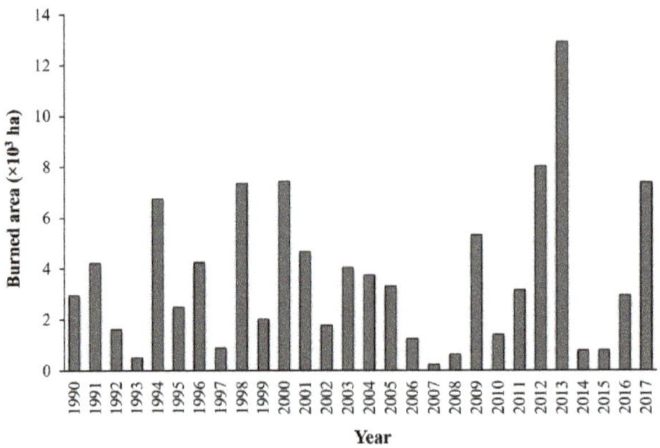

Figure 4. The area burned between 1990 and 2017 in the Sabor river basin.

Table 1. Maps and records were used in the construction of the hydrological model, in the evaluation of Land Use and Land Cover (LULC) changes between 1990 and 2008, as well as in the analysis of the afforestation and wildfire scenarios. All data types used have reference to the purpose, their owner institution, and the source of data.

Data Type	Purpose	Owner Institution	URL of Internet Website
Construction of the hydrological model			
Digital Elevation Model	Calculation of the elevation and slope	Directorate-General of Territory	http://www.dgterritorio.pt/
Drainage network	Definition of the stream network and delineation of the watershed	Portuguese Water Institute	http://geo.snirh.pt/AtlasAgua/
Corine Land Cover 1990	Creation of a land use grid for assessing vegetation cover type and calculation of curve number (CN)	European Environmental Agency	http://www.eea.europa.eu/
Soils	Creation of a soil grid for assessment of soil types and calculation of CN	Directorate-General of Territory	http://scrif.igeo.pt/
Records of weather stations	Construction of the hydrological model to estimate the water balance at the sub-basin level	Portuguese Water Institute	http://snirh.apambiente.pt/
Records of hydrometric stations	Calibration of the hydrological model	Portuguese Water Institute	http://snirh.apambiente.pt/
Evaluation of the Land Use and Land Cover changes between 1990 and 2008			
CORINE Land Cover in 2000 and 2006	Assessment of the LULC changes between 1990 and 2008	European Environmental Agency	http://www.eea.europa.eu/
Analysis of the afforestation and wildfires scenarios			
Regional Forestry Management Plan of Douro and Nordeste to 2045	Assessment of the impact of the afforestation projection on the water balance	Portuguese Institute for the Conservation of Nature and Forests	http://www.icnf.pt/
Cartography of burned areas between 1990 and 2017	Assessment of the impact of the wildfires scenario on the water balance	Portuguese Institute for the Conservation of Nature and Forests	http://www.icnf.pt/

2.3. Conceptual Framework Model

The conceptual framework model was developed to assess the effects of LULC changes and the forecast of afforestation and the occurrence of wildfires on water balance (Figure 5). SWAT was used to construct the hydrological model of the Sabor river basin with the following set of data: digital elevation model, land use, soil types, and weather station (Table 1). With these data, the drainage network, the delimitation of the basin and sub-basins, and the hydrologic response units (HRU) were defined. After that, the model was calibrated by SWAT-CUP with streamflow discharge data on a daily basis. Then, the hydrological model was simulated with the values of the parameters obtained in the calibration procedure. This model was used in the diagnostic phase and in future scenarios. In the diagnostic phase, the model was used to assess the effects of LULC changes between 1960 and 2008 on water balance components with the CORINE Land Cover in 1990, 2000, and 2006. In future scenarios, the model was used to evaluate the effects of the forecast of afforestation and the occurrence of wildfires on water balance components.

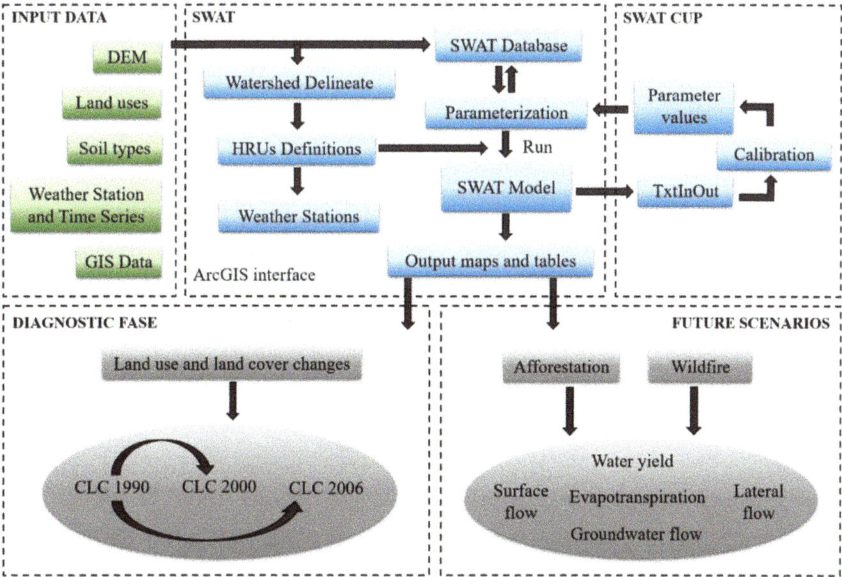

Figure 5. The conceptual framework model for constructing and modelling the water resources in the SWAT (Soil and Water Assessment Tool).

2.4. SWAT Model

The SWAT is based on a physical process to be simulated in a watershed [49]. In a watershed, SWAT continuously simulates the hydrological pattern processes, water balance, water quality, nutrients, and sediment exportation [49,50]. For simulations, information collected from various sources (e.g., topography, land cover) is necessary, and this is then assimilated into the database (e.g., soil characteristics, meteorological data). For modelling purposes, the SWAT model divided the watershed into a sub-basin linked in a cascade by the stream network [62]. In turn, the land area in a sub-basin is divided into HRUs. Each HRU is a portion of a sub-basin that comprises unique land cover, soil, and slope attributes [49]. The SWAT treats the HRU as a homogeneous unit of land use, management techniques, and soil properties and then quantifies the relative impacts of vegetation, management, soil, and climate change within each HRU [63]. The output of the hydrological model (e.g., runoff, sediments, nutrients) is calculated separately from each HRU and then added together to determine the total loading from a sub-basin. The advantage of HRUs is the increase in accuracy which adds to the prediction of loading from a sub-basin [49].

The SWAT uses a modified Soil Conservation Service Curve Number (SCS CN) methodology, and the Penman–Monteith equation to calculate the runoff and evapotranspiration, respectively [50]. Curve numbers have been developed and published for a wide range of land cover types and uses and can be found in [64]. The Penman–Monteith equation requires daily values of precipitation (mm), maximum and minimum temperature (°C), solar radiation (MJ/m^2 day), relative humidity (%), and wind speed (ms^{-1}), but this observed data is usually available with gaps and thus may limit the performance and results of the model [49,62]. To fill the gaps, SWAT includes the WXGEN stochastic weather generator model to generate climatic data [63,65]. The WXGEN generates daily weather information that is missing from the monthly average data summarized over a number of years [49].

2.5. SWAT Input Data

The 2012 version of ArcSWAT was used to build a hydrological model of the Sabor river basin. ArcSWAT is an ArcGIS extension and graphic user input interface for SWAT, which is used worldwide and is continuously under development [66]. We selected ArcSWAT because it has been used in numerous hydrologic, decision-making, and environmental applications [62,65,67], and the authors have experience with using ArcGIS for the processing, overlay, and combination of multiscale and multi-type spatial data in thematic surveys or projects focused on the collection and interpretation of spatial data [34,68–73].

The input data used to construct the hydrological model were (i) the topography of Trás-os-Montes and Alto Douro to generate a digital elevation model and slope, (ii) the drainage network of the Sabor river basin to define the stream and delineate the basins and sub-basins, (iii) the CORINE Land Cover to create a land use grid, (iv) the soil types of Trás-os-Montes and Alto Douro to create a soil grid, and (v) weather data of stations located inside or near the basin to simulate climatic data (Figure 6, Table 1).

A precipitation dataset was compiled between 1957 and 2008 by the various climatologic stations located inside and near the basin (Figure 6b). The gaps in the daily precipitation time series were calculated by inverse distance weighted interpolation. This method is frequently used in climatic predictions and has already been proven to provide good results [17,62]. The time series of weather information insert in the WXGEN weather generator model was provided by the SNIRH meteorological station of Folgares (06N/01C) (Figure 6b). The WXGEN filled the daily missing data based on an average of 46 years of weather data.

A total of 37 sub-basins were defined in the basin area (Figure 1a), with an average area and standard deviation of 86 and 69 km^2, respectively. A total of about 523 HRUs were defined within sub-basins, with a threshold value of 10% for the land use, soil, and slope classes. The SWAT model was executed on a daily basis from 1957 to 1999 with a warm-up period of 3 years. A warm-up period is recommended to initialize the simulation process with the objective of ensuring the establishment of basic flow conditions and hydrologic processes equilibrium as well as to help to minimize the model values for the initial hydrological conditions [62,67].

The land use grid used in the construction of the hydrological model was the CORINE Land Cover 1990 (CLC) (Table 1), but this land cover does not provide information that corresponds to the SWAT land cover classes. Therefore, the CLC classes were reclassified, which led to several CLC classes having the same SWAT land cover classes (Figure 6c) [62,65]. For example, all classes of the heterogeneous agricultural areas of CLC classes were generalized into the SWAT land cover class agricultural land—generic. Among the different SWAT land cover classes, the most appropriate reclassification for the burned areas of CLC classes was barren (BARR). The reclassification is presented in the Supplementary Material (worksheet 1).

The State Soil and Geographic (STATSGO) is the soil database integrated into the SWAT [74]. These soil categories are unavailable in Portugal, and a match could not be found between those categories and the available ones for the research area. The soil map available in the Sabor river basin was extracted from the digital soil map of the Trás-os-Montes and Alto Douro region (Table 1). The soil categories were lithosols, cambisols, alisols, anthrosols, fluvisols, and urban land (Figure 6d). Therefore, data, including the soil component parameters and soil layer parameters, were inserted into the SWAT database for each soil type, except for urban land, which already existed.

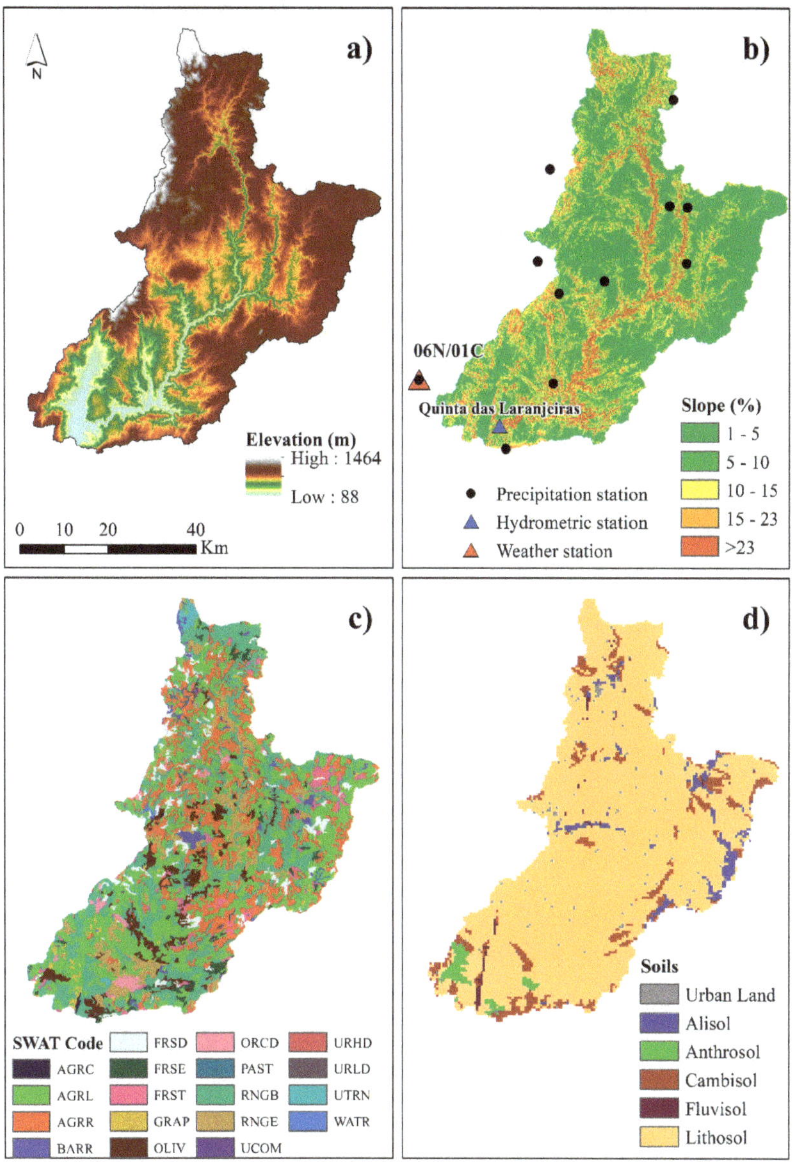

Figure 6. The SWAT (Soil and Water Assessment Tool) input data used for constructing the hydrological model were (**a**) the elevation; (**b**) the slope, meteorological, and hydrometrical data; (**c**) the SWAT codes of land cover classes; and (**d**) the soil types. The SWAT codes are agricultural land—close-grown (AGRC), agricultural land—generic (AGRL), agricultural land—row crops (AGRR), barren (BARR), forest—deciduous (FRSD), forest—evergreen (FRSE), forest—mixed (FRST), vineyard (GRAP), olives (OLIV), orchard (ORCD), pasture (PAST), range—brush (RNGB), range—grasses (RNGE), commercial (UCOM), residential—high density (URHD), residential—low density (URLD), transportation (UTRN), and water (WATR).

The hydrological model was calibrated from 1960 to 1999 and validated from 2000 to 2008. A hydrometric station called Quinta das Laranjeiras (Figure 6b) was used to calibrate and validate the streamflow on a daily basis.

2.6. SWAT-CUP Model Calibration

The computer program used for calibrating the SWAT 2012 models was the SWAT-CUP 2012 (Calibration and Uncertainty Procedures) [75]. The SWAT-CUP 2012 consists of five different calibration procedures, including functionalities for validation and sensitivity analysis. The calibration procedure SUFI-2 (Sequential Uncertainty Fitting) was used in this work. The SUFI-2 algorithm is quite efficient for large-scale, time-consuming models [75,76]. In SUFI-2, the uncertainty analysis is based on the discrepancy assessment between observed and simulated values, taking into account potential sources of uncertainty, like observed data, the conceptual model, and parameters. The degree of uncertainty is quantified by the 95% prediction uncertainty (95PPU), measured by the P-factor and R-factor. The P-factor is the percentage of measured data bracketed by the 95PPU and varying from 0 to 1, wherein 1 indicates a 100% bracketing fit of the observed values. The R-factor measures the calibration quality and indicates the thickness of the 95PPU. A P-factor of 1 and an R-factor of 0 mean a perfect fit for the observed and calibrated values [75].

SUFI-2 has several criteria and quantitative statistical methods to evaluate the outcome simulation values from SWAT, as compared with the observed data [75]. The objective function selected as the calibrated parameter set was the coefficient of determination (R^2). The statistical methods used to assess the model performance were the following: the Nash–Sutcliffe efficiency (NS), the percent bias (PBIAS), the ratio of the root mean square error to the standard deviation of measured data (RSR), and the coefficient of determination (R^2). The model performance is considered satisfactory whenever R^2 and NS are greater than 0.5, RSR is less than 0.7, and PBIAS is less than ±25% for streamflow [77,78].

2.7. Diagnostic Phase of Land Use and Land Cover Changes

In the diagnostic phase, CLC 1990, CLC 2000, and CLC 2006 were used to determine the effects of LULC changes on the water balance components of the Sabor river basin. The SWAT hydrological model was constructed and calibrated with CLC 1990. In forthcoming sections, this hydrological model will be referred to as the reference model. After that, two more hydrological models were constructed, one with the CLC 2000 and another one with the CLC 2006. The values of the calibration parameters of the reference model were inserted into these two hydrological models. Using the same parameter values in all hydrological models ensures that changes in the streamflow are exclusively due to LULC changes. Thus, the study of the LULC changes consisted of comparing the values of the water balance components between the reference model and both the CLC 2000 hydrological model and the CLC 2006 hydrological model. The reference model was simulated between 2000 and 2008, the CLC 2000 hydrological model was simulated between 2000 and 2005, and the CLC 2006 hydrological model was simulated between 2006 and 2008.

2.8. Future Scenarios of Afforestation and Wildfires

In the Sabor river basin, the afforestation scenario was based on the Regional Plan for Forest Management (RPFM) of Douro [79] and Northeast [80], and the wildfires scenario was based on the burned area between 1990 and 2007 [61]. The Regional Plan for Forest Management (Portuguese Regulate Decree no. 3/2007, published on 17 January 2007) includes a plan to be accomplished until 2045, and the main objectives are to reduce the risk of wildfire occurrence, and to adjust the proportions of resinous and deciduous species based on the application of correct forest management models. In the RPFM, the Sabor river basin occupies eleven homogeneous sub-regions in the Douro and Northeast regions, which are referenced in the Supplementary Material (worksheet 2).

In order to create a model of the afforestation scenario, a map was created with the coniferous and broad-leaved areas proposed by RPFM until 2045. Afforestation consisted of counting the areas of coniferous, broad-leaved, and mixed forest in CLC 1990 and calculating the missing areas until reaching the percentage of afforestation proposed by the RPFM for each sub-region (Figure 7a). According to this technical report, the area of afforestation of agricultural land will include 40% coniferous forest and 60% broad-leaved forest, and the afforestation of semi-natural areas will include 70% coniferous forest and 30% broad-leaved forest [79,80]. To make the map, the afforestation of agricultural land was carried out in the following order of CLC classes: agro-forestry areas, land mainly occupied by agriculture, with significant areas of natural vegetation and complex cultivation patterns. The afforestation of semi-natural areas was done in the following order of CLC classes: transitional woodland-shrub, sclerophyllous vegetation, moors and heathland, and burned areas. A map of the afforested area is presented in Figure 7b, and calculation of the respective areas is found in the Supplementary Material (worksheet 2). The map of the afforestation scenario inserted in the SWAT was the CLC 1990 updated with afforestation areas. Also, the map of the wildfires scenario inserted in the SWAT was the CLC 1990 updated with all burned areas in the Sabor river basin between 1990 and 2017 (Figure 7d).

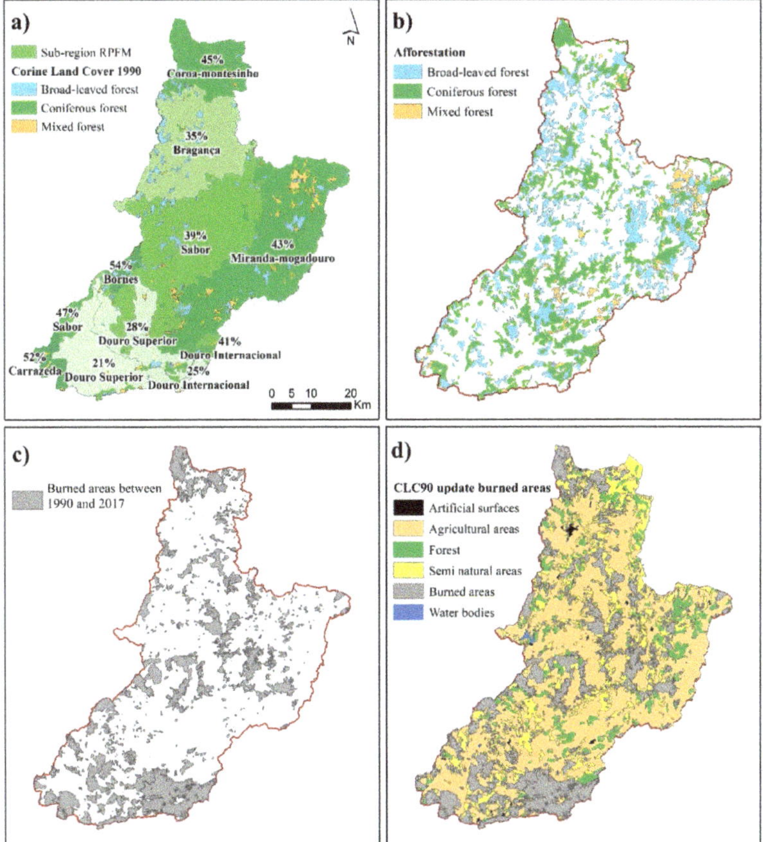

Figure 7. The maps of afforestation and wildfires scenarios inserted in the SWAT for constructing the hydrological models were (**a**) broad-leaved, coniferous, and mixed forest from the CORINE Land Cover 1990 and the projected percentages of afforestation per sub-region for 2045, proposed by the Regional Plan for Forest Management; (**b**) the afforestation projection for 2045; (**c**) areas burned between 1990 and 2017; and (**d**) the CORINE Land Cover 1990 updated with areas burned between 1990 and 2017.

After creating the maps, two hydrological models were constructed, one with the map of the afforestation scenario and another with the map of the wildfires scenario. The values of the calibration parameters of the reference model were inserted into these two hydrological models. Once again, the use of the same parameters in both hydrological models allowed the scenarios to be compared with the reference data. In other words, the factors used to compare the afforestation and wildfires scenarios with the reference model included the water yield surface flow, evapotranspiration, lateral flow, and groundwater flow.

3. Results

3.1. Calibration and Validation of the Streamflow

The streamflow was calibrated for a 39-year period (1960–1999) and validated for a 9-year period (2000–2008) both on a daily basis. The parameters and their respective values resulting from the calibration are shown in Table 2. The graphics in Figure 8a,b show good agreement between the observed and simulated streamflow for both calibration and validation. The calibration and validation, illustrated in Figure 8a,b, respectively, were performed on a daily basis, but for best visualization, they are represented on a monthly basis. The goodness-of-fit indicators for the streamflow calibration (Table 3), based on the R^2, RSR, and NS show satisfactory performances (with values of 0.63 and 0.62) and PBIAS shows a very good performance (2.7%) [50,51]. The same goodness-of-fit indicators were obtained for the validation with a very good performance for R^2 (with 0.8), and satisfactory performances for RSR, NS, and PBIAS with 0.63%, 0.61%, and −24%, respectively (Table 3). The negative value of PBIAS indicates a model overestimation bias [77].

Table 2. The parameters used in the calibration procedure of streamflow. In the legend of methods, R is relative and V is the replacement value.

Method and Parameter	Description	Units	Minimum Value	Maximum Value	Fitted Value
R_CN2.mgt	Curve number for moisture condition II	-	−0.08	0.10	0.05
V_ALPHA_BF.gw	Base flow alpha factor	days	0.53	0.82	0.59
V_GW_DELAY.gw	Flow delay time for aquifer recharge	days	−80.89	99.30	14.84
V_GWQMN.gw	Flow threshold depth of water in the shallow aquifer	mm	0.72	1.81	1.26
V_REVAPMN.gw	Threshold depth of water in the shallow aquifer	mm	256.09	383.16	295.00
V_GW_REVAP.gw	Groundwater re-evaporation coefficient	-	0.09	0.19	0.12
V_RCHRG_DP.gw	Flow deep aquifer percolation coefficient	-	−0.05	0.69	−0.03
V_SHALLST.gw	Initial depth of water in the shallow aquifer	mm	−23476.17	9946.85	−5302.40
V_CH_N2.rte	Manning's "n" value in the main channel	-	0.08	0.16	0.12
V_CH_K2.rte	Effective hydraulic conductivity in the main channel	-	56.47	213.88	147.97
V_ALPHA_BNK.rte	Baseflow alpha factor for bank storage	-	1.30	2.01	1.91
V_ESCO.hru	Soil evaporation compensation factor	-	0.53	0.90	0.88
V_EPCO.hru	Plant uptake compensation factor	-	0.08	0.58	0.39
V_SLSUBBSN.hru	Average slope length	-	171.37	248.23	214.12
R_SOL_AWC (1).sol	Soil available water capacity (soil 1st layer)	-	0.02	0.18	0.18
R_SOL_K (1).sol	Saturated hydraulic conductivity	-	301.52	633.36	411.44
V_SURLAG.bsn	Surface runoff lag coefficient	-	2.68	9.03	6.53
V_CH_K1.sub	Effective hydraulic conductivity in the tributary channel alluvium	-	108.63	194.72	178.04

Table 3. Goodness-of-fit indicators for daily calibration between 1960 and 1999 and validation of streamflow between 2000 and 2008 in the Sabor river basin.

Measure	Calibration	Acceptable Ranges
Calibration		
R^2 (coefficient of determination)	0.63	> 0.5 acceptable [51]
RSR (standardized RMSE)	0.62	Satisfactory [50]
NS (Nash–Sutcliffe coefficient)	0.62	Satisfactory [50]
PBIAS (percent bias)	2.7%	Very good [50]
Validation		
R^2 (coefficient of determination)	0.80	> 0.75 very good [51]
RSR (standardized RMSE)	0.63	Satisfactory [50]
NS (Nash-Sutcliffe coefficient)	0.61	Satisfactory [50]
PBIAS (percent bias)	−24%	Satisfactory [50]

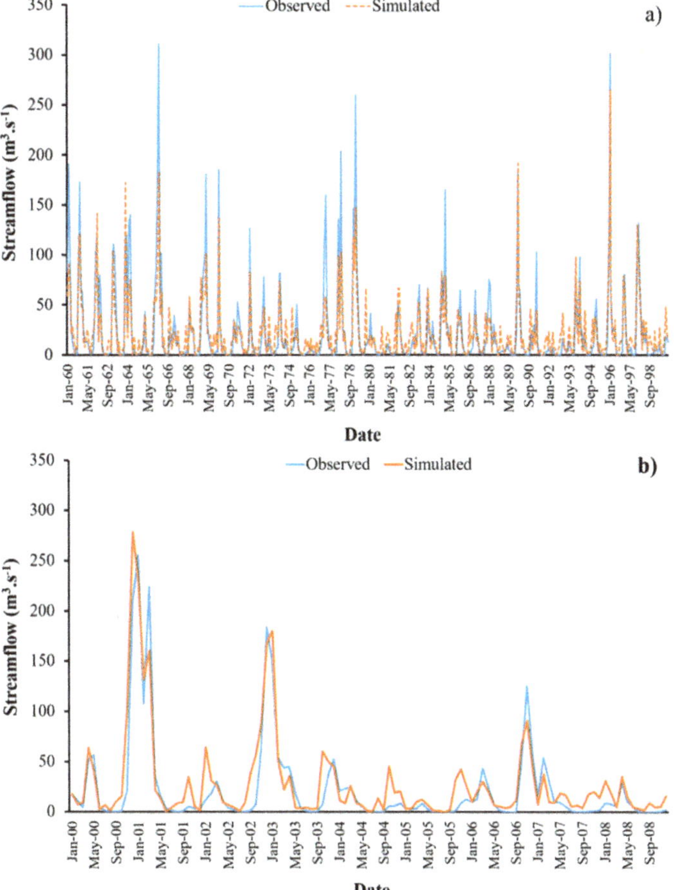

Figure 8. The comparison of observed and simulated streamflow during (**a**) the calibration (between 1960 and 1999); and (**b**) validation (between 2000 and 2008) in the Sabor river basin. The simulation of the streamflow was executed on a daily basis, but for best visualization was present on a monthly basis.

3.2. Land Cover and Land Use Changes

Figure 9 shows the graphics of the water balance components of the reference model (with the CLC 1990, simulated between 2000 and 2008) the hydrological model of the CLC 2000 (simulated between 2000 and 2005), and the hydrological model of the CLC 2006 (simulated between 2006 and 2008). The results show that the LULC changes that occurred in 2000 and 2006 led to a decrease in the water yield and an increase in evapotranspiration (Figure 9a,b). The water yield decreased by an average of 91 and 52 mm·year^{-1} for the LULC changes in 2000 and 2006, respectively. The evapotranspiration increased by an average of 90 and 55 mm·year^{-1} for the LULC changes in 2000 and 2006, respectively. The values of surface flow and groundwater flow decreased, while the lateral flow increased (Figure 9c,e). On average, for the LULC changes in 2000 and 2006, the surface flow decreased by 28 and 23 mm·year^{-1}, and the groundwater flow decreased by 91 and 50 mm·year^{-1}, respectively. The lateral flow increased by 10 and 5 mm·year^{-1} for the LULC changes in 2000 and 2006, respectively.

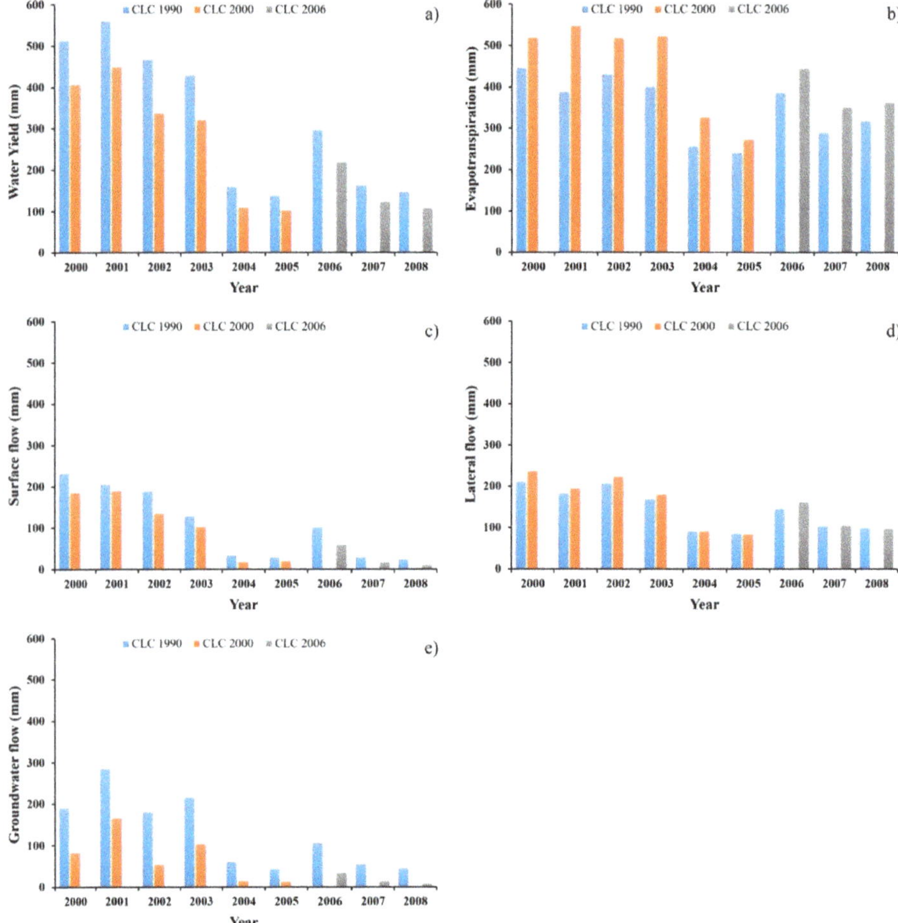

Figure 9. Quantity of water (mm) (**a**) in the water yield; (**b**) evapotranspiration; (**c**) surface flow; (**d**) lateral flow; and (**e**) groundwater flow. The blue bars are the reference model (simulated between 2000 and 2008 with the CLC 1990), the orange bars are the CLC 2000 model (simulated between 2000 and 2005), and the grey bars are the CLC 2006 model (simulated between 2006 and 2008).

3.3. *Afforestation and Wildfires Scenarios*

The values of the water balance components were calculated on a monthly basis between the reference model (CLC90) and both the afforestation and wildfires scenarios. The results are illustrated in Figures 10 and 11, which represent the monthly values and spatial distribution, respectively. The graphics of Figure 10a,b show that, in both scenarios and in every month of the year, the water yield decreased and the evapotranspiration increased. The decrease in the water yield was more pronounced in the rainy season (autumn and winter) than in the dry season (spring and summer) and in the afforestation scenario compared with the wildfires scenario. In the afforestation scenario, these large decreases varied from 28 hm^3·year^{-1} in October to 62 hm^3·year^{-1} in January, while in the wildfires scenario they were somewhat smaller, varying from 15 hm^3·year^{-1} in October to 49 hm^3·year^{-1} in January. The month of August registered much lower water yield decreases: 3 hm^3·year^{-1} in the afforestation scenario and 2 hm^3·year^{-1} in the wildfires scenario. The evapotranspiration increase was more pronounced in January, February, and March, with values ranging from 46 to 82 hm^3·year^{-1} for the afforestation scenario and from 33 to 61 hm^3·year^{-1} for the wildfires scenario. In the afforestation scenario, the lowest increase in evapotranspiration was registered in August (3 hm^3·year^{-1}), while in the wildfires scenario, a small decrease in evapotranspiration was registered in September (0.4 hm^3·year^{-1}).

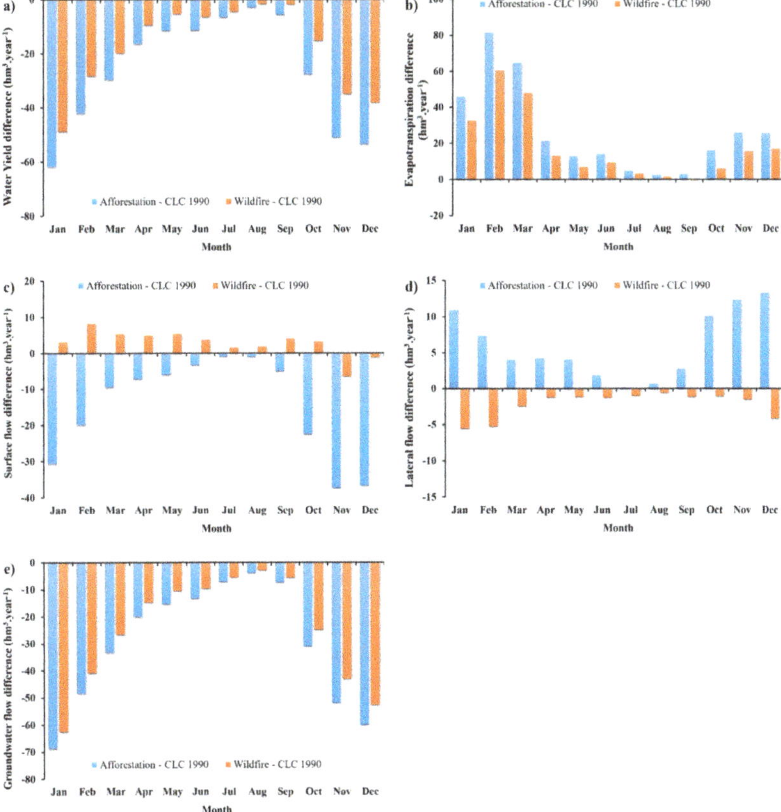

Figure 10. Quantity of water (hm^3·year^{-1}) (**a**) in the water yield; (**b**) in evapotranspiration; (**c**) in surface flow; (**d**) in lateral flow; and (**e**) in groundwater flow. The blue bars show the difference between the reference model and the afforestation scenario model, and the orange bars show the difference between the reference model and the wildfires scenario model.

The water available for surface flow and lateral flow showed different behaviors in both scenarios. In the afforestation scenario, the surface flow decreased and the lateral flow increased, and the opposite occurred for the wildfires scenario (Figure 10c,d). In the afforestation scenario, the major decrease in the surface flow occurred in January, November, and December, with values ranging from 31 to 37 $hm^3 \cdot year^{-1}$, while the smallest decrease occurred in July with 0.8 $hm^3 \cdot year^{-1}$. The major increase in the lateral flow occurred between October and January with values ranging from 10 $hm^3 \cdot year^{-1}$ in October to 13 $hm^3 \cdot year^{-1}$ in December, while the smallest increase occurred in July (0.2 $hm^3 \cdot year^{-1}$). In the wildfires scenario, the surface flow increased between January and October with values ranging between 2 $hm^3 \cdot year^{-1}$ in July and 8 $hm^3 \cdot year^{-1}$ in February, and it decreased in November and December with values of 1 and 6 $hm^3 \cdot year^{-1}$, respectively. The lateral flow decreased in every month of the year, with values ranging from 1 $hm^3 \cdot year^{-1}$ in August to 6 $hm^3 \cdot year^{-1}$ in January. The major decrease in the groundwater flow occurred in the rainy season in both scenarios (Figure 10e). In the afforestation scenario, these large decreases varied from 31 $hm^3 \cdot year^{-1}$ in October to 69 $hm^3 \cdot year^{-1}$ in January, while in the wildfires scenario, they were somewhat smaller, varying from 25 $hm^3 \cdot year^{-1}$ in October to 62 $hm^3 \cdot year^{-1}$ in January. The lowest groundwater flow was registered in August, with 4 $hm^3 \cdot year^{-1}$ in the afforestation scenario and 3 $hm^3 \cdot year^{-1}$ in the wildfires scenario.

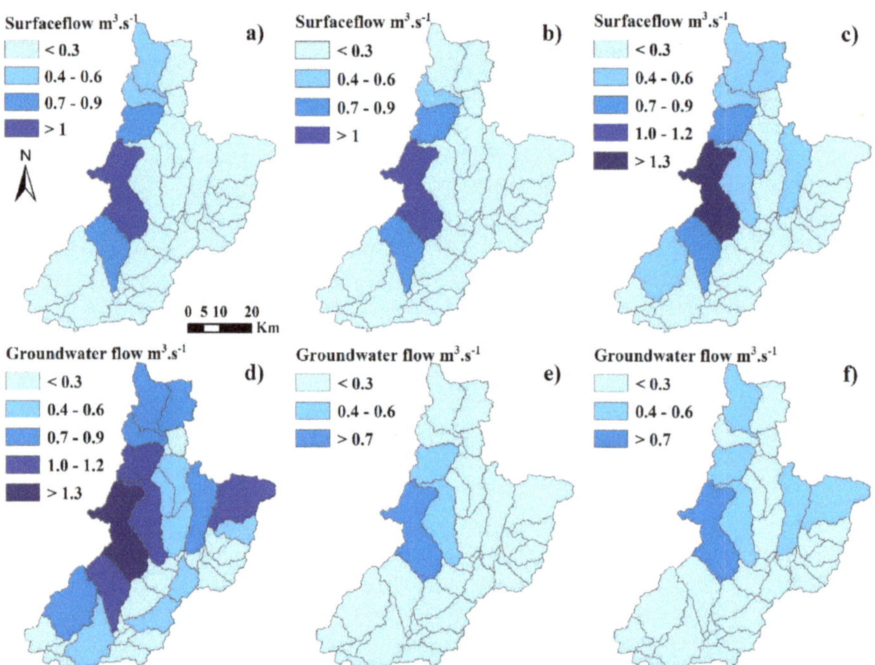

Figure 11. SWAT output maps of (**a–c**) the surface flow and (**d–f**) the groundwater flow. The surface flow matches (**a–c**) the reference model, the afforestation scenario, and the wildfires scenario, respectively. The groundwater flow matches (**d–f**) the reference model, the afforestation scenario, and the wildfires scenario, respectively.

SWAT output maps of the surface flow and the groundwater flow for the reference model and scenarios are illustrated in Figure 11. Figure 11a–c shows the surface flow of the reference model, for the afforestation scenario, and for the wildfires scenarios, respectively. Figure 11d–f shows the groundwater flow of the reference model, for the afforestation scenario, and for the wildfires scenario, respectively. The maps show that the highest surface flow and groundwater flow values in the reference model (Figure 11a,d) and scenarios (Figure 11b,c,e,f) are located in the sub-basins of the Northwest. When comparing the maps of the reference model with the scenarios, it can be seen that the surface flow and groundwater flow values decrease. However, these decreases are major in the afforestation scenario.

4. Discussion

4.1. Performance of Streamflow and TP Calibration

The performance statistics of the model for the daily streamflow during calibration and validation periods were satisfactory and good, respectively. Similar results were achieved by other authors [81,82] using the SWAT for streamflow simulation in both the calibration and validation periods in the Alto Sabor river basin (located in the upstream part of the Sabor river basin). The results of the performance statistics indicated that SWAT was able to capture and reproduce the average flows and seasonal variations of the Sabor river basin (Figure 8a). However, the SWAT also shown had difficulty reproducing the observed extreme values of the streamflow, as shown by the regular underestimation of the most important peak events for the calibration period. Many studies have also detected the inability of the SWAT models to reproduce the highly complex hydrological processes during the high flow periods [52,53,81,82]. Major high flows or extreme conditions occurred in the winter of the years 1960, 1966, 1969, 1985, and between 1976 and 1979 (Figure 8a). Several authors analyzed how the rainfall-runoff translation occurs in reality compared with that simulated by the model [52]. These authors verified that the same amount of annual rainfall did not correspond to the same amount of observed streamflow. In contrast, the model presented a similar streamflow for the same amount of rainfall. The authors stated that the curve number method in SWAT uses the average daily rainfall and does not take into account the intensity and duration of rainfall. This is a factor that can explain why peak flows were underestimated by the model. One example of the high intensity of rainfall was registered in four basins of the North of Portugal, between 2000 and 2006, which had a considerable number of days that exceeded 20 mm of precipitation [68].

Despite the goodness-of-fit indicators for the streamflow calibration being considered satisfactory, there are some potential factors which compromise the performance of the model. We identified the following potential factors: (i) the climate stations cannot represent the orographic effect on the precipitation rates, (ii) the meteorological data are considerably affected by orographic variations, (iii) the human activities across the catchment may not have been accounted for in the model (e.g., artesian wells and boreholes, irrigation, and domestic consumption), (iv) the observed flow values are not 100% accurate. The precipitation gauges are located between 441 and 905 m of elevation with an average altitude of 659 m, and they cannot represent the orographic effect which varies between 88 and 1464 m of elevation. The orographic variations have a considerable effect on meteorological data; for example, the vertical temperature gradient, the depressions, and the shade in the valley result in lower minimum temperatures. These local variations are not entirely covered by weather stations. The last factor that influences the effectiveness of calibration is the problem that the hydrometric gauges faced during data collection. The discharge flow was estimated from hydrometric levels which were transformed into streamflow by a state discharge curve, but it was only valid within a range of hydrometric heights. Thus, the calculation used was not able to capture the actual hydrological pattern within extreme events.

4.2. Effects of LULC Changes and Future Scenarios for Water Resources

A major change in the LULC was the replacement of range—brush (scrub and herbaceous vegetation associations) with forest (essentially coniferous and deciduous) from CLC 1990 to CLC 2000 and the replacement of range—grasses (natural grasslands), agricultural land—generic (heterogeneous agricultural areas), forest (coniferous, deciduous, and mixed) and burned area in range—brush from CLC 1990 to CLC 2006 (Figure 3). These LULC changes are a result of a complex process of plant recolonization of shrubs and forests as a consequence of farmland abandonment. This natural reforestation has the advantage of limiting soil erosion. However, the authors of [56,83] showed that many Mediterranean areas evolved into simplified and continuous landscapes with plenty of flammable wood, which led to fire occurrences. The wildfires increased soil erosion and progressively reduced the potential for recolonization of plants, leading to stony soils. This explains the many degraded panoramic landscapes in Mediterranean areas [16].

The results showed that the LULC changed from CLC 1990 to 2000, and from CLC 1990 to CLC 2006, causing decreases in the water yield, surface flow, and groundwater flow and increases in evapotranspiration and lateral flow. The major changes were observed in groundwater flow and evapotranspiration. The decrease in the groundwater flow in 200 and 2006 was, on average, 91 and 50 mm·year^{-1}, respectively (Figure 9a). The increase in evapotranspiration in 2000 and 2006 was, on average, 90 and 55 mm·year^{-1}, respectively (Figure 9b). The same results were obtained with the afforestation projection scenario proposed by the Regional Plan for Forest Management to 2045. In both the afforestation scenario and LULC changes, the decreases in the water yield, surface flow, and groundwater flow were due to a large amount of water being lost by evapotranspiration.

The increase in the lateral flow suggests that the forest provides an increase of water infiltration into the soil, but the high transpiration rate of the trees removes the water from the groundwater into the atmosphere. The increase in soil water infiltration by afforestation was confirmed by Ilstedt et al. [12]. The authors applied a meta-analysis based on several articles and concluded that the infiltration capacity increased, on average, by approximately three-fold after afforestation in agricultural fields. Another study [11] selected 20 studies to compare water flows in tropical watersheds under natural or planted forests and non-forest lands to provide useful results for valuing the watershed ecosystem services. The planted forests were Pines and Eucalyptus and lowland natural forests. The main results showed significantly lower total flows or water yields and groundwater flows under planted forests than under non-forest land uses. These results were explained by the high transpiration rates of Eucalyptus. According to these results, the authors argue that the effect of natural forests on base flow results from two competing processes: the high infiltration under the forest contributes to soil water recharge and the base flow increase, while the high transpiration of the trees contributes to the base flow decrease. Furthermore, the infiltration of soil water may be higher under planted forests than under non-forest land uses, but may not be sufficient to offset the loss of water by transpiration [12]. The compilation of paired-watershed results, based on a total of 137 basins, showed that reforestation results in a decrease in water yield and deforestation results in an increase [3]. Several authors have observed that reforestation could reduce the amount of sediment entering streams [84], as vegetation reduces the streamflow and increases infiltration [8,12]. The results of Serrano-Muela et al. [85] confirmed that forest conservation in the Central Spanish Pyrenees reduces floods and soil erosion, particularly on steep slopes. In this study, the major decreases in water yield, surface flow and groundwater flow, as well as the increases in lateral flow and evapotranspiration, were obtained in the wet season. For example, the decrease in water yield varied between 28 hm^3·year^{-1} in October and 62 hm^3·year^{-1} in January. Moreover, an increase in the lateral flow occurred between October and January with values ranging from 10 hm^3·year^{-1} in October to 13 hm^3·year^{-1} in December. The evapotranspiration was also higher in winter, because there was more water available to evaporate, but the highest values were obtained in February and March (with values ranging between 46 and 82 hm^3·year^{-1}). The reason for this is the combination of the rainfall and the increase of temperature that support evaporation and the growth of the canopy trees, which increases transpiration.

In contrast, in the wildfires scenario, increases in the surface flow and evapotranspiration and decreases in the lateral flow and groundwater flow were observed (Figure 10b–e). Similarly to the afforestation scenario, the major increases or decreases of water in the wildfires scenario mainly occurred in winter. For example, the water yield varied between 15 hm^3·year^{-1} in October and 49 hm^3·year^{-1} in January, and the increase in evapotranspiration was more pronounced in January, February, and March, with values ranging between 33 and 61 hm^3·year^{-1}.

Wildfires are agents of change that can dramatically alter evaporation and transpiration rates, and their short-term effects on surface evaporation rates may be complex. The increase in runoff after the fire has been attributed to destroyed vegetation as well as to reduced evapotranspiration [16,86–88]. The study by Miranda et al. [89] showed that immediately after a fire, evapotranspiration rates decrease. However, this situation is reversed with the first rains, because the burned area rapidly becomes a stronger sink for CO_2 and has higher evapotranspiration rates than nearby unburned areas. This difference persists throughout the wet season and is attributable to the greater physiological activity of the growing vegetation in the burned area. Similar rates of evaporation immediately after rainfall for both burned and unburned plots were observed in Campo Sujo, near Brazil [90]. However, within a few days of the soil drying, lower evaporation rates from the burned plot were observed. About 1 month after the fire, evaporation rates from the burned plot were typically as much as 80% that of the unburned control early in the wet season [90]. This was attributed to greatly enhanced rates of soil evaporation in the burned plot, especially immediately after the rain. Nagra et al. [91] also obtained similar results and justified the increase in the evaporation rates post-fire as a result of low albedo and reduced vegetation cover (lack of shading).

The increase in surface flow and the decreases in lateral flow and groundwater flow were due to the reduced infiltration capacity of the soil caused by the removal of vegetation by fire. Previous studies [19,92,93] showed that changes in vegetation cover and topsoil (composed of organic matter) have important impacts on the hydrological regime. A review of the literature developed in the European Mediterranean showed that wildfires make the soil more susceptible to removal by water erosion, less likely to allow infiltration, and more likely to promote surface flow [16]. The increase in surface flow, as well as soil erosion caused by fires, were proven in several studies in Portugal [19,92,93] and in the Mediterranean region [16,94,95]. For example, in burnt Eucalyptus and Pinus forest plantations in the Águeda Basin, North-Central Portugal, the surface flow on burned plots was 5%–25% higher than on unburnt ones [96]. For burned pine plantations in a small catchment located in Central Portugal, a runoff of up to 48.5% was found (1.1 km^2) [19]. In the Arbúcies basin, North-East Spain, there was an increase of 30% in flood runoff [94]. In the Rimbaud catchment in South-East France, an increase in the annual runoff of 30% was also verified during the first year after the fire [86].

A reduction in the infiltration capacity and the absence of vegetation improved the flood peak and soil erosion [16,93]. Nunes et al. [1] reported that during the study period (2005 and 2006), more than 80% of the total rainfall fell in autumn and winter. Consequently, the monthly rainfall erosivity, based on the Modified Fournier Index, increased considerably during these months. This result suggests that the energy available for erosion and transport was highly concentrated in this season. They also verified that the significantly larger amounts of runoff caused high soil loss. The overland flow was three times higher in 2006 than in 2005 which resulted in a sediment yield twice as high. However, the soil losses fundamentally depended on the amount and intensity of rainfall. The stability of the soil structure and aggregates is usually thought to be reduced by fire, producing more easily eroded soil [16].

Even so, modest post-fire soil losses could be important for soil longevity in some areas, because of soil organic matter and nutrient losses in solution or adsorbed onto eroded sediment particles [16,97]. Much of the post-fire nutrient content is in the form of ash, which is prone to removal by wind and water [19]. However, it can also benefit the soil quality through nutrient-rich ash becoming incorporated into the soil [16]. The high fire frequency and thin soils in a nutrient-poor ecosystem, typical of many fire-prone Mediterranean areas causes the risk of soil fertility depletion to be high [16,19].

4.3. *Overview of the Effects of the LULC Changes on Water Resources*

Changes in the LULC of the Sabor river basin have had impacts on the water balance. The main changes were the increases in forest and shrubs between 2000 and 2008. These changes caused decreases in the water yield and groundwater flow and an increase in evapotranspiration as well as the afforestation scenario. The decrease in water yield was due to the reductions of surface flow and groundwater flow. The decrease in the surface flow was due to the increase in soil water infiltration by forests. Additionally, the decrease in groundwater flow can be partly attributed to the high transpiration rate of the trees. Despite forests causing substantial reductions in water yield, they also improve the water quality, reduce erosion, and benefit biodiversity. According to Francis et al. [10], a low to intermediate tree cover can improve soil hydraulic properties by up to 25 m from its canopy edge, which means that the hydrologic gains can be proportionally higher than the additional losses from the increased transpiration.

In the afforestation and wildfires scenarios, a decrease in the water yield was observed as well as an increase in evapotranspiration. The increase in evapotranspiration in the wildfires scenario was due to the evaporation of the water as a result of low albedo and a reduction in the vegetation cover. The comparison of the two scenarios showed that the trees were responsible for the reduction of groundwater flow. However, they contributed to the decrease in sediment yield and nutrients in surface flow. On the other hand, the wildfires scenario increased the surface flow as well as the sediment yield and the concentration of nutrients in the surface flow. The recurrence of fires has long-term implications because of the increase in soil erosion and the loss of nutrients, making it difficult for plants to recolonize. The water quality was affected by the increase in nutrient concentration due to eutrophication of the aquatic environment when associated with the increase in temperatures in the summer. The construction of two dams in the Sabor river for the production of electricity has increased the problem of eutrophication. According to the technical report [98] and the article [99], the changes in water quality caused by dam construction and the consequent stream water impoundment were significant. The increases in temperature and electric conductivity and accumulation of phosphorus and nitrogen in the reservoirs triggered the growth of algae, the increase of chlorophyll a, and a drop in the transparency of the water. The consequences of water deterioration for the aquatic fauna were severe, marked by abrupt declines of native fish species and the invasion of exotic species.

5. Conclusions

In this study, the SWAT2012 model was applied to assess the spatiotemporal dynamics of the water balance in the Sabor river basin. First of all, the LULC changes between 1960 and 2008 were analyzed with the CLC 1990, CLC 2000, and CLC 2006. Secondly, two scenarios were created, afforestation and wildfires. The afforestation scenario used was the projection proposed by the Regional Plan for Forest Management until 2045. The wildfires scenario used was based on areas burned in the Sabor river basin between 1990 and 2017.

The overall results of LULC changes between 1960 and 2008 show that the increase in forest and shrubs is a consequence of farmland abandonment in a basin where the population density is less than 20 inhabitants per km^2. The changes caused decreases in the water yield, surface flow, and groundwater flow and increases in evapotranspiration and lateral flow. The major changes occurred in the water yield and evapotranspiration. The water yield decreased in 2000 and 2006, on average, by 91 and 52 $mm \cdot year^{-1}$, respectively. Evapotranspiration increased in 2000 and 2006, on average, by 90 and 55 $mm \cdot year^{-1}$, respectively. The decrease in groundwater flow and increase in evapotranspiration were attributed to the high transpiration rate of the trees. Similar results were obtained for the afforestation scenario: decreases in the water yield and groundwater flow and an increase in evapotranspiration. The major decrease occurred in winter for groundwater flow with values varying between 28 $hm^3 \cdot year^{-1}$ in October and 62 $hm^3 \cdot year^{-1}$ in January. The increase in evapotranspiration was more pronounced in January, February, and March, with values ranging from 46 to 82 $hm^3 \cdot year^{-1}$.

In contrast, in the wildfires scenario, increases in the surface flow and evapotranspiration and decreases in the lateral flow and groundwater flow were observed. Similarly to the afforestation scenario, the major increases or decreases of water in the wildfires scenario occurred in winter. The major decrease in groundwater flow occurred in winter, for which values varied between 25 $hm^3 \cdot year^{-1}$ in

October and 62 $hm^3 \cdot year^{-1}$ in January. The increase in evapotranspiration was more pronounced in January, February, and March, with values ranging from 33 to 61 $hm^3 \cdot year^{-1}$.

The afforestation projection, proposed by the Regional Plan for Forest Management, will cause greater decreases in water yield, surface flow and groundwater than the wildfires scenario. A decrease in groundwater has implications for streamflow, because it is this flow that feeds the rivers in the dry season. However, it generates an increase in lateral flow, with more available water for irrigation. This is an important factor, because the basin is essentially agroforestry. With this study, we have shown that afforestation cannot simultaneously maximize all environmental benefits, but any form of afforestation should provide environmental improvements over agricultural land. Reductions in water yields could be minimized by planting trees with low water use at low densities and by avoiding landscape positions with access to the groundwater. An improvement in water quality could be achieved by using the contrasting strategies.

Supplementary Materials: The following are available online at http://www.mdpi.com/2073-4441/11/7/1464/s1, The Supplementary Materials comprise the following worksheets: Worksheet 1—Reclassification of CORINE Land Cover classes into SWAT counterparts; Worksheet 2—Forest occupation and predicted afforestation until 2045, in the eleven homogeneous sub-regions in the Douro and Northeast regions.

Author Contributions: Conceptualization, R.M.B.S. and F.A.L.P.; methodology, R.M.B.S. and F.A.L.P.; software, R.M.B.S.; validation, F.A.L.P.; formal analysis, L.F.S.F.; investigation, R.M.B.S.; resources, L.F.S.F. and R.M.V.C.; data curation, R.M.B.S., F.A.L.P., L.F.S.F., and R.M.V.C.; writing—original draft preparation, R.M.B.S.; writing—review and editing, F.A.L.P.; visualization, R.M.B.S.; supervision, F.A.L.P. and L.F.S.F.; project administration, R.M.V.C.; funding acquisition, R.M.V.C.

Funding: This research was funded by the INTERACT project "Integrated Research in Environment, Agro-Chain and Technology", no. NORTE-01-0145-FEDER-000017, in the line of research entitled BEST "Bio-economy and Sustainability", and co-financed by the European Regional Development Fund (ERDF) through NORTE 2020 (the North Regional Operational Program 2014/2020). For authors at the CITAB research centre, this research was further financed by the FEDER/COMPETE/POCI—Operational Competitiveness and Internationalization Programme under Project POCI-01-0145-FEDER-006958, and by National Funds of FCT—Portuguese Foundation for Science and Technology under the project UID/AGR/04033/2019. For the author in the CQVR, this research was additionally supported by National Funds of FCT—Portuguese Foundation for Science and Technology under the project UID/QUI/00616/2019.

Conflicts of Interest: The authors declare no conflict of interest. The funders had no role in the design of the study; in the collection, analyses, or interpretation of data; in the writing of the manuscript or in the decision to publish the results.

References

1. Nunes, A.N.; de Almeida, A.C.; Coelho, C.O.A. Impacts of land use and cover type on runoff and soil erosion in a marginal area of Portugal. *Appl. Geogr.* **2011**, *31*, 687–699. [CrossRef]
2. García-Ruiz, J.M.; Nadal-Romero, E.; Lana-Renault, N.; Beguería, S. Erosion in Mediterranean landscapes: Changes and future challenges. *Geomorphology* **2013**, *198*, 20–36. [CrossRef]
3. Andréassian, V. Waters and forests: From historical controversy to scientific debate. *J. Hydrol.* **2004**, *291*, 1–27. [CrossRef]
4. Pitman, A.J.; Narisma, G.T.; Pielke, R.A.; Holbrook, N.J. Impact of land cover change on the climate of southwest Western Australia. *J. Geophys. Res. Atmos.* **2004**, *109*, 1–12. [CrossRef]
5. Liu, M.; Tian, H.; Chen, G.; Ren, W.; Zhang, C.; Liu, J. Effects of land-use and land-cover change on evapotranspiration and water yield in China during 1900–2000. *J. Am. Water Resour. Assoc.* **2008**, *44*, 1193–1207. [CrossRef]
6. Serpa, D.; Nunes, J.P.; Santos, J.; Sampaio, E.; Jacinto, R.; Veiga, S.; Lima, J.C.; Moreira, M.; Corte-Real, J.; Keizer, J.J.; et al. Impacts of climate and land use changes on the hydrological and erosion processes of two contrasting Mediterranean catchments. *Sci. Total Environ.* **2015**, *538*, 1–14. [CrossRef] [PubMed]
7. Meyfroidt, P.; Lambin, E.F. Global Forest Transition: Prospects for an end to deforestation. *Annu. Rev. Environ. Resour.* **2011**, *36*, 343–371. [CrossRef]
8. Cunningham, S.C.; Mac Nally, R.; Baker, P.J.; Cavagnaro, T.R.; Beringer, J.; Thomson, J.R.; Thompson, R.M. Balancing the environmental benefits of reforestation in agricultural regions. *Perspect. Plant Ecol. Evol. Syst.* **2015**, *17*, 301–317. [CrossRef]

9. Carvalho-Santos, C.; Sousa-Silva, R.; Gonçalves, J.; Honrado, J.P. Ecosystem services and biodiversity conservation under forestation scenarios: Options to improve management in the Vez watershed, NW Portugal. *Reg. Environ. Chang.* **2016**, *16*, 1557–1570. [CrossRef]
10. Francis, C.F.; Thornes, J.B. Trees, forests and water: Cool insights for a hot world. *Glob. Environ. Chang.* **2017**, *43*, 51–61.
11. Locatelli, B.; Vignola, R. Managing watershed services of tropical forests and plantations: Can meta-analyses help? *For. Ecol. Manag.* **2009**, *258*, 1864–1870. [CrossRef]
12. Ilstedt, U.; Malmer, A.; Verbeeten, E.; Murdiyarso, D. The effect of afforestation on water infiltration in the tropics: A systematic review and meta-analysis. *For. Ecol. Manag.* **2007**, *251*, 45–51. [CrossRef]
13. Desir, G.; Marín, C. Factors controlling the erosion rates in a semi-arid zone (Bardenas Reales, NE Spain). *Catena* **2007**, *71*, 31–40. [CrossRef]
14. European Environment Agency. *CORINE Soil Erosion Risk and Important Land Resources-In the Southern Regions of the European Community*; European Environmental Agency: Copenhagen, Denmark, 1994; ISBN 92-826-2545-1.
15. Grimm, M.; Jones, R.; Montanarella, L. *Soil Erosion Risk in Europe*; European Commission Joint Research Centre/Institute for Environment and Sustainability/European Soil Bureau: Ispra, Italy, 2001; Volume 19939.
16. Shakesby, R.A. Post-wildfire soil erosion in the Mediterranean: Review and future research directions. *Earth-Sci. Rev.* **2011**, *105*, 71–100. [CrossRef]
17. Pereira, M.G.; Fernandes, L.S.; Carvalho, S.; Santos, R.B.; Caramelo, L.; Alencoão, A. Modelling the impacts of wildfires on runoff at the river basin ecological scale in a changing Mediterranean environment. *Environ. Earth Sci.* **2016**, *75*, 1–14. [CrossRef]
18. DaCamara, C.; Trigo, R.M.; Pinto, M.M.; Nunes, S.A.; Trigo, I.F. The tragic fire event of 17 June 2017 in Portugal: The meteorological perspective. In Proceedings of the American Geophysical Union Fall Meeting 2017, New Orleans, LA, USA, 11–15 December 2017.
19. Ferreira, A.J.D.; Coelho, C.O.A.; Ritsema, C.J.; Boulet, A.K.; Keizer, J.J. Soil and water degradation processes in burned areas: Lessons learned from a nested approach. *Catena* **2008**, *74*, 273–285. [CrossRef]
20. Smith, H.G.; Sheridan, G.J.; Lane, P.N.J.; Nyman, P.; Haydon, S. Wildfire effects on water quality in forest catchments: A review with implications for water supply. *J. Hydrol.* **2011**, *396*, 170–192. [CrossRef]
21. Pacheco, F.A.L. Regional groundwater flow in hard rocks. *Sci. Total Environ.* **2015**, *506–507*, 182–195. [CrossRef]
22. Santos, R.M.B.; Sanches Fernandes, L.F.; Moura, J.P.; Pereira, M.G.; Pacheco, F.A.L. The impact of climate change, human interference, scale and modeling uncertainties on the estimation of aquifer properties and river flow components. *J. Hydrol.* **2014**, *519*, 1297–1314. [CrossRef]
23. Pacheco, F.A.L.; Van der Weijden, C.H. Role of hydraulic diffusivity in the decrease of weathering rates over time. *J. Hydrol.* **2014**, *512*, 87–106. [CrossRef]
24. Pacheco, F.A.L.; Van der Weijden, C.H. Mineral weathering rates calculated from spring water data: A case study in an area with intensive agriculture, the Morais Massif, northeast Portugal. *Appl. Geochem.* **2002**, *17*, 583–603. [CrossRef]
25. Van der Weijden, C.H.; Pacheco, F.A.L. Hydrogeochemistry in the Vouga River basin (central Portugal): Pollution and chemical weathering. *Appl. Geochem.* **2006**, *21*, 580–613. [CrossRef]
26. Pacheco, F.A.L.; Sousa Oliveira, A.; Van Der Weijden, A.J.; Van Der Weijden, C.H. Weathering, biomass production and groundwater chemistry in an area of dominant anthropogenic influence, the Chaves-Vila Pouca de Aguiar region, north of Portugal. *Water Air Soil Pollut.* **1999**, *115*, 481–512. [CrossRef]
27. Pacheco, F.A.L.; Van der Weijden, C.H. Integrating topography, hydrology and rock structure in weathering rate models of spring watersheds. *J. Hydrol.* **2012**, *428–429*, 32–50. [CrossRef]
28. Pacheco, F.A.L.; Van der Weijden, C.H. Weathering of plagioclase across variable flow and solute transport regimes. *J. Hydrol.* **2012**, *420–421*, 46–58. [CrossRef]
29. Pacheco, F.A.L.; Van der Weijden, C.H. Modeling rock weathering in small watersheds. *J. Hydrol.* **2014**, *513*, 13–27. [CrossRef]
30. Pacheco, F.A.L.; Szocs, T. "Dedolomitization reactions" driven by anthropogenic activity on loessy sediments, SW Hungary. *Appl. Geochem.* **2006**, *21*, 614–631. [CrossRef]
31. Pacheco, F.A.L.; Landim, P.M.B. Two-Way Regionalized Classification of Multivariate Datasets and its Application to the Assessment of Hydrodynamic Dispersion. *Math. Geol.* **2005**, *37*, 393–417. [CrossRef]

32. Bellu, A.; Sanches Fernandes, L.F.; Cortes, R.M.V.; Pacheco, F.A.L. A framework model for the dimensioning and allocation of a detention basin system: The case of a flood-prone mountainous watershed. *J. Hydrol.* **2016**, *533*, 567–580. [CrossRef]
33. Salgado Terêncio, D.P.; Sanches Fernandes, L.F.; Vitor Cortes, R.M.; Moura, J.P.; Leal Pacheco, F.A.; Salgado Terêncio, D.P.; Sanches Fernandes, L.F.; Vitor Cortes, R.M.; Moura, J.P.; Leal Pacheco, F.A. Can Land Cover Changes Mitigate Large Floods? A Reflection Based on Partial Least Squares-Path Modeling. *Water* **2019**, *11*, 684. [CrossRef]
34. Terêncio, D.P.S.; Sanches Fernandes, L.F.; Cortes, R.M.V.; Pacheco, F.A.L. Improved framework model to allocate optimal rainwater harvesting sites in small watersheds for agro-forestry uses. *J. Hydrol.* **2017**, *550*, 318–330. [CrossRef]
35. Pacheco, F.A.L.; Fallico, C. Hydraulic head response of a confined aquifer influenced by river stage fluctuations and mechanical loading. *J. Hydrol.* **2015**, *531*, 716–727. [CrossRef]
36. Terêncio, D.P.S.; Sanches Fernandes, L.F.; Cortes, R.M.V.; Moura, J.P.; Pacheco, F.A.L. Rainwater harvesting in catchments for agro-forestry uses: A study focused on the balance between sustainability values and storage capacity. *Sci. Total Environ.* **2018**, *613–614*, 1079–1092. [CrossRef] [PubMed]
37. Pacheco, F.A.L. Hydraulic diffusivity and macrodispersivity calculations embedded in a geographic information system. *Hydrol. Sci. J.* **2013**, *58*, 930–944. [CrossRef]
38. Sanches Fernandes, L.F.; Santos, C.; Pereira, A.; Moura, J. Model of management and decision support systems in the distribution of water for consumption: Case study in North Portugal. *Eur. J. Environ. Civ. Eng.* **2011**, *15*, 411–426. [CrossRef]
39. Fernandes, L.F.S.; Marques, M.J.; Oliveira, P.C.; Moura, J.P. Decision support systems in water resources in the demarcated region of Douro-case study in Pinhão river basin, Portugal. *Water Environ. J.* **2014**, *28*, 350–357. [CrossRef]
40. Modesto Gonzalez Pereira, M.J.; Sanches Fernandes, L.F.; Barros Macário, E.M.; Gaspar, S.M.; Pinto, J.G. Climate Change Impacts in the Design of Drainage Systems: Case Study of Portugal. *J. Irrig. Drain. Eng.* **2015**, *141*, 05014009. [CrossRef]
41. Fonseca, A.R.; Sanches Fernandes, L.F.; Fontainhas-Fernandes, A.; Monteiro, S.M.; Pacheco, F.A.L. The impact of freshwater metal concentrations on the severity of histopathological changes in fish gills: A statistical perspective. *Sci. Total Environ.* **2017**, *599–600*, 217–226. [CrossRef]
42. Fonseca, A.R.; Sanches Fernandes, L.F.; Fontainhas-Fernandes, A.; Monteiro, S.M.; Pacheco, F.A.L. From catchment to fish: Impact of anthropogenic pressures on gill histopathology. *Sci. Total Environ.* **2016**, *550*, 972–986. [CrossRef]
43. Pacheco, F.A.L. Application of Correspondence Analysis in the Assessment of Groundwater Chemistry. *Math. Geol.* **1998**, *30*, 129–161. [CrossRef]
44. Valle Junior, R.F.; Varandas, S.G.P.; Sanches Fernandes, L.F.; Pacheco, F.A.L. Multi Criteria Analysis for the monitoring of aquifer vulnerability: A scientific tool in environmental policy. *Environ. Sci. Policy* **2015**, *48*, 250–264. [CrossRef]
45. Ferreira, A.R.L.; Sanches Fernandes, L.F.; Cortes, R.M.V.; Pacheco, F.A.L. Assessing anthropogenic impacts on riverine ecosystems using nested partial least squares regression. *Sci. Total Environ.* **2017**, *583*, 466–477. [CrossRef] [PubMed]
46. Pacheco, F.A.L. Finding the number of natural clusters in groundwater data sets using the concept of equivalence class. *Comput. Geosci.* **1998**, *24*, 7–15. [CrossRef]
47. Hughes, S.J.; Cabecinha, E.; Andrade dos Santos, J.C.; Mendes Andrade, C.M.; Mendes Lopes, D.M.; da Fonseca Trindade, H.M.; dos Santos Cabral, J.A.F.A.; dos Santos, M.G.S.; Lourenço, J.M.M.; Marques Aranha, J.T.; et al. A predictive modelling tool for assessing climate, land use and hydrological change on reservoir physicochemical and biological properties. *Area* **2012**, *44*, 432–442. [CrossRef]
48. Álvarez, X.; Valero, E.; Santos, R.M.B.; Varandas, S.G.P.; Sanches Fernandes, L.F.; Pacheco, F.A.L. Anthropogenic nutrients and eutrophication in multiple land use watersheds: Best management practices and policies for the protection of water resources. *Land Use Policy* **2017**, *69*, 1–11. [CrossRef]
49. Neitsch, S.; Arnold, J.; Kiniry, J. *Soil and Water Assessment Tool: Theoretical Documentation*; Version 2009; Texas A&M University System: College Station, TX, USA, 2011; pp. 1–647.

50. Arnold, J.G.; Moriasi, D.N.; Gassman, P.W.; Abbaspour, K.C.; White, M.J.; Srinivasan, R.; Santhi, C.; Harmel, R.D.; van Griensven, A.; van Liew, M.W.; et al. SWAT: Model use, calibration, and validation. *Trans. ASABE* **2012**, *55*, 1491–1508. [CrossRef]
51. Abbaspour, K.C.; Rouholahnejad, E.; Vaghefi, S.; Srinivasan, R.; Yang, H.; Kløve, B. A continental-scale hydrology and water quality model for Europe: Calibration and uncertainty of a high-resolution large-scale SWAT model. *J. Hydrol.* **2015**, *524*, 733–752. [CrossRef]
52. Shrestha, M.K.; Recknagel, F.; Frizenschaf, J.; Meyer, W. Assessing SWAT models based on single and multi-site calibration for the simulation of flow and nutrient loads in the semi-arid Onkaparinga catchment in South Australia. *Agric. Water Manag.* **2016**, *175*, 61–71. [CrossRef]
53. Saha, P.P.; Zeleke, K.; Hafeez, M. Streamflow modeling in a fluctuant climate using SWAT: Yass River catchment in south eastern Australia. *Environ. Earth Sci.* **2014**, *71*, 5241–5254. [CrossRef]
54. DGT Directorate-General of Territory. Available online: http://www.dgterritorio.pt/ (accessed on 28 November 2018).
55. AHBS-Aproveitamento Hidroelétrico do Baixo Sabor. *Relatório de Conformidade Ambiental do Projeto de Execução (RECAPE). Volume 1—Sumário Executivo*, Technical report developed by Ecossistema and Agri Pro Ambiente Consultores; Lisbon, Portugal, 2006; 1–33.
56. SNIRH-National System of Water Resources Information. Available online: https://snirh.apambiente.pt/ (accessed on 18 June 2018).
57. Cuba Agroconsultores. *Carta de Solos, Carta do Uso Actual da Terra e Carta de Aptidão da Terra do Nordeste de Portugal*; Version 2009; Engenharia de Recursos Agrários; Coba-Consultores de Engenharia e Ambiente; University of Trás-os-Montes and Alto Douro: Vila Real, Portugal, 1991; pp. 1–311.
58. EEA-European Environmental Agency. Corine Land Cover. Available online: https://www.eea.europa.eu/publications/COR0-landcover (accessed on 20 June 2018).
59. APA-Agência Portuguesa do Ambiente. *Plano de Gestão da Região Hidrográfica do Douro–RH3 Relatório de Base, Parte 2–Caracterização e Diagnóstico da Região Hidrográfica*; Agência Portuguesa do Ambiente: Lisbon, Portugal, 2016; pp. 1–213.
60. INE-Instituto Nacional de Estatística. Available online: https://www.ine.pt (accessed on 21 September 2018).
61. ICNF-Institute for Conservation of Nature and Forestry. Available online: https://www.icnf.pt/ (accessed on 21 September 2018).
62. Rocha, J.; Roebeling, P.; Rial-Rivas, M.E. Assessing the impacts of sustainable agricultural practices for water quality improvements in the Vouga catchment (Portugal) using the SWAT model. *Sci. Total Environ.* **2015**, *536*, 48–58. [CrossRef]
63. Baker, T.J.; Miller, S.N. Using the Soil and Water Assessment Tool (SWAT) to assess land use impact on water resources in an East African watershed. *J. Hydrol.* **2013**, *486*, 100–111.
64. Rawls, W.J.; Ahuja, L.R.; Brakensiek, D.L.; Shirmohammadi, A. Infiltration and soil water movement. In *Handbook of Hydrology*; McGraw-Hill: New York, NY, USA, 1993.
65. Koulov, B.; Zhelezov, G. *Sustainable Mountain Regions: Challenges and Perspectives in Southeastern Europe*; Springer: New York, NY, USA, 2016.
66. SWAT-Soil & Water Assessment Tool. Available online: https://swat.tamu.edu/ (accessed on 15 May 2018).
67. Li, Z.; Shao, Q.; Xu, Z.; Cai, X. Analysis of parameter uncertainty in semi-distributed hydrological models using bootstrap method: A case study of SWAT model applied to Yingluoxia watershed in northwest China. *J. Hydrol.* **2010**, *385*, 76–83. [CrossRef]
68. Santos, R.M.B.; Sanches Fernandes, L.F.; Pereira, M.G.; Cortes, R.M.V.; Pacheco, F.A.L. A framework model for investigating the export of phosphorus to surface waters in forested watersheds: Implications to management. *Sci. Total Environ.* **2015**, *536*, 295–305. [CrossRef] [PubMed]
69. Pacheco, F.A.L.; Varandas, S.G.P.; Sanches Fernandes, L.F.; Valle Junior, R.F. Soil losses in rural watersheds with environmental land use conflicts. *Sci. Total Environ.* **2014**, *485–486*, 110–120. [CrossRef] [PubMed]
70. Pacheco, F.A.L.; Santos, R.M.B.; Sanches Fernandes, L.F.; Pereira, M.G.; Cortes, R.M.V. Controls and forecasts of nitrate yields in forested watersheds: A view over mainland Portugal. *Sci. Total Environ.* **2015**, *537*, 421–440. [CrossRef] [PubMed]
71. Fernandes, L.F.S.; Terêncio, D.P.; Pacheco, F.A. Rainwater harvesting systems for low demanding applications. *Sci. Total Environ.* **2015**, *529*, 91–100.

72. Fernandes, L.S.; Fernandes, A.C.P.; Ferreira, A.R.L.; Cortes, R.M.V.; Pacheco, F.A.L. A partial least squares–Path modeling analysis for the understanding of biodiversity loss in rural and urban watersheds in Portugal. *Sci. Total Environ.* **2018**, *626*, 1069–1085. [CrossRef] [PubMed]
73. Santos, R.M.B.; Sanches Fernandes, L.F.; Pereira, M.G.; Cortes, R.M.V.; Pacheco, F.A.L. Water resources planning for a river basin with recurrent wildfires. *Sci. Total Environ.* **2015**, *526*, 1–13. [CrossRef]
74. Peschel, J.M.; Haan, P.K.; Lacey, R.E. Influences of Soil Dataset Resolution on Hydrologic Modeling. *J. Am. Water Resour. Assoc.* **2007**, *42*, 1371–1389. [CrossRef]
75. Atkinson, H.D.E.; Johal, P.; Falworth, M.S.; Ranawat, V.S.; Dala-Ali, B.; Martin, D.K. Adductor tenotomy: Its role in the management of sports-related chronic groin pain. *Arch. Orthop. Trauma Surg.* **2010**, *130*, 965–970.
76. Yang, J.; Reichert, P.; Abbaspour, K.C.; Xia, J.; Yang, H. Comparing uncertainty analysis techniques for a SWAT application to the Chaohe Basin in China. *J. Hydrol.* **2008**, *358*, 1–23. [CrossRef]
77. Moriasi, D.N.; Arnold, J.G.; Van Liew, M.W.; Bingner, R.L.; Harmel, R.D.; Veith, T.L. Model Evaluation guidelines. *Trans. ASABE* **2007**, *50*, 885–900.
78. Santhi, C.; Arnold, J.G.; Williams, J.R.; Dugas, W.A.; Srinivasan, R.; Hauck, L.M. Validation of the SWAT model on a large river basin with point and nonpoint sources. *J. Am. Water Resour. Assoc.* **2001**, *37*, 1169–1188. [CrossRef]
79. Machado, C.; Bento, J. *Plano Regional de Ordenamento Florestal do Douro*; Fase 2–Proposta de Plano; Technical report developed by Directorate-General of Forest Resources; University of Trás-os-Montes and Alto Douro and Nordeste Rural: Lisbon, Portugal, 2006; p. 266.
80. Machado, C.; Bento, J. *Plano Regional de Ordenamento Florestal do Nordeste*; Fase 2–Proposta de Plano; Technical report developed by Directorate-General of Forest Resources; University of Trás-os-Montes and Alto Douro and Nordeste Rural: Lisbon, Portugal, 2006; p. 242.
81. Carvalho-Santos, C.; Monteiro, A.T.; Azevedo, J.C.; Honrado, J.P.; Nunes, J.P. Climate Change Impacts on Water Resources and Reservoir Management: Uncertainty and Adaptation for a Mountain Catchment in Northeast Portugal. *Water Resour. Manag.* **2017**, *31*, 3355–3370. [CrossRef]
82. Rua, J.C.P. Establishment of a Hydrological Model in a Subbasin of the Sabor River. Ph.D. Thesis, Instituto Politécnico de Bragança, Escola Superior Agrária, Bragança, Portugal, 2012; pp. 1–40.
83. Úbeda, X.; Lorca, M.; Outeiro, L.R.; Bernia, S.; Castellnou, M. Effects of prescribed fire on soil quality in Mediterranean grassland (Prades Mountains, north-east Spain). *Int. J. Wildland Fire* **2005**, *14*, 379. [CrossRef]
84. Francis, C.F.; Thornes, J.B. Runoff hydrographs from three Mediterranean vegetation cover types. In *Vegetation and Erosion*; Processes and environments; Thornes, J.B., Ed.; John Wiley and Sons Ltd.: Bristol, UK, 1990; pp. 363–384. ISBN 0471926302.
85. Serrano-Muela, M.P.; Lana-Renault, N.; Nadal-Romero, E.; Regüés, D.; Latron, J.; Martí-Bono, C.; García-Ruiz, J.M. Forests and Their Hydrological Effects in Mediterranean Mountains. *Mt. Res. Dev.* **2008**, *28*, 279–285. [CrossRef]
86. Lavabre, J.; Torres, D.S.; Cernesson, F. Changes in the hydrological response of a small Mediterranean basin a year after a wildfire. *J. Hydrol.* **1993**, *142*, 273–299. [CrossRef]
87. Shakesby, R.A.; Doerr, S.H. Wildfire as a hydrological and geomorphological agent. *Earth-Sci. Rev.* **2006**, *74*, 269–307. [CrossRef]
88. Seibert, J.; McDonnell, J.J.; Woodsmith, R.D. Effects of wildfire on catchment runoff response: A modelling approach to detect changes in snow-dominated forested catchments. *Hydrol. Res.* **2010**, *41*, 378–390. [CrossRef]
89. Miranda, A.C.; Lloyd, J.; Santos, A.J.B.; Silva, G.T.D.A.; Miranda, H.S. Effects of fire on surface carbon, energy and water vapour fluxes over campo sujo savanna in central Brazil. *Funct. Ecol.* **2003**, *17*, 711–719.
90. Dunin, F.X.; Miranda, H.S.; Miranda, A.C.; Lloyd, J. Evapotranspiration responses to burning of campo sujo savanna in Central Brasil. In Proceedings of the In Bushfire'97: Australian Bushfire Conference, CSIRO Tropical Ecosystems Research Centre, Darwin, Australia, 8–10 July 1997; pp. 146–151.
91. Nagra, G.; Treble, P.C.; Andersen, M.S.; Fairchild, I.J.; Coleborn, K.; Baker, A. A post-wildfire response in cave dripwater chemistry. *Hydrol. Earth Syst. Sci.* **2016**, *20*, 2745–2758. [CrossRef]
92. Coelho, C.D.O.A.; Ferreira, A.J.D.; Boulet, A.K.; Keizer, J.J. Overland flow generation processes, erosion yields and solute loss following different intensity fires. *Q. J. Eng. Geol. Hydrogeol.* **2004**, *37*, 233–240. [CrossRef]

93. Ferreira, A.J.D.; Coelho, C.O.A.; Boulet, A.K.; Leighton-Boyce, G.; Keizer, J.J.; Ritsema, C.J. Influence of burning intensity on water repellency and hydrological processes at forest and shrub sites in Portugal. *Aust. J. Soil Res.* **2005**, *43*, 327–336. [CrossRef]
94. López, R.; Batalla, R.J. Análisis del comportamiento hidrológico de la cuenca mediterránea de Arbúcies antes y después de un incendio forestal. In Proceedings of the III Congreso Forestal Español, Granada, Espanã, 25–28 de Septiembre 2001; pp. 1–6.
95. Mayor, A.G.; Bautista, S.; Llovet, J.; Bellot, J. Post-fire hydrological and erosional responses of a Mediterranean landscpe: Seven years of catchment-scale dynamics. *Catena* **2007**, *71*, 68–75. [CrossRef]
96. Walsh, R.P.D.; Boakes, D.; Coelho, C.D.O.; Gonçalves, A.J.B.; Shakesby, R.A.; Thomas, A. Impact of fire-induced hydrophobicity and post-fire forest litter on overland flow in northern and central Portugal. In Proceedings of the Second International Conference on Forest Fire Research, Coimbra, Portugal, 21–24 November 1994; pp. 1149–1159.
97. Shakesby, R.A.; Bento, C.P.M.; Ferreira, C.S.S.; Ferreira, A.J.D.; Stoof, C.R.; Urbanek, E.; Walsh, R.P.D. Impacts of prescribed fire on soil loss and soil quality: An assessment based on an experimentally-burned catchment in central Portugal. *Catena* **2015**, *128*, 278–293. [CrossRef]
98. AHBS-Aproveitamento Hidroelétrico do Baixo Sabor. *Relatório de Monitorização do Estado das Águas Superficiais (PMEAS), Relatório Anual de 2013*, Technical report developed by Ecovisão-Tecnologias do Meio Ambiente; Lisbon, Portugal, 2014; 1–180.
99. Sanches Fernandes, L.F.; Pacheco, F.A.L.; Cortes, R.M.V.; Jesus, J.J.B.; Varandas, S.G.P.; Santos, R.M.B. Integrative assessment of river damming impacts on aquatic fauna in a Portuguese reservoir. *Sci. Total Environ.* **2017**, *601–602*, 1108–1118.

Article

Effects of Human Activities on Hydrological Components in the Yiluo River Basin in Middle Yellow River

Xiujie Wang [1], Pengfei Zhang [1,2], Lüliu Liu [3,*], Dandan Li [1] and Yanpeng Wang [1]

1 State Key Laboratory of Hydraulic Engineering Simulation and Safety, Tianjin University, Tianjin 300350, China; Wangxiujie@tju.edu.cn (X.W.); m18763822967@163.com (P.Z.); 18832012027@163.com (D.L.); wangyanpeng0512@163.com (Y.W.)

2 Tianjin University Frontier Technology Research Institute, No.1 Xinxing Road, Wuqing Development Area, Wuqing District, Tianjin 301700, China

3 National Climate Center, China Meteorological Administration, Beijing 100081, China

* Correspondence: liull@cma.gov.cn; Tel.: +86-10-58995906

Received: 4 March 2019; Accepted: 29 March 2019; Published: 3 April 2019

Abstract: Land use and land cover change (LUCC) and water resource utilization behavior and policy (WRUBAP) affect the hydrological cycle in different ways. Their effects on streamflow and hydrological balance components were analyzed in the Yiluo River Basin using the delta method and the Soil and Water Assessment Tool (SWAT). The multivariable (runoff and actual evapotranspiration) calibration and validation method was used to reduce model uncertainty. LUCC impact on hydrological balance components (1976–2015) was evaluated through comparison of simulated paired land use scenarios. WRUBAP impact on runoff was assessed by comparing natural (simulated) and observed runoff. It showed that urban area reduction led to decreased groundwater, but increased surface runoff and increased water area led to increased evaporation. LUCC impact on annual runoff was found limited; for instance, the difference under the paired scenarios was <1 mm. Observed runoff was 34.7–144.1% greater than natural runoff during November–June because of WRUBAP. The effect of WRUBAP on wet season runoff regulation was limited before the completion of the Guxian Reservoir, whereas WRUBAP caused a reduction in natural runoff of 21.6–35.0% during the wet season (July–October) after its completion. The results suggest that WRUBAP has greater influence than LUCC on runoff in the Yiluo River Basin. Based on existing drought mitigation measures, interbasin water transfer measures and deep groundwater exploitation could reduce the potential for drought attributable to predicted future climate extremes. In addition to reservoir regulation, conversion of farmland to forestry in the upstream watershed could also reduce flood risk.

Keywords: changes in hydrological components; effects of human activities; LUCC; WRUBAP; Yiluo River

---

1. Introduction

The climate and human activities are two factors that can affect the hydrological cycle in different ways. Climate change alters hydrological systems by inducing both spatiotemporal variations of regional precipitation and changes in temperature [1–3]. Compared with climate change, human activities are more controllable; thus, alteration of human activities constitutes the principal measure for dealing with the potential impacts of climate change on hydrological systems. With recent developments of society and technology, human activities have gradually increased and their consequential impacts on the hydrological cycle on different spatiotemporal scales, such as river basins, have become widely recognized. In general, human activities in river basins can be divided into land use and land cover change (LUCC), e.g., vegetation degradation, deforestation, and urbanization, and water resource utilization behavior and policy (WRUBAP), e.g., agricultural irrigation [4], reservoir regulation [5–7], deep groundwater extraction, and interbasin water diversion.

Numerous studies have shown that LUCC can have considerable impact on watershed hydrology in terms of water quantity and water quality. For example, vegetation degradation causes decline in the water storage capacity, which ultimately causes changes in the hydrological balance components (e.g., evapotranspiration, infiltration, and baseflow), or vice versa [8,9]. Moreover, changes in surface albedo, surface aerodynamic roughness, leaf area, and rooting depth caused by changing land use can influence the hydrological cycle via different processes. Land use conflicts have impacts on concentration or yield of nitrates, phosphorus, and sulphates, etc. in both surface water and ground water [10–12]. Hence, understanding the hydrological response of a watershed to LUCC constitutes an important step toward sustainable water resources management. Reasonable WRUBAP is another focus of global attention, particularly in semiarid and subhumid agricultural regions where appropriate watershed management is extremely important in relation to the prevention of droughts/flooding. WRUBAP can affect runoff directly via water intake, water transfer, and reservoir operation and indirectly via other components of the water balance, e.g., groundwater, actual evapotranspiration (ET) and infiltration. The mechanisms via which LUCC and WRUBAP affect hydrological processes are different. Therefore, it is highly important to separate the impacts of LUCC and WRUBAP for sustainable water resources utility and water management, which could ultimately ensure manageable agricultural development and ecological environment protection [13–15].

The paired catchments approach [16–18], statistical analyses, and other modeling approaches [19,20] are methods commonly adopted to further the understanding of how human activities might influence basin hydrology. The paired catchments approach has certain limitations attributable to the sizes and characteristics of the watersheds. Analyses based on statistical methods represent a data-driven approach and thus are not based on physical processes. Distributed hydrological models are physically based models that have been used widely to simulate hydrological responses to human activities [13,21,22]. The Soil and Water Assessment Tool (SWAT) [23–25], which is one of the models used most commonly in hydrological response studies, can be used to assess the influence of human activities on the hydrological balance. The delta approach is often applied in association with the SWAT both to analyze the impacts of LUCC through comparison of hydrological responses under two different land use scenarios for the same time frame, and to evaluate the impacts of WRUBAP by comparing observed values with simulated values that are considered representative of natural runoff.

The Yiluo River Basin in China is located in a region where a semiarid area borders a subhumid area. It supplies water for an important grain-producing region in Henan Province. Regional agricultural production and food security are frequently threatened by serious flooding and droughts attributable to the monsoon climate. Increasingly severe problems regarding water resources are expected in the future because drier springs and increasingly severe flooding over long return periods (25 and 50 years) have been projected under the 1.5 and 2 °C global warming scenarios [26]. However, changes in regional land use have been detected since the 1990s, e.g., the area of cultivated land has decreased and the area of urbanized land has increased [27]. The effects of such changes in land use on hydrological processes have been investigated in previous studies [21,28]. For example, the change characteristics of intra-annual and interannual runoff, as well as the relative contributions of climate change and human activities on the decrease in annual runoff, have been explored in earlier research [29,30]. However, their effects on other hydrological variables have rarely been investigated.

This study had two primary objectives: (1) to assess the impacts of LUCC on hydrological balance components, and to determine the relative contributions of individual land use types and (2) to analyze the impact of WRUBAP on streamflow during different periods. The remainder of this paper is organized as follows. Section 2 provides an overview of the study river basin, data, and methods. Section 3 presents the results of both the calibration and validation of the hydrological model and the responses of the hydrological components to LUCC and WRUBAP. The approaches adopted to reduce the uncertainty of model parameters and the effects of LUCC and WRUBAP on the hydrological processes are discussed in Section 4. Finally, the main conclusions are summarized in Section 5.

2. Materials and Method

2.1. Study Area

The Yiluo River Basin (33°–35° N, 109°–113° E) covers an area of around 18,563 km^2 (Figure 1). The river, which is 449 km long, runs from Shaanxi Province through Henan Province and into the Yellow River in the city of Luoyang. The Heishiguan hydrological station is located at the outlet of the Yiluo River in the northeast of the basin. The Yiluo River has two principal tributaries: the Yihe River and the Luohe River, on each of which is a large reservoir. The Lushun Reservoir, which is located in the middle reaches of the Yihe River, was completed in 1965. The Guxian Reservoir, located in the middle reaches of the Luohe River, was completed in 1994. There are ten types of soil within the Yiluo River watershed: Acrisols, Alisols, Andosols, Arenosols, Anthrosols, Cambisols, Fluvisols, Gleysols, Solonchaks, and Vertiaols (Figure 2). The major land use classes of the region are cultivation (AGRR), barren (BARR), forest (FRST), grassland (RNGE), urban (URBN), and water (WATR) (Figure 3).

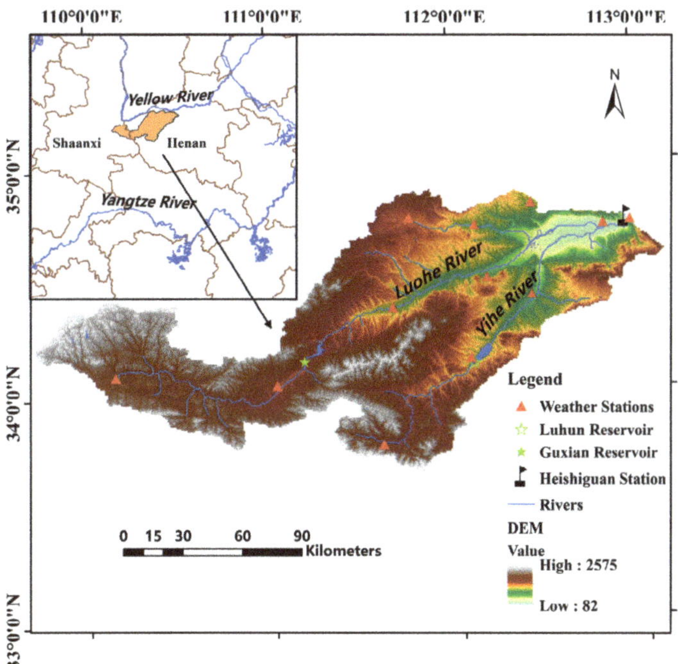

Figure 1. Study area with weather stations, reservoirs, and watershed outlets.

Dominated by a typical continental monsoon climate, precipitation is concentrated in the wet season, and about 60% of the total precipitation falls during June–September [30]. The annual mean temperature is about 12 °C, and the annual averaged precipitation is in the range 414–1066 mm (1960–2016). The fluctuation of precipitation causes corresponding changes in the hydrological components. The main source of water supply in the basin is local surface runoff, shallow groundwater and deep groundwater. During the 1980s, local surface runoff accounted for approximately 58–69% of the total. Local surface runoff is lower in dry years than in normal years, while water demand increases in dry years because of the requirements of agriculture; consequently, such conditions lead to water shortages [31]. Generally, water consumption increases with socioeconomic development and an increasing tendency has been observed during the past 20 years in China from Water Resources Bulletins from 1997–2017 (www.mwr.gov.cn).

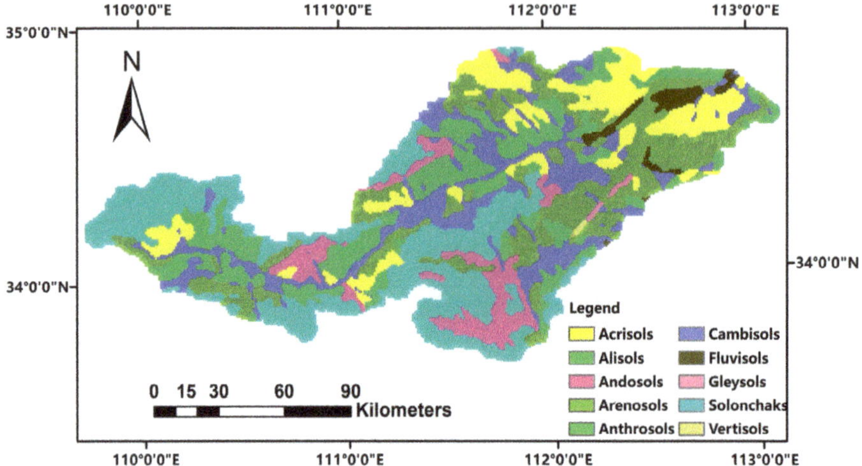

Figure 2. Soil types within the study area.

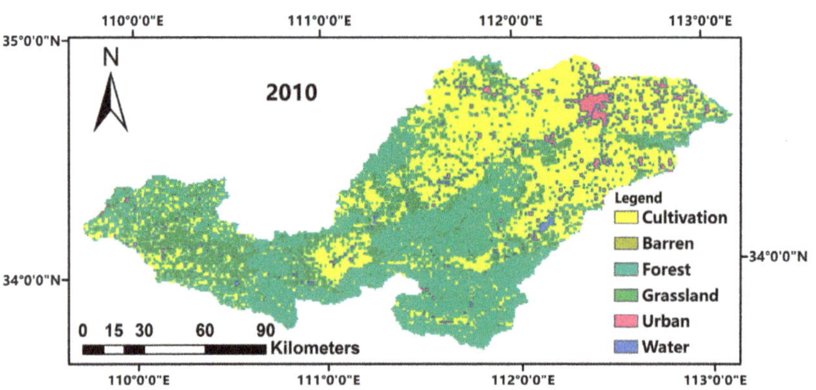

Figure 3. Map of land use maps in 2010 in the Yiluo River Basin (maps for 1980, 1990, and 2000 not shown).

2.2. Dataset

A digital elevation model (DEM) maps of soil types and properties, LUCC, meteorological variables, and river discharge data were used in this study. The digital elevation model with 90-m resolution was obtained from the Geospatial Data Cloud of China (http://www.gscloud.cn). The LUCC maps with 1-km resolution for 1980, 1990, 2000, and 2010 were acquired from the Resource and Environment Data Cloud Platform (http://www.resdc.cn). These were used to analyze the changes in land use and to investigate the effects of such changes on the water balance. Soil maps with 1-km resolution were obtained from the Harmonized World Soil Database v1.1 (http://www.fao.org/soils-portal/soil-survey/en). Soil parameters used in the hydrological model were estimated using the Soil–Plant–Air–Water budgeting tool.

Daily minimum temperature, maximum temperature, precipitation, relative humidity, wind speed, and sunshine duration at 12 meteorological stations and daily evaporation at the Guxian hydrological station during 1969–2015 were obtained from the National Meteorological Information Center of the China Meteorological Administration, as shown in Figure 1. Evaporation was monitored using E601B pan.

Monthly data of observed streamflow through the Heishiguan hydrological station during 1971–2015 were obtained from the Hydrological Yearbook of the Yellow River. These data were used both to calibrate and validate the hydrological model and to analyze the effects of WRUBAP on runoff.

2.3. Hydrological Modeling

2.3.1. Model Introduction

SWAT [23,32] is a semidistributed hydrological model that is used widely to simulate the effects of varying soils, land use, and management practices on hydrological conditions, sediment loading, and contamination on the basin scale [32,33]. This model was applied in this study to the Yiluo River Basin to assess the impacts of LUCC and WRUBAP on hydrological balance components: surface runoff, evapotranspiration, return flow, and lateral flow. The daily water-balance equation can be expressed as follows:

$$SW_t = SW_0 + \sum_{i=1}^{t}\left(R_{day} - Q_{surf} - ET - W_{seep} - Q_{gw}\right) \quad (1)$$

where SW_t is the final soil moisture content on day t (mm), SW_0 is the initial soil moisture content (mm), R_{day} is the amount of precipitation on day i (mm), Q_{surf} is the amount of surface runoff on day i (mm), ET is the amount of actual evapotranspiration on day i (mm), W_{seep} is the amount of water transferred from the soil profile into the gas zone on day i (mm), and Q_{gw} is the return flow of day i (mm).

2.3.2. Model Setup

The Yiluo River Basin was divided into 29 sub-basins (Figure 4) and the hydrological response units were based on the land uses, soil types, and slope classes. The modified Soil Conservation Service Curve Number method [34,35], Penman–Monteith method [36,37], and Muskingum routing method were used to simulate the hydrological components.

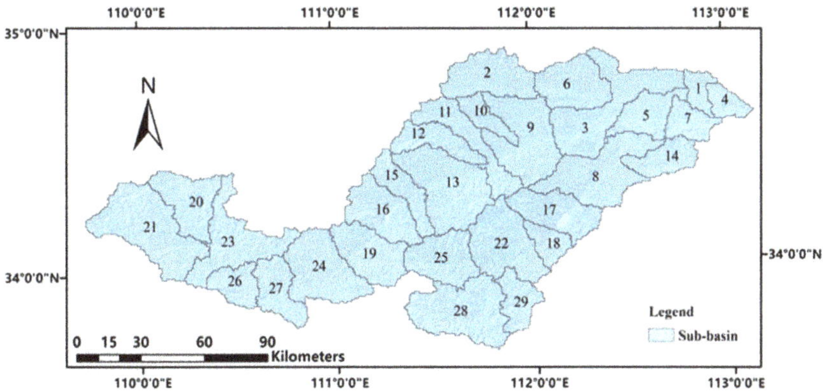

Figure 4. Sub-basins within the study area.

2.3.3. SWAT Calibration and Validation

The SWAT model was calibrated using observed monthly runoff and evaporation data series from 1971–1982, and it was validated using monthly historical data from 1983–1985. The 1985 cutoff date was based on the recognition that natural runoff has been changed by human activities since 1986, because a change point in the annual runoff series was detected around 1986, whereas no change point was observed in the annual precipitation series [30].

To reduce model parameter uncertainty and equifinality, the multivariable (runoff and actual evapotranspiration) calibration and validation method (MCVM) was used in this study. Parameters were auto-calibrated using Particle Swarm Optimization (PSO) [38] in the SWAT-CUP (SWAT Calibration and Uncertainty Programs) package [39], based on monthly observed discharge through the Heishiguan hydrological station and the set of 20 initial sensitive parameters listed in Table 1. The initial two years of the simulated outputs were used as a warm-up period [40]. After autocalibration, the performance of ET was improved further by manual adjustment of EPCO (Plant uptake compensation factor) and ESCO (Soil evaporation compensation factor), which are parameters related to soil evaporation and plant transpiration [41]. The formula for calculation of ET is as follows:

$$ET = K \times ET_{pan} \tag{2}$$

where ET is the actual evapotranspiration (mm), ET_{pan} is the observed evaporation monitored using an E601B pan and K is a conversion coefficient of evaporation, the value of which is 0.81 according to [42].

Three indices including the Nash–Sutcliffe efficiency (NSE) [43], coefficient of determination (R^2), and percent bias ($PBIAS$) were used to evaluate the model's performance. The performance of the SWAT model can be judged satisfactory if R^2 and NSE are both >0.5 and $PBIAS$ is <±25% [44].

$$NSE = 1 - \frac{\sum_{i=1}^{n}\left(Q_i^{sim} - Q_i^{obs}\right)^2}{\sum_{i=1}^{n}\left(Q_i^{obs} - \overline{Q^{obs}}\right)^2} \tag{3}$$

$$R^2 = \frac{\left[\sum_{i=1}^{n}\left(Q_i^{obs} - \overline{Q^{obs}}\right)\left(Q_i^{sim} - \overline{Q^{sim}}\right)\right]^2}{\sum_{i=1}^{n}\left(Q_i^{obs} - \overline{Q^{obs}}\right)^2 \sum_{i=1}^{n}\left(Q_i^{sim} - \overline{Q^{sim}}\right)^2} \tag{4}$$

$$PBIAS = \frac{\sum_{i=1}^{n}\left(Q_i^{obs} - Q_i^{sim}\right) \times 100}{\sum_{i=1}^{n}\left(Q_i^{obs}\right)} \tag{5}$$

where n is the total number of sample pairs, Q_i^{obs} is the observed value, $\overline{Q^{obs}}$ is the mean of the observed values, Q_i^{sim} is the simulated value and $\overline{Q^{sim}}$ is the mean of the simulated values.

Table 1. Optimal parameter values calibrated for the Yiluo River Basin.

Parameters	Description	Range	Value
a_CN2	Curve number	±20	−10.65
v_GWQMN	Threshold depth of water in the shallow aquifer required for return flow to occur (mm)	50–100	59.68
v_GW_REVAP	"Revap" coefficient	0–0.2	0.22
v_EPCO*	Plant uptake compensation factor	0.01–1	0.47
v_CH_N2	Manning's "n" value for the main channel	0–0.3	0.01
v_CH_K2	Effective hydraulic conductivity in main channel (mm/h)	0–180	219.56
r_SOL_AWC(1)	Available water capacity of the soil layer	±100%	−21.10%
r_SOL_K(1)	Saturated hydraulic conductivity (mm/h)	±100%	−66.74%
v_ ALPHA_BF	Base-flow recession constant	0.1–0.8	0.67
v_RCHRG_DP	Deep aquifer percolation fraction	0.1–0.8	0.32
v_REVAPMN	Threshold depth of water in the shallow aquifer for "Revap" to occur (mm)	50–450	174.59
v_CANMX	Maximum canopy storage	3	12.55
v_SURLAG	surface runoff lag time	1–24	31.47
v_ALPHA_BNK	Baseflow alpha factor for bank storage	0.1–0.8	0.60
v_SFTMP	Snowfall temperature (°C)	±5	0.51
v_SMTMP	Snowmelt base temperature (°C)	±5	2.42
v_SMFMX	Melt factor for snow on June 21 (mm H_2O/°C-day)	0–10	4.30
v_SMFMN	Melt factor for snow on December 21 (mm H_2O/°C-day)	0–10	1.33
V_TIMP	Snow pack temperature lag factor	0.1–0.8	0.83
v_ESCO*	Soil evaporation compensation coefficient	0.01–1	0.54

v__ means the current parameter value is to be replaced by a given value, a__ means a given value is added to the current parameter value, and r__ means the current parameter value is multiplied by (1 + a given value). *means the optimal parameter value was finally determined by manual calibration based on actual evaporation (ET).

2.4. Evaluation Effects of LUCC and WRUBAP

The delta approach was applied to assess the impact of LUCC. First, four scenarios (S1, S2, S3, and S4) were defined based on available LUCC data for 1980, 1990, 2000, and 2010 and the contemporaneous climate, which was taken as the ten-year mean during 1976–1985, 1986–1995, 1996–2005, and 2006–2015, respectively. Among them, S1 was set as the baseline period. The runoff simulated under the four actual LUCC scenarios approximately emulated the natural runoff on the monthly scale. Then, three LUCC scenarios (S2*, S3*, and S4*) were defined using the same climate series as in S2, S3, and S4 and LUCC data for 1980 (Table 2). The effects of LUCC were evaluated by comparing S2 against S2*, S3 against S3*, and S4 against S4* using the following Equation:

$$\text{Deviation}_{\text{lucc}} = W^{S}_{sim} - W^{S*}_{sim} \tag{6}$$

where $\text{Deviation}_{\text{lucc}}$ is the change of the water balance term under the actual scenario relative to the baseline scenario, W^{S*}_{sim} is the water balance term under the baseline scenario and W^{S}_{sim} is the water balance term under the actual scenario.

The effects of WRUBAP on river runoff were evaluated by comparing the simultaneously simulated and observed runoff under each of the S1, S2, S3, and S4 scenarios using the following Equation:

$$\text{Deviation}_{\text{WRUBAP}} = Q_{sim} - Q_{obs} \tag{7}$$

where $\text{Deviation}_{\text{WRUBAP}}$ is the difference between the simulated and observed runoff, Q_{sim} is the simulated runoff under the actual LUCC and Q_{obs} is the observed runoff.

Table 2. Scenario setting based on climate and land use and land cover change (LUCC).

Scenarios		Climate Data	LUCC Data
Baseline	S1	1976–1985	1980
Actual LUCC	S2	1986–1995	1990
	S3	1996–2005	2000
	S4	2006–2015	2010
Baseline LUCC	S2*	1986–1995	1980
	S3*	1996–2005	1980
	S4*	2006–2015	1980

3. Result

3.1. Model Calibration and Validation

The *NSE* value for monthly runoff during the calibration periods was 0.85, the R^2 value was 0.88 and the *PBIAS* value was 13.37. The *NSE* for yearly *ET* during calibration was 0.84, the R^2 value was 0.87 and the *PBIAS* value was −1.36, indicating good agreement between the monthly simulated values and the observation values during the calibration period. Moreover, satisfactory results were also obtained during the validation, with *NSE*, R^2, and *PBIAS* values of 0.72, 0.75, and 12.01, respectively, in the runoff validation and 0.80, 0.81, and −6.06, respectively, in the *ET* validation. The results of the peak value were reasonably accurate, but the base flow was underestimated (Figure 5a). Similarly, the peaks of $ET > 150$ mm/month were also underestimated (Figure 5b). Overall, the simulations for water discharge and *ET* indicated satisfactory agreement between the observed and simulated values; thus, the parameters were considered credible and the model deemed suitable for simulating the natural water balance of the Yiluo River Basin.

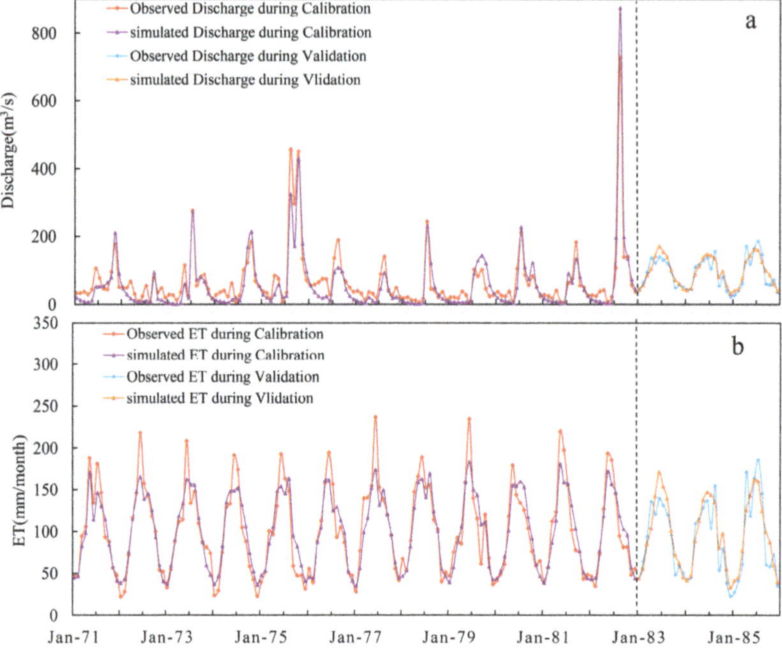

Figure 5. Simulated and observed monthly (**a**) discharge and (**b**) actual evapotranspiration (*ET*) during calibration period (1971–1982) and validation period (1983–1985).

3.2. Analysis of LUCC

The areas of AGRR, FRST, RNGE, URBN, WATR, and BARR in 1980, 1990, 2000, and 2010, as well as their changes relative to 1980, are listed in Table 3. In 1980, the main land use types were AGRR, RNGE, and FRST, which accounted for 94.8% of the total watershed area. Through urban centralization and the "Returning farmland to grass" policy, sporadic areas of URBN were converted to AGRR, while some AGRR land was converted to RNGE during 1980–1990. The RNGE area increased by 115 km^2 from 3038 km^2, but the area of AGRR, URBN, and FRST decreased by 59, 48, and 7 km^2, respectively, in 1990 relative to 1980. After 1990, because of reservoir construction and enhanced urbanization, the areas of WATR and URBN both increased, while the areas of the other four main land use types decreased. The WATR area increased by 35 km^2 in 2000 and by 38 km^2 in 2010, URBN increased by 121 km^2 in 2000 and by 187 km^2 in 2010, but AGRR decreased by 72 and 148 km^2 in 2000 and 2010, respectively, relative to 1980.

Table 3. LUCC in 1980, 1990, 2000, and 2010 relative to 1980.

LUCC	Area(km^2)					
	AGRR	**FRST**	**RNGE**	**URBN**	**WATR**	**BARR**
1980	8166	6481	3038	664	297	9
1990	8107	6474	3153	616	297	7
2000	8094	6459	2976	785	332	8
2010	8018	6469	2977	851	335	8
1990–1980	−59	−7	115	−48	0	−2
2000–1980	−72	−22	−62	121	35	−1
2010–1980	−148	−12	−61	187	38	−1

3.3. Responses of Hydrological Components and Runoff to LUCC

Table 4 summarizes the annual rainfall, *ET*, surface runoff (SURQ), return flow (GWQ), and runoff depth (RD) under each scenario and their difference between paired scenarios. Comparatively, the values of ET and GWQ are higher by 0.2 and 0.1 mm but SURQ and RD are lower by 0.3 and 0.4 mm, respectively, under scenario S2 relative to S2*. The values of ET, SURQ, and RD are higher by 1.0, 0.5, and 0.2 mm, respectively, and GWQ is lower by 0.5 mm under scenario S3 relative to S3*. The values of *ET* and SURQ are higher by 0.9 and 0.7 mm, respectively, and GWQ is lower by 0.3 mm under scenario S4 relative to S4*.

Table 4. Water balance components and the difference between paired scenarios.

Scenario	Rainfall (mm)	*ET* (mm)	SURQ (mm)	GWQ (mm)	RD (mm)
S1	698.6	574.8	60	51.9	133.4
S2	611.4	563	27.2	22.4	61.5
S2*	611.4	562.8	27.5	22.3	61.9
S2–S2*	0	0.2	−0.3	0.1	−0.4
S3	669.7	562.4	53	44.7	116.7
S3*	669.7	561.4	52.5	45.2	116.5
S3–S3*	0	1.0	0.5	−0.5	0.2
S4	645.8	571.9	41.4	30.2	88.8
S4*	645.8	571	40.7	30.5	88.8
S4–S4*	0	0.9	0.7	−0.3	0

Generally, ET occurs in three forms: vegetation transpiration (VT), soil evaporation (SE), and water evaporation (WE) [45]. In the efficient consumption of ET, vegetation land uses (AGRR, FRST, and RNGE) possess greater VT and SE than URBN land use. The total amount of WE from WATR is much higher than from the other five LUCC types. Under scenario S2, the increase in ET was caused mainly by the reduction of the URBN area and the expansion of RNGE. It also caused a slight decrease of RD. The URBN and RNGE under scenarios S3 and S4 presented an opposite trend to scenario S2 However, the trend of increase in ET was enhanced, which could be explained by the obvious expansion of the WATR area. In addition, plant root systems change the physical properties of soil, improve infiltration and GWQ and ultimately decrease SURQ. This suggests that the expansion of vegetation land uses leads to increase of both GWQ and ET but to decrease of SURQ, which consequently leads to higher water-holding capacity. These effects can be seen under scenario S2 relative to S2*. The effects of URBN expansion are opposite to those of vegetation expansion because URBN does not allow much soil evaporation or water infiltration; thus, most precipitation within an URBN area flows into rivers in the form of SURQ. The continuous urbanization after 1990 in the study area has played a positive role in the increase of SURQ.

3.4. Response of Runoff to WRUBAP

This study analyzed the effect of WRUBAP on RD only because of data availability. In this part, the simulated RD under the baseline scenario (S1) and the actual LUCC scenarios (S2, S3, and S4) are assumed to represent natural RD.

Scenario S1 (1976–1985): At Hesihiguan station, the observed monthly runoff under scenario S1 was obviously higher than the natural runoff in the first seven months of the year but lower in the final three months (Figure 6a). The main cause was reservoir regulation performed to meet the needs of agricultural irrigation during April–June. Consistency between the dry season and the irrigation period in the study area requires reservoir regulation. Water storage in the reservoir during October–December in most cases caused the observed runoff to be lower than the natural runoff during this period. To supply crop growth with sufficient water during April–June, the watershed management commission would release reservoir storage. That action resulted in observed runoff (78.1 mm) being higher than the natural RD (56.7 mm) during the dry season, and WRUBAP accounted for about 37.7% of the natural RD (Table 5). In addition, because of flood prevention, reservoir storage capacity was reserved by discharging water in the early wet season; therefore, observed runoff was slightly higher than natural runoff during July. Ultimately, the effect of WRUBAP on runoff was reasonably small during the wet season under scenario S1.

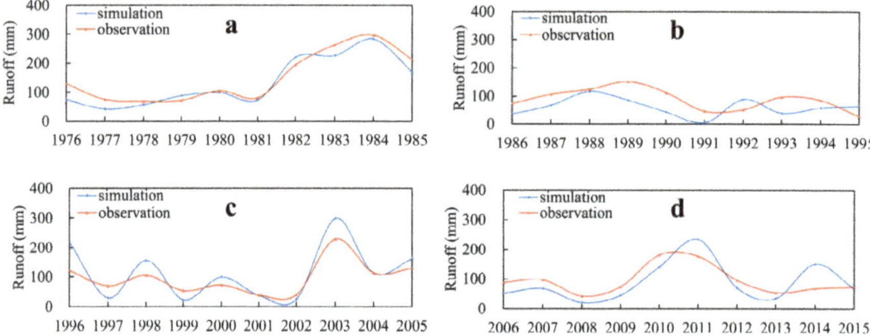

Figure 6. Comparison between the simulated and observed annual runoff values for scenarios (**a**) S1, (**b**) S2, (**c**) S3, and (**d**) S4.

Table 5. Comparison between the simulated (natural) and observed mean discharge during the dry season and the wet season under scenarios S1, S2, S3, and S4.

Scenario	Time Frame	RD (mm)		Deviation (mm)	Deviation Rate (%)
		Simulation	Observation		
S1	annual	133.4	149.7	16.3	12.2
	Wet season	286.8	292.7	5.9	2.1
	Dry season	56.7	78.1	21.4	37.7
S2	annual	61.5	88.6	27.1	44.1
	Wet season	125.0	121.1	−3.9	−3.1
	Dry season	29.7	72.4	42.7	144.1
S3	annual	116.7	97.3	−19.4	−16.6
	Wet season	254.2	165.2	−89.0	−35.0
	Dry season	47.0	63.3	16.3	34.7
S4	annual	88.8	95.4	6.6	7.5
	Wet season	185.7	145.5	−40.1	−21.6
	Dry season	40.8	70.4	29.6	72.6

Notes: Deviation = observation − simulation; Deviation rate = Deviation/simulation×100%. Dry season: January–June, November and December, Wet season: July–October.

Scenario S2 (1986–1995): Observed annual runoff under scenario S2 was greater than natural runoff in most years except 1992 and 1995, as shown in Figure 6b. Moreover, the observed monthly runoff was greater than the natural runoff in both the dry season and the early wet season (July), as shown in Figure 7b. This might be attributable to reservoir regulation and to irrigation through the extraction of deep groundwater. Storing water in the reservoir upstream of the Heishiguan station during August and September led to the observed runoff being smaller than the natural runoff. Conversely, discharging water during the dry season led to the observed runoff being greater than the natural runoff. The extraction of deep groundwater to supply irrigation increased the base flow, which ultimately led to the observed RD (72.4 mm) being higher by 144.1% than the natural RD (29.7 mm) during the dry season (Table 5). Consequently, the impact of WRUBAP on runoff in the wet season under scenario S2 was reasonably small.

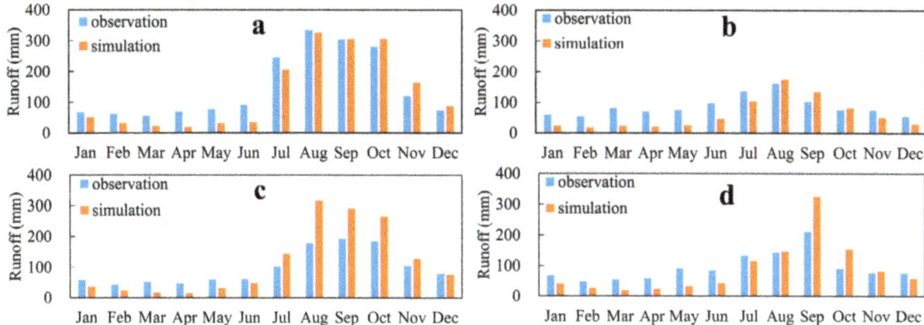

Figure 7. Simulated and observed monthly runoff under scenarios (**a**) S1, (**b**) S2, (**c**) S3, and (**d**) S4.

Scenario S3 (1996–2005): As shown in Figure 6c, annual observed runoff was smaller in the relatively wet years (1996, 1998, 2000, 2003, and 2005) but greater in the relatively dry years (1997, 1999, and 2002) than natural runoff. This might be due to stronger interannual water resource management than under scenarios S1 and S2 to prevent floods during the wet years and to prevent droughts during the dry years. The effects of regulation can also be seen in Figure 7c and Table 5. The observed RD (165.2 mm) was smaller by 35.0% than the natural RD (254.2 mm) during the wet season, but the observed RD (63.3 mm) was higher by 34.7% than the natural RD (47.0 mm) during the dry season. It indicates that the effect of WRUBAP on runoff under scenario S3 was enhanced during the wet season.

Scenario S4 (2006–2015): Similar to scenario S3, observed annual runoff was smaller in relatively wet years but greater in relatively dry years (Figure 7d) than natural runoff. Natural annual RD under scenario S3 (116.7 mm) was higher by 27.9 mm than under scenario S4 (88.8 mm). However, annual observed RD under scenario S3 (97.3 mm) was almost equal to that under scenario S4 (95.4 mm) (Table 5). Observed monthly RD (145.5 mm) was lower by 21.6% than natural RD (185.7 mm) during the wet season but higher by 34.7% than natural RD (40.8 mm) during the wet season. It suggests the effect of WRUBAP under scenario S4 alleviates the unbalanced temporal distribution of the water resource.

4. Discussion

Parameter uncertainty and equifinality can influence numerical model performance. At the basin scale, multiobjective optimization algorithms [46–48], multiperiod parameterization [49,50], and the MCVM are methods commonly adopted to reduce uncertainty. Generally, the MCVM is based on observed discharge at multiple hydrological stations. In this study, the two variables of discharge and *ET* were used to calibrate/validate a numerical model to reduce parameter uncertainty and to improve *ET* performance. Although human activities, especially WRUBAP, were not accounted for in the calibration and validation of SWAT model, the model performance is satisfactory. Simulation, not observation, during the baseline period was compared with other scenarios to explore the effects of human activities.

In this study, the CN value of FRST was lower than other five land use types, which indicates that FRST has a stronger ability of water storage and is likely to reduce more SURQ. It has previously been found that Grassland and cultivation have similar influences on hydrological balance components in this area [14], and the mutual transformation between them has little effect on water yield within the Yiluo River basin. Compared with herbaceous plants (grassland and cultivation), woody plants have more developed root systems and canopies; thus, conversion of herbaceous plants into forests would enhance the capacity of catchment to conserves soil and water, and reduce water yield [51,52], which consequently affects the yield of phosphorus and nitrates [12,53]. Considering that the upstream regions of the Yiluo River Basin are mountainous areas and are not the main crop region, returning farmland to forests is conducive to preventing land degradation, increasing infiltration, and thus reducing flood risk [54,55]. However, quantitative analysis has not been performed in this study. This will be carried out in further study.

Hydrological changes have important impacts on the local use of natural water [56]. However, compared with LUCC, WRUBAP has more dynamic and greater impact on hydrological processes, particularly in monsoon climate zones. It is noticed that the Guxian Reservoir greatly improved the capacity of runoff regulation in the wet season. Therefore, reservoirs could be used to respond to climate variety, which increases runoff in the dry season while decreasing runoff in the wet season. A recent study made similar conclusions, indicating that large and small reservoirs can have equally large effects on runoff [57].

5. Conclusions

In this study, the SWAT model and the MCVM were used to assess the impact of human activities (LUCC and WRUBAP) on the catchment hydrology of the Yiluo River Basin during 1976–2015. Based on the analysis, the following conclusions were drawn.

(1) Increased areas of urban land and decreased areas of vegetation land resulted in decrease of both groundwater and ET, but increase of average surface runoff, and vice versa. Although the expansion of water areas resulted in an increase of ET, it had little effect on groundwater and surface runoff in the studied region. It was found that LUCC had limited effect on annual runoff, i.e., the difference in annual runoff under paired scenarios was <1 mm.

(2) WRUBAP has greater influence than LUCC on runoff. Affected by WRUBAP, the observed RD varied in the range 63.3–78.1 mm during the dry season, which was 34.7%–144.1% greater than natural RD. After completion of the Guxian Reservoir, the observed RD was found to be 21.6%–35.0% less than natural RD.

(3) Returning farmland to forestry in upstream regions and adopting interbasin water transfer and deep groundwater exploitation could be useful supplementary measures for reducing the risk of extreme hydrological events in the future.

Author Contributions: Conceptualization, X.W. and L.L; data curation, P.Z.; Methodology L.L. and P.Z.; Resources, X.W. and L.L.; Software, P.Z. and L.L.; Supervision, Z.W. and L.L; Writing—Original Draft, L.L., X.W. and P.Z.; Manuscript revision/review D.L. and Y.W.

Funding: This study was supported by the National Key R&D Program of China (Grant Nos. 2017YFA0605004, 2016YFE0102400, and 2018YFC1508403), and National Natural Science Foundation of China (51579173).

Acknowledgments: The authors wish to acknowledge the assistance of all the editors and reviewers. We thank James Buxton MSc from Liwen Bianji, Edanz Group China (www.liwenbianji.cn./ac), for editing the English text of this manuscript.

Conflicts of Interest: The authors declare no conflict of interest.

References

1. Woldesenbet, T.A.; Elagib, N.A.; Ribbe, L.; Heinrich, J. Catchment response to climate and land use changes in the Upper Blue Nile sub-basins, Ethiopia. *Sci. Total Environ.* **2018**, *644*, 193–206. [CrossRef] [PubMed]
2. Silva, V.d.P.R.; Campos, J.H.; Silva, M.T.; Azevedo, P.V. Impact of global warming on cowpea bean cultivation in northeastern Brazil. *Agric. Water Manag.* **2010**, *97*, 1760–1768. [CrossRef]
3. Pereira, D.d.R.; Martinez, M.A.; da Silva, D.D.; Pruski, F.F. Hydrological simulation in a basin of typical tropical climate and soil using the SWAT Model Part II: Simulation of hydrological variables and soil use scenarios. *J. Hydrol. Reg. Stud.* **2016**, *5*, 149–163. [CrossRef]
4. Zhang, X.; Ren, L.; Kong, X. Estimating spatiotemporal variability and sustainability of shallow groundwater in a well-irrigated plain of the Haihe River basin using SWAT model. *J. Hydrol.* **2016**, *541*, 1221–1240. [CrossRef]
5. Buytaert, W.; Célleri, R.; De Bièvre, B.; Cisneros, F.; Wyseure, G.; Deckers, J.; Hofstede, R. Human impact on the hydrology of the Andean páramos. *Earth Sci. Rev.* **2006**, *79*, 53–72. [CrossRef]

6. Shen, Z.Y.; Gong, Y.W.; Li, Y.H.; Hong, Q.; Xu, L.; Liu, R.M. A comparison of WEPP and SWAT for modeling soil erosion of the Zhangjiachong Watershed in the Three Gorges Reservoir Area. *Agric. Water Manag.* **2009**, *96*, 1435–1442. [CrossRef]
7. Wang, Y.; Zhang, N.; Wang, D.; Wu, J.; Zhang, X. Investigating the impacts of cascade hydropower development on the natural flow regime in the Yangtze River. *China Sci. Total Environ.* **2018**, *624*, 1187–1194. [CrossRef] [PubMed]
8. Descheemaeker, K.; Nyssen, J.; Poesen, J.; Raes, D.; Haile, M.; Muys, B.; Deckers, S. Runoff on slopes with restoring vegetation: A case study from the Tigray highlands, Ethiopia. *J. Hydrol.* **2006**, *331*, 219–241. [CrossRef]
9. Pacheco, F.; Varandas, S.G.P.; Fernandes, L.S.; Junior, R.V. Soil losses in rural watersheds with environmental land use conflicts. *Sci. Total Environ.* **2014**, *485*, 110–120. [CrossRef]
10. Junior, R.V.; Varandas, S.G.P.; Fernandes, L.S.; Pacheco, F.A.L. Groundwater quality in rural watersheds with environmental land use conflicts. *Sci. Total Environ.* **2014**, *493*, 812–827.
11. Junior, R.F.V.; Varandas, S.G.; Pacheco, F.A.; Pereira, V.R.; Santos, C.F.; Cortes, R.M.; Fernandes, L.F.S. Impacts of land use conflicts on riverine ecosystems. *Land Use Policy* **2015**, *43*, 48–62. [CrossRef]
12. Pacheco, F.; Santos, R.M.B.; Fernandes, L.S.; Pereira, M.G.; Cortes, R.M.V. Controls and forecasts of nitrate yields in forested watersheds: A view over mainland Portugal. *Sci. Total Environ.* **2015**, *537*, 421–440. [CrossRef] [PubMed]
13. Woldesenbet, T.A.; Elagib, N.A.; Ribbe, L.; Heinrich, J. Hydrological responses to land use/cover changes in the source region of the Upper Blue Nile Basin, Ethiopia. *Sci. Total Environ.* **2017**, *575*, 724–741. [CrossRef] [PubMed]
14. Li, Y.; Chang, J.; Wang, Y.; Jin, W.; Bai, X. Spatiotemporal responses of runoff to land use change in Wei River Basin. *Trans. Chin. Soc. Agric. Eng.* **2016**, *32*, 232–238.
15. Song, W.; Deng, X.; Yuan, Y.; Wang, Z.; Li, Z. Impacts of land-use change on valued ecosystem service in rapidly urbanized North China Plain. *Ecol. Model.* **2015**, *318*, 245–253. [CrossRef]
16. Bi, H.; Liu, B.; Wu, J.; Yun, L.; Chen, Z.; Cui, Z. Effects of precipitation and land use on runoff during the past 50 years in a typical watershed in Loess Plateau, China. *Int. J. Sediment Res.* **2009**, *24*, 352–364. [CrossRef]
17. Zhang, Y.-K.; Schilling, K. Increasing streamflow and baseflow in Mississippi River since the 1940 s: Effect of land use change. *J. Hydrol.* **2006**, *324*, 412–422. [CrossRef]
18. Zeng, S.; Xia, J.; Du, H. Separating the effects of climate change and human activities on runoff over different time scales in the Zhang River basin. *Stoch. Environ. Res. Risk Assess.* **2014**, *28*, 401–413. [CrossRef]
19. Christensen, N.S.; Wood, A.W.; Voisin, N.; Lettenmaier, D.P.; Palmer, R.N. The Effects of Climate Change on the Hydrology and Water Resources of the Colorado River Basin. *Clim. Chang.* **2004**, *62*, 337–363. [CrossRef]
20. Rostamian, R.; Jaleh, A.; Afyuni, M.; Mousavi, S.F.; Heidarpour, M.; Jalalian, A.; Abbaspour, K.C. Application of a SWAT model for estimating runoff and sediment in two mountainous basins in central Iran. *Hydrol. Sci. J.* **2008**, *53*, 977–988. [CrossRef]
21. Liu, Q.; Yang, Z.; Cui, B.; Sun, T. Temporal trends of hydro-climatic variables and runoff response to climatic variability and vegetation changes in the Yiluo River basin, China. *Hydrol. Process. Int. J.* **2009**, *23*, 3030–3039. [CrossRef]
22. Wang, Q.; Liu, R.; Men, C.; Guo, L. Application of genetic algorithm to land use optimization for non-point source pollution control based on CLUE-S and SWAT. *J. Hydrol.* **2018**, *560*, 86–96. [CrossRef]
23. Kiniry, J.R.; Williams, J.R.; King, K.W. Soil and Water Assessment Tool Theoretical Documentation (Version 2005). *Comput. Speech Lang.* **2011**, *24*, 289–306.
24. Douglas-Mankin, K.R.; Srinivasan, R.; Arnold, A.J. Soil and Water Assessment Tool (SWAT) Model: Current Developments and Applications. *Trans. ASABE* **2010**, *53*, 1423–1431. [CrossRef]
25. Arnold, J.G.; Fohrer, N. SWAT2000: Current capabilities and research opportunities in applied watershed modelling. *Hydrol. Process.* **2005**, *19*, 563–572. [CrossRef]
26. Liu, L.; Xu, H.; Wang, Y.; Jiang, T. Impacts of 1.5 and 2 °C global warming on water availability and extreme hydrological events in Yiluo and Beijiang River catchments in China. *Clim. Chang.* **2017**, *145*, 145–158. [CrossRef]
27. Li, S.; Hu, C.; Wang, H.; Zhang, Z.; Li, X. Pattern Analysis on Land Use/Cover Changes in Yiluo Drainage Basin. *Meteorol. Environ. Sci.* **2015**, *38*, 108–113.
28. Wang, W.; Qian, L. Spatial distribution and seasonal variation of evapotranspiration in Yiluo River basin on MODIS data. *Resour. Sci.* **2012**, *34*, 1582–1590.

29. He, R.; Wang, G.; Zhang, J. Impacts of environmental change on runoff in the Yiluohe River Basin of the Middle Yellow River. *Res. Soil Water Conserv.* **2007**, *2*, 297–301.
30. Liu, X.; Dai, X.; Zhong, Y.; Li, J.; Wang, P. Analysis of changes in the relationship between precipitation and streamflow in the Yiluo River, China. *Theor. Appl. Climatol.* **2013**, *114*, 183–191. [CrossRef]
31. Guo, J.; Zheng, J. *Yiluo River Basin Annals*; China Science & Technology Press: Henan, China, 1995.
32. Neitsch, S.L. *Soil and Water Assessment Tool Theoretical Documentation Version 2009*; Texas Water Resources Institute: Texas, TX, USA, 2011.
33. Arnold, J.G.; Srinivasan, R.; Muttiah, R.S.; Williams, J.R. Large area hydrologic modeling and assessment part I: Model development 1. *JAWRA J. Am. Water Resour. Assoc.* **1998**, *34*, 73–89. [CrossRef]
34. Lyon, S.W.; Walter, M.T.; Gérard-Marchant, P.; Steenhuis, T.S. Using a topographic index to distribute variable source area runoff predicted with the SCS curve-number equation. *Hydrol. Process.* **2004**, *18*, 2757–2771. [CrossRef]
35. Mishra, S.K.; Singh, V.P. *Soil Conservation Service Curve Number (SCS-CN) Methodology*; Springer Science & Business Media: Berlin, Germany, 2013; Volume 42.
36. Allen, R.G.; Jensen, M.E.; Wright, J.L.; Burman, R.D. Operational estimates of reference evapotranspiration. *Agron. J.* **1989**, *81*, 650–662. [CrossRef]
37. Monteith, J.L. Evaporation and environment. *Symp. Soc. Exp. Biol.* **1965**, *19*, 205–234. [PubMed]
38. Shi, Y.; Eberhart, R. A modified particle swarm optimizer. In Proceedings of the 1998 IEEE International Conference on Evolutionary Computation Proceedings, IEEE World Congress on Computational Intelligence, Anchorage, AK, USA, 4–9 May 1998.
39. Arnold, J.G.; Moriasi, D.N.; Gassman, P.W.; Abbaspour, K.C.; White, M.J.; Srinivasan, R.; Harmel, R.D.; van Griensven, A.; Van Liew, M.W.; Kannan, N. SWAT: Model use, calibration, and validation. *Trans. ASABE* **2012**, *55*, 1491–1508. [CrossRef]
40. Neupane, R.P.; White, J.D.; Alexander, S.E. Projected hydrologic changes in monsoon-dominated Himalaya Mountain basins with changing climate and deforestation. *J. Hydrol.* **2015**, *525*, 216–230. [CrossRef]
41. Saharia, A.M.; Sarma, A.K. Future climate change impact evaluation on hydrologic processes in the Bharalu and Basistha basins using SWAT model. *Nat. Hazards* **2018**, *92*, 1463–1488. [CrossRef]
42. Jing-Huan, T.; Cui, Q.; Xu, J.; Zhou, X. Surface -evaporation of large and middle reservoirs affects the cunount of water resource in the yellow river valley. *J. Shandong Agric. Univ.* **2005**, *36*, 391–394.
43. Nash, J.E.; Sutcliffe, J.V. River flow forecasting through conceptual models part I—A discussion of principles. *J. Hydrol.* **1970**, *10*, 282–290. [CrossRef]
44. Moriasi, D.N.; Arnold, J.G.; Van Liew, M.W.; Bingner, R.L.; Harmel, R.D.; Veith, T.L. Model evaluation guidelines for systematic quantification of accuracy in watershed simulations. *Trans. ASABE* **2007**, *50*, 885–900. [CrossRef]
45. Ling, Z.; Karthikeyan, R.; Bai, Z.; Srinivasan, R. Analysis of streamflow responses to climate variability and land use change in the Loess Plateau region of China. *CATENA* **2017**, *154*, 1–11.
46. Chen, Y.; Chen, X.; Xu, C.Y.; Zhang, M.; Liu, M.; Gao, L. Toward Improved Calibration of SWAT Using Season-Based Multi-Objective Optimization: A Case Study in the Jinjiang Basin in Southeastern China. *Water Resour. Manag.* **2018**, *32*, 1193–1207. [CrossRef]
47. Deb, K.; Pratap, A.; Agarwal, S.; Meyarivan, T.A.M.T. A fast and elitist multiobjective genetic algorithm: NSGA-II. *IEEE Trans. Evol. Comput.* **2002**, *6*, 182–197. [CrossRef]
48. Vrugt, J.A.; Gupta, H.V.; Bastidas, L.A.; Bouten, W.; Sorooshian, S. Effective and efficient algorithm for multiobjective optimization of hydrologic models. *Water Resourc. Res.* **2003**, *39*. [CrossRef]
49. Kim, H.; Lee, S. Assessment of a seasonal calibration technique using multiple objectives in rainfall–runoff analysis. *Hydrol. Process.* **2014**, *28*, 2159–2173. [CrossRef]
50. Kim, H. Potential improvement of the parameter identifiability in ungauged catchments. *Water Resour. Manag.* **2016**, *30*, 3207–3228. [CrossRef]
51. Haddeland, I.; Skaugen, T.; Lettenmaier, D. Hydrologic effects of land and water management in North America and Asia: 1700? 1992. *Hydrol. Earth Syst. Sci. Discuss.* **2007**, *11*, 1035–1045.
52. Giertz, S.; Junge, B.; Diekkrüger, B. Assessing the effects of land use change on soil physical properties and hydrological processes in the sub-humid tropical environment of West Africa. *Phys. Chem. Earthparts A/B/C* **2005**, *30*, 485–496. [CrossRef]

53. Qiu, G.Y.; Yin, J.; Tian, F.; Geng, S. Effects of the "Conversion of Cropland to Forest and Grassland Program" on the water budget of the Jinghe River catchment in China. *J. Environ. Qual.* **2011**, *40*, 1745–1755. [CrossRef] [PubMed]
54. Jiang, G.; Gao, P.; Mu, X.; Chai, X. Effect of Conversion of Farmland to Forestland or Grassland on the Change in Runoff and Sediment in the Upper Reaches of Beiluo River. *Res. Soil Water Conserv.* **2015**, *22*, 1–6.
55. Santos, R.; Fernandes, L.S.; Pereira, M.G.; Cortes, R.M.V.; Pacheco, F.A.L. A framework model for investigating the export of phosphorus to surface waters in forested watersheds: Implications to management. *Sci. Total Environ.* **2015**, *536*, 295–305. [CrossRef] [PubMed]
56. Santos, R.; Fernandes, L.S.; Moura, J.P.; Pereira, M.G.; Pacheco, F.A.L. The impact of climate change, human interference, scale and modeling uncertainties on the estimation of aquifer properties and river flow components. *J. Hydrol.* **2014**, *519*, 1297–1314. [CrossRef]
57. Dong, N.; Yang, M.; Meng, X.; Liu, X.; Wang, Z.; Wang, H.; Yang, C. CMADS-Driven Simulation and Analysis of Reservoir Impacts on the Streamflow with a Simple Statistical Approach. *Water* **2019**, *11*, 178. [CrossRef]

Article

Flood Vulnerability, Environmental Land Use Conflicts, and Conservation of Soil and Water: A Study in the Batatais SP Municipality, Brazil

Anildo Monteiro Caldas [1], Teresa Cristina Tarlé Pissarra [2], Renata Cristina Araújo Costa [2], Fernando Cartaxo Rolim Neto [1], Marcelo Zanata [3], Roberto da Boa Viagem Parahyba [4], Luis Filipe Sanches Fernandes [5] and Fernando António Leal Pacheco [6,*]

1. Departamento de Tecnologia Rural, Universidade Federal Rural de Pernambuco, Rua Manuel de Medeiros, s/n-Dois Irmãos, Recife-PE 52171-900, Brazil; monteiro.dtr.ufrpe@gmail.com (A.M.G.); fernandocartaxo@yahoo.com.br (F.C.R.N.)
2. Departamento de Engenharia Rural, Universidade Estadual Paulista, Faculdade de Ciências Agrárias e Veterinárias, Via de Acesso Prof.Paulo Donato Castellane s/n, Jaboticabal-SP 14884-900, Brazil; teresap1204@gmail.com (T.C.T.P.); renata.criscosta@gmail.com (R.C.A.C.)
3. Instituto Florestal, Rua do Horto, 931, São Paulo-SP 02377-000, Brazil; marcelozanata@netsite.com.br
4. Empresa Brasileira de Pesquisa Agropecuária, Centro Nacional de Pesquisa de Solos, Unidade de Execução de Pesquisa e Desenvolvimento de Recife, Rua Antônio Falcão, 402-Boa Viagem, Recife-PE 51020-240, Brazil; rbvparahyba@gmail.com
5. Centro de Investigação e Tecnologias Agroambientais e Biológicas, Universidade de Trás-os-Montes e Alto Douro, Ap 1013, Vila Real 5001-801, Portugal; lfilipe@utad.pt
6. Centro de Química de Vila Real, Universidade de Trás-os-Montes e Alto Douro, Ap 1013, Vila Real 5001-801, Portugal

* Correspondence: fpacheco@utad.pt; Tel.: +35-191-751-9833

Received: 12 September 2018; Accepted: 28 September 2018; Published: 29 September 2018

Abstract: In many regions across the planet, flood events are now more frequent and intense because of climate change and improper land use, resulting in risks to the population. However, the procedures to accurately determine the areas at risk in regions influenced by inadequate land uses are still inefficient. In rural watersheds, inadequate uses occur when actual uses deviate from land capability, and are termed environmental land use conflicts. To overcome the difficulty to evaluate flood vulnerability under these settings, in this study a method was developed to delineate flood vulnerability areas in a land use conflict landscape: the Batatais municipality, located in the State of São Paulo, Brazil. The method and its implementation resorted to remote sensed data, geographic information systems and geo-processing. Satellite images and their processing provided data for environmental factors such as altitude, land use, slope, and soil class in the study area. The importance of each factor for flood vulnerability was evaluated through the analytical hierarchy process (AHP). According to the results, vast areas of medium to high flood vulnerability are located in agricultural lands affected by environmental land use conflicts. In these areas, amplified flood intensities, soil erosion, crop productivity loss and stream water deterioration are expected. The coverage of Batatais SP municipality by these vulnerable areas is so extensive (60%) that preventive and recovery measures were proposed in the context of a land consolidation–water management plan aiming flood control and soil and water conservation.

Keywords: flood vulnerability; land use conflicts; water management; soil conservation; spatial multi-criteria analysis; geographic information system

1. Introduction

Natural floods occur when stream flow exceeds the river channel's capacity, and consequently spills over the natural or artificial embankments. Floods are particularly related with extreme precipitation events and specific geomorphologic settings. When floods reach the marginal urban and rural areas they can cause significant economic damage and even human deaths [1,2].

Land uses, which are forms of land occupation and refer to the management and modification of a natural environment, influence the intensity of processes such as infiltration and runoff. The areas of greater impermeability, such as settlements, favor the runoff process and, hence, cause a substantial accumulation of rainwater in drainage channels, especially during periods of intense rainfall. The same holds for intensively tilled grounds, such as arable fields or pastures, as well as for managed woods. Therefore, the spatial distributions of urban and agricultural areas at the neighborhood of a river channel, and especially their dynamic changes over time can be the cause of frequent and often unexpected overflows with crop and property losses [3–12].

The frequency and magnitude of floods can increase where environmental land use conflicts have developed. In general, a land use conflict occurs when there are conflicting views on land use policies, such as when an increasing population creates competitive demands for the use of the land, causing a negative impact on other land uses nearby [13]. However, a different definition has been presented to describe land use conflicts in rural watersheds mostly occupied by agriculture, pastures, and forests [14,15]. In that context, an environmental land use conflict develops where the current land use differs from a natural use set up on the basis of specific morphometric parameters, namely drainage density and hill slope. In the same watersheds, conflicts were also associated with changes in permanent preservation areas, such as destruction of riparian vegetation. The consequences of conflicts for floods are apparent. For example, in the long term the conversion of forest into agricultural land reduces soil permeability, because the use of heavy machines over many years increases compaction of topsoil layers and deteriorates the soil structure [4,16,17]. The stability of the soil influences its infiltration capacity [3,4] and its ability to preserve structure under the impact rain drops [18,19]. The overall result is increased runoff and soil erosion. Thus, flood prevention as a priority measure should be emphasized to avoid environmental land use conflicts. To accomplish this goal, it is important to improve design solutions for natural drainage by enhancing and promoting innovative forms of land use structures, as well as the correction of riverbeds and specific sectors along the river, especially riparian and floodplain areas [20].

A pre-requisite for the proposition and implementation of flood control solutions and conservation of soil and water measures, is the evaluation and mapping of flood vulnerability and land uses [21,22], besides environmental land use conflicts [14,15] if the focus problem is to assess human-induced impacts on flood prone agro-systems triggered by land use changes. However, if the analysis of flood vulnerability and environmental land use conflicts, taken separately, is relatively abundant, the potential amplifying factor of environmental land use conflicts on flood frequency and intensity, as well as the feedback of amplified floods on agro-systems has barely been discussed. This study aims contributing to fill in this gap. To accomplish this general goal the study established the following specific objectives, using a geographic information system (GIS) to implement them: (1) prepare a flood vulnerability map for the studied area, using multi-criteria analysis based on relevant flood-control attributes (e.g., terrain slope, land use); (2) prepare a map of environmental land use conflicts based on comparisons between land capabilities (natural uses) and actual uses [23]; (3) cross-tabulate the data on flood vulnerability and environmental land use conflicts to verify if flood prone regions extend beyond the areas in environmental land use conflict and estimate the overlapping areas; and (4) Using Pareto diagrams, identify the regions requiring priority action as regards flood control/soil and water conservation and propose preventive or recovery methods.

The studied area comprises the Batatais SP municipality, located in Brazil, which is a flood prone region with severe environmental land use conflicts caused by replacement of natural vegetation with sugar cane plantations. The local population and government have the right and the duty to perceive the risks they are vulnerable to, to plan flood control and adapt land uses in the sequel. It is expected that the identification of areas susceptible to flooding and areas of conflicts, offered in this study, subsidize decision-making in a context of land consolidation–water management.

2. Material and Methods

2.1. Study Area

The study area is located in the northeast of São Paulo State, Brazil (Figure 1), in the municipality of Batatais, at the latitude 20°53′28″ S, longitude 47°35′06″ W of Greenwich, and average altitude of 862 meters. The Batatais municipality covers an area of approximately 851 km^2 and has a population of 56,476 inhabitants [24]. The climate is considered Cwa in the classification of Köppen-Geiger, described as tropical (mild) with dry winter. The rainy period runs from November to March. The average precipitation (period 1970–2017) is 1972 mm/year and 100.4 mm/day. The average annual temperature is 21.9 °C (http://www.inmet.gov.br). In the past century precipitation has steadily increased on the annual and daily basis. Both trends are displayed in Figure 2 and point to an average increase of 10 mm/year and 0.7 mm/day. Within this period, most years after 1995 were characterized by precipitation above the average, while before that most years received rainfall below the average. Relief comprises the structural plateaus of Ribeirão Preto, characterized by 500–700 m of altitude and shallow incised stream valleys, and the residual plateaus of Franca/Batatais installed at higher altitudes (800–1100 m) and characterized by deep incised valleys (http://www.abagrp.cnpm.embrapa.br). According to the Prefeitura Municipal da Estância Turística de Batatais (https://www.batatais.sp.gov.br), in 2014 land use was largely dominated by sugar cane plantations (49,065 hectares; 57.7% of municipality area) and brachiaria pastures (10,437 hectares; 12.3%). Other agriculture activities were much less represented, such as coffee (2625 hectares; 3.1%), corn (2247 hectares; 2.6%) or soya (1349 hectares; 1.6%). The predominant type of vegetation cover is the Cerrado, composed of forest savanna, wooded savanna, park savanna and gramineous-woody savanna. The Batatais municipality hosts a Conservation Unit called the State Forest of Batatais, which occupies an area of approximately 1353.3 hectares (1.6%). Other forested areas are occupied by pine (976 hectares; 1.1%) or eucalyptus (419 hectares; 0.5%), while the urban centers occupy 751 hectares (0.9%). The dominant soil type is distroferric red latosols with sandy to sandy-loam texture.

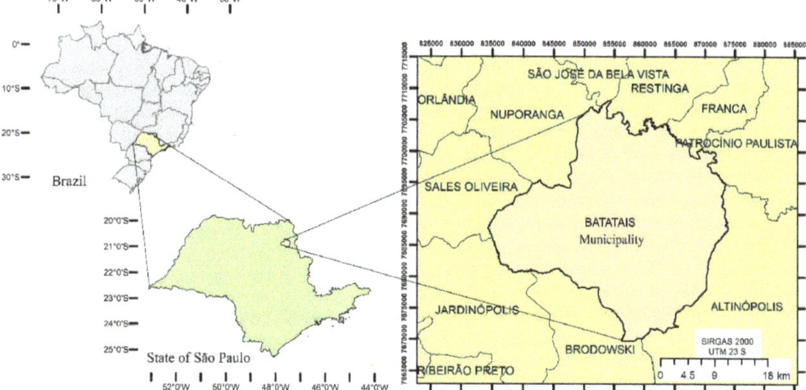

Figure 1. Study area, municipality of Batatais, São Paulo State, Brazil.

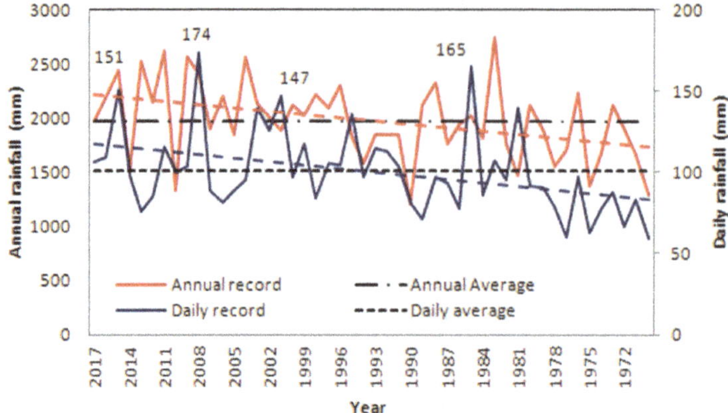

Figure 2. Precipitation records of Batatais State Forest Station relative to period 1970–2017 (red and blue lines), and associated increase trends (dashed black lines). The labels close to blue line peaks refer to precipitations at reported major flood events. The daily precipitation of 174 mm occurred in 29 January 2008 and provoked a severe flood in Batatais.

The scientific studies on floods are not abundant for the Batatais municipality, but several videos on the consequences of urban and rural flood events can be viewed through Brazilian television websites (e.g., the Globo Tv). Based on comparisons between daily precipitation records (e.g., Figure 2) and reported flood events, a precipitation >140 mm/day can be used as reference for the occurrence of a severe flood in Batatais. Figure 2 points to the occurrence of four major events in the 1970–2017 period (labeled peaks), which represents approximately one major event per decade. Figure 3 illustrates the severe flood occurred in 29 January 2008, when the daily precipitation reached 174 mm, the largest value registered in the Batatais Forest Station in the 1970–2017 period. Besides the four events highlighted in Figure 2, some local newspapers have reported the occurrence of floods in 20 January 2014, 9 January 2016, or 29 December 2016, presumably at lower daily precipitations. This is indication that moderate floods can occur at rather short return periods (e.g., 1–2 years).

Figure 3. Illustration of 29 January 2008 severe flood in the Batatais town.

2.2. *Raw Data and Software*

The image dataset comprises radar images from the SRTM (Shuttle Radar Topography Mission), with a spatial resolution of 30 m, referring to the WGS84-Zone 23K (SF-23-VA and SF-23-VC); and a mosaic of images from the Landsat 8 optical sensor, with a spatial resolution of 30 m and temporal resolution of 16 days, referred to Datum WGS84. These datasets were used to prepare all the maps and to extract data on environmental attributes. The images were obtained using public data sources and processed in ArcGIS software from ESRI, licensed to Universidade Estadual Paulista (UNESP), Departamento de Engenharia Rural, Campus Jaboticabal. The public data sources comprised the websites of EMBRAPA—Empresa Brasileira de Pesquisa Agropecuária (http://www.relevobr.cnpm.embrapa.br) and of INPE—Instituto Nacional de Pesquisas Espaciais (http://www.dgi.inpe.br/CDSR).

2.3. Preparation of Workable Data

The flood risk map was calculated and drawn for the municipality of Batatais SP, based on four variables, called environmental attributes, represented by altitude, slope class, land use/occupation and soil type. The maps of all environmental attributes were prepared in raster format as described in the next paragraph.

The altitude and slope classes were developed from a 30 m resolution digital elevation model (DEM), following Brazilian principles and rules on cartography. The DEM images were first corrected for relief depressions and then submitted to the slope routine of ArcGIS that calculated and projected the slopes. The map of land use was drawn with four classes: agriculture, pasture, forest and urban. It was generated from the Landsat 8 images using the supervised classification method, whereby statistical algorithms are run for the recognition of spectral patterns that subsequently are labeled as land uses. The map of soil units was obtained from [25] in paper format and then scanned, geo-referenced, vectorized, and finally transformed into the required raster format. The four maps are illustrated in Figure 4.

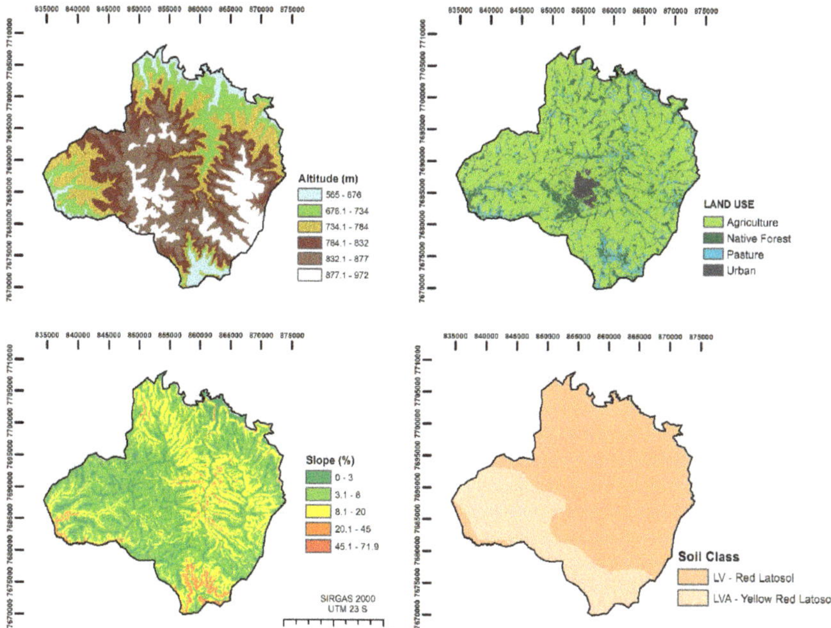

Figure 4. Maps of environmental attributes within the municipality of Batatais SP.

2.4. Flood Vulnerability

The mapping of flood vulnerability involved a weighted combination of the environmental attribute maps (Figure 4) operated in ArcGIS using map algebra tools. The selection of attributes was based on the judgment of experts who know the region and could empirically relate historical flood events to these variables. The processing of attributes to produce the map of vulnerable areas followed a conventional multi-criteria analysis (MCA), whereby the original variables are first harmonized into a common dimensionless scale (rated), then are allocated a specific importance (weight), and finally are aggregated into a global vulnerability index. The MCA is used in many environmental applications [26,27].

The first step in the MCA comprised the reclassification of environmental attributes into a dimensionless scale ranging from 1 to 10, where "1" means least susceptible to floods and "10" most susceptible to floods. This common range allows equivalence among attributes that otherwise would not be comparable given the different scales. The range 1–10 is arbitrary but is regularly used in MCA analyses. The reclassification scheme is depicted in Table 1 and resulted from bibliographical surveys and debates within a multidisciplinary team composed of agronomists, forestry engineers, biologists, lawyers, and geographers.

Table 1. Reclassification of environmental attributes according to their flood susceptibility. The source for the soil classes is [28].

Altitude Attribute		
Classes (m)	**Rating**	**Rank**
565–676	10	7
676–734	7	
734–784	6	
784–832	4	
832–877	3	
877–972	1	
Land Use Attribute		
Classes	**Rating**	**Rank**
Urban	10	5
Pasture	9	
Agriculture	8	
Native Forest	1	
Slope Attribute		
Classes (%)	**Rating**	**Rank**
0–3	9	3
3–8	8	
8–20	7	
20–45	5	
45–75	1	
> 75	1	
Soil Class Attribute		
Classes	**Rating**	**Rank**
Red latosols	4	1
Yellow-red latosols	3	

As regards the altitude attribute, the lower elevations were given the highest susceptibility (rated as 10) because they receive runoff from the higher altitudes (rated as susceptibility 1) and therefore are more prone to flooding. The attribution of susceptibility ratings to the land use attribute was based on potential infiltration capacity of land covers, assessed through curve numbers (https://swat.tamu.edu). The larger the curve number, the larger was rated the susceptibility, because big curve numbers are in favor of big overland flows. In that context, the urban areas were ascribed the largest possible susceptibility (10) because these areas comprise large portions of impermeable surface (e.g., street pavements, commercial and industrial areas), characterized by very large curve numbers (>80 in a maximum of 100). The low susceptibility end-member (rated as susceptibility 1) was the forested areas because vegetation has the capability to retain water in the canopy promoting infiltration and reducing runoff downstream. These areas are frequently characterized by low curve numbers (e.g., between 30 and 60 for forests and bushes located on permeable soils). The occupations by agriculture or pasture were given high susceptibility ratings because characteristic curve numbers in these areas are intermediate to high (60–75). The distribution of susceptibility ratings per slope classes considered the effect of topography in the share of overland flow between exportation and accumulation. The steepest slopes (>75%) were attributed the lowest rating (1) because overland flow in these areas is mostly exported under the action o gravity towards the plains where the surface is leveled promoting deceleration of water velocity, accumulation and floods in the sequel. For the opposite reasons, the low slope areas (slope < 10%) were assigned a high susceptibility rating (8–9). The susceptibility ratings of soil classes took into account the expected permeability of dominant soil types, deduced from soil texture. The textures of red and yellow-red latosols are mostly sandy or sandy-loam, which means the soils are permeable. For that reason, the susceptibility ratings of both soil types were small (3–4).

The analytic hierarchy process AHP; 29 was used to weight the importance of environmental attributes as regards flood susceptibility. This method is based on pairwise comparisons among the attributes according to the fundamental scale of absolute numbers (Table 2). Firstly, a hierarchy of attributes is defined according to expert's judgment. In this study, ranks were assigned to the four environmental attributes as indicated in the last column of Table 1. The altitude attribute was given the largest relative importance (7), because altitude determines where the overland flow is likely to accumulate producing floods. The smallest importance (1) was attributed to soils, because soil types in the studied region are relatively homogeneous as regards permeability. A similar rationale was used to rate the slope class parameter. In this case, the assignment of a rating 3 was related to the observation that the Batatais municipality is generally a flat to undulated region, incised by deep valleys only in specific sectors (e.g., the south limit). An assignment of a large relative importance to these steady attributes would tend to smooth the final susceptibility map masking important contributions from the other attributes. Land use was given the second largest relative importance (5), because the role of vegetation cover is prominent in the control of overland flow, but also because land use/occupation is a dynamic attribute that can change overtime. A large relative importance assigned to this parameter highlights the impact of a land use change, and therefore is likely to improve the efficacy of decision support systems focused on the flood control versus land use management issue.

Fourthly, attribute weights (w) are obtained from the main eigenvector of matrix B as follows (last column of Table 4):

$$w_i = \frac{\sum_j b_{ij}}{\sum_i \sum_j b_{ij}} \text{ with } i \text{ and } j = 1, 2, \ldots, p \quad (2)$$

with:

$$\sum_i w_i = 1 \text{ with } i = 1, 2, \ldots, p \quad (3)$$

Finally, a consistency ratio (CR) is computed to check whether the elements of matrix B have been randomly generated, in which case the attribute weights are not statistical significant and for that reason are invalid within the MCA. The consistency ratio ranges from 0 and 1 and invalidate the weights when $CR \geq 0.1$ [29]. The details on the calculation of CR are beyond the scope of this paper and can be consulted elsewhere [30]. For the weights depicted in Table 4, a $CR = 0.043$ was estimated, which means the weights are valid.

Having validated the weights the vulnerability index is calculated using the following aggregation equation:

$$\textit{Flood susceptibility} = 0.06 \times \textit{soil} + 0.12 \times \textit{slope} + 0.26 \times \textit{land use} + 0.57 \times \textit{altitude} \quad (4)$$

where the numeric coefficients represent the attribute weights (last column of Table 4).

Table 2. The fundamental scale of absolute numbers, according to [29].

Importance	Definition	Explanation
1	Equal importance	Two activities contribute equally to the objective
2	Weak or slight	
3	Moderate importance	Experience and judgment slightly favor one activity over another
4	Moderate plus	
5	Strong importance	Experience and judgment strongly favor one activity over another
6	Strong plus	
7	Very strong or demonstrated importance	An activity is favored very strongly over another; its dominance demonstrated in practice
8	Very, very strong	
9	Extreme importance	The evidence favoring one activity over another is of the highest order of affirmation
Reciprocals of above	If the activity *i* has one of the above non-zero numbers assigned to it when compared with activity *j*, then *j* has the reciprocal value when compared with *i*	
1.1–1.9	If the activities are very close	May be difficult to assign the best value but when compared with other contrasting activities the size of the small numbers would not be too noticeable, yet they can still indicate the relative importance of the activities.

Secondly, the ranks are assembled in a pairwise comparison matrix, denoted matrix A of a_{ij} elements (Table 3). Thirdly, matrix A is normalized as matrix B with elements b_{ij} (Table 4), as follows:

$$b_{ij} = \frac{a_{ij}}{\sum_i a_{ij}} \text{ with } i \text{ and } j = 1, 2, \ldots, p \tag{1}$$

Table 3. Pairwise comparison matrix based on the attribute ranks listed in the last column of Table 1.

Attributes	Soil	Slope	Land Use	Altitude
Soil	1	1/3	1/5	1/7
Slope	3	1	1/3	1/5
Land Use	5	3	1	1/3
Altitude	7	5	3	1

Table 4. AHP to obtain the attribute weights. The values under the heading of each attribute represent b_{ij} scores calculated by Equation (1). In these columns, the nominator values to the left of the equal sign are a_{ij} scores transported from Table 3, while the corresponding denominator values represent the sum of nominator values. The attribute weights (w) are depicted in the last column. They result from application of Equation (2) to the b_{ij} scores calculated previously.

Attributes	Soil	Slope	Land Use	Altitude	Weight (w)
Soil	1/16 = 0.0625	0.33/9.33 = 0.0357	0.20/4.53 = 0.0441	0.14/1.68 = 0.0852	0.0553
Slope	3/16 = 0.1875	1/9.33 = 0.1075	0.33/4.53 = 0.0735	0.20/1.68 = 0.1193	0.1175
Land Use	5/16 = 0.3125	3/9.33 = 0.3214	1/4.53 = 0.2206	0.33/1.68 = 0.1988	0.2622
Altitude	7/16 = 0.4375	5/9.33 = 0.5357	3/4.53 = 0.6618	1/1.68 = 0.5966	0.5650

2.5. Environmental Land Use Conflicts

Environmental land use conflicts relate to uses of the land that ignore land capability. In this study, land capability was assessed using the ruggedness number concept ($RN = Sl \times D_d$) based on the morphometric parameters hill slope (Sl) and drainage density (D_d). The RN is easily estimated from a digital elevation model (e.g., Figure 4; top-left map) using the adequate terrain modeling tools within the GIS software, namely the slope and density functions of ArcGIS. The relationship between the RN and land capability can be described as follows [31]: sectors of a watershed with a low RN are assumed capable for the practicing of cropping agriculture as they correspond to undulated low dissected areas. When the RN is high the sectors are found proper for an occupation by forests because they are sloping and considerably dismembered. Finally, sectors of the basin with intermediate RN values are adequate for livestock pasturing or for a mosaic of natural pastures and forests. The boundaries of each RN class are set up in four consecutive steps: (a) a number of sub-basins is identified and delineated across the watershed in study; (b) the mean Sl and D_d of each sub-basin are estimated, from which the corresponding RN is obtained; (c) the RN amplitude (maxRN–minRN) is estimated, where maxRN is the highest and minRN the lowest RN, and then divided by a pre-defined number of land capability classes (n), for example the four classes used in this study (n = 4): 1—Agriculture, 2—Pastures for livestock production, 3—Pastures for livestock production/Forestry and 4—Forestry; (d) the sub-basins are assigned to these classes, based on their individual RN scores, from which a land capability map is prepared.

The environmental land use conflict is the divergence of an actual use (e.g., Figure 4; top-right image) from land capability. This conflict can be graduated to express the departure between the natural and actual land uses. The steps required to accomplish this goal were thoroughly described in the study of [15]. Firstly, land capabilities (labeled N) and actual uses (labeled A) were ranked in ascending order of their resemblance with agriculture, and given the codes N = A = 1 (agriculture), 2 (livestock production), 3 (mixed livestock production and forestry), and 4 (forestry). Secondly, the conflict was equated to the difference between codes of capability ($1 \leq N \leq 4$) and actual land use ($1 \leq A \leq 4$). According to this method, (a) the no conflict areas are represented by regions where $N - A \leq 0$, with the negative values representing areas with potential for a sustainable expansion of agriculture or livestock pasturing; (b) areas where the value of the difference between capability and actual use is equal to 1 or 2 are classified as Class 1 (N – A = 1) or Class 2 (N – A = 2) conflict areas, respectively; (c) areas with potential for forestry (N = 4) but occupied with a farm (A = 1) are referred to as Class 3 (N – A = 3). According to [23], the risks and limitations allocated to conflict classes 1 to 3 are: (Class 1) these lands represent a low risk for surface water contamination and riverine ecosystem degradation when used for annual crops or pastures. The attenuation of the risks may be accomplished by the implementation of soil conservation measures resorting to vegetative and(or) mechanical techniques; (Class 2) these lands represent a medium risk for surface water contamination and riverine ecosystem degradation and, hence, are incapable for the practice of agriculture, although being adapted for livestock pasturing, reforestation or environmental preservation; (Class 3) these lands represent a high risk for surface water contamination and riverine ecosystem degradation and therefore are not capable of being used for cropping or livestock pasturing; nonetheless, they are still capable of being reforested or used for environmental preservation.

2.6. Combined Analysis of Flood Vulnerability, Land Use, and Land Use Conflicts

The potential impact of floods on agricultural systems, with or without eventual amplification caused by environmental land use conflicts, was assessed through a combined analysis of flood vulnerability, land use and environmental land use conflicts, using Pareto diagrams. A Pareto chart, also called a Pareto distribution diagram, is a vertical bar graph in which values are plotted in decreasing order of relative frequency from left to right. A point-to-point graph, which shows the cumulative relative frequency, may be superimposed on the bar graph. Pareto charts are extremely useful for analyzing what problems need attention first because the taller bars on the chart, which represent frequency, clearly illustrate which variables have the greatest cumulative effect on a given system. The Pareto chart provides a graphic depiction of the Pareto principle, a theory maintaining that 80% of the output in a given situation or system is produced by 20% of the input (https://whatis.techtarget.com).

3. Results

The vulnerability to floods in the Batatais SP municipality is represented in Figure 5 and the associated areas quantified in Table 5. The areas with higher vulnerability (levels high and very high) are concentrated in sectors around the municipal boundary where altitudes are lower, namely the west, south, and northeast sectors. The highest vulnerability level (very high, displayed in dark blue in Figure 5) occupies an area of 39.45 km^2, which represents 4.64% of municipal territory. These areas are located along the water course margins where the relief is smooth. They represent a serious problem concerning human, structural, or material losses, should an anomalous natural phenomenon occur, namely an intense and very prolonged rainfall event.

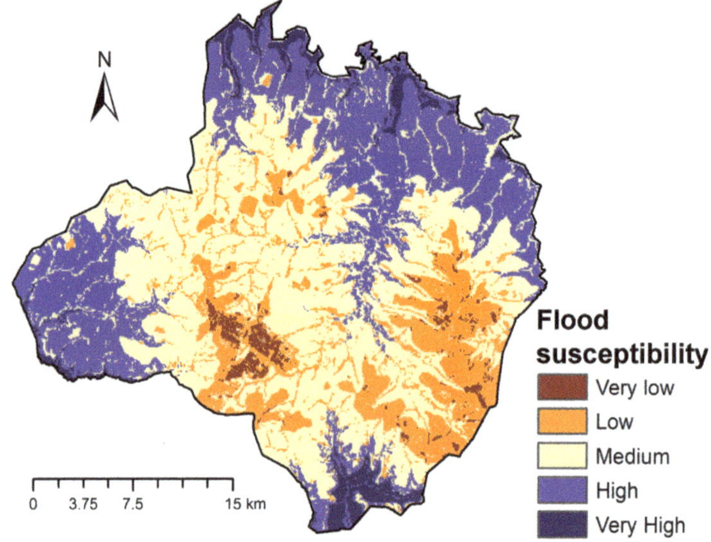

Figure 5. Flood vulnerability map of Batatais SP municipality.

Table 5. Quantification of flood vulnerability areas within the municipality of Batatais SP, as function of vulnerability class. The spatial incidence of vulnerability classes is represented in Figure 5.

Vulnerability Class	Area of Incidence (km^2)	Percentage of Municipal Area (%)
Very low	20.31	2.39
Low	152.77	17.95
Medium	388.92	45.70
High	249.57	29.33
Very High	39.45	4.64
Total	851	100

The very low (20.31 km^2; 2.39%) and low (152.77 km^2; 17.95%) vulnerability levels occupy the southwest and southeast sectors of Batatais SP municipality. Vulnerability in these areas is generally low (sometimes very low) because these regions occupy areas of high altitude where accumulation of runoff less likely. The southwest region comprises a particularly low vulnerability sector (the dark brown area), because vegetation cover (Cerrado), permeable forest soil, and low slope gradients all potentiate infiltration increase and runoff decrease, which combined with the high altitudes further reduce flood vulnerability. The medium vulnerability level occupies the largest area (388.92 km^2) representing 45.7% of municipal area. This outcome is strongly influenced by the MCA options, whereby altitude and land use were ranked as most relevant factors in the flood vulnerability assessment while agriculture and medium altitudes were ascribed relatively large ratings (Tables 3 and 4). Therefore, a large portion of medium vulnerability areas occupies agricultural areas located at medium altitudes. It is worth mentioning that the urban area of Batatais SP municipality is also described as medium vulnerability area. This is enough reason for civil defense agencies to implement monitoring programs, as well as to adopt prevention measures to minimize the potential losses of floods and the transmission of communicable diseases, such as water-borne (e.g., typhoid fever, cholera) and/or vector-borne (e.g., malaria, dengue) diseases.

Overall, the flood vulnerability analysis reveals a flood-prone municipality where the total area susceptible to floods is much larger than the less susceptible areas, since the regions of medium and high flood vulnerability account for 75.03% of all municipal area. Additionally, the most vulnerable regions, although representing a small area, are located in the vicinity of drainage channels requiring special attention from local authorities.

The map describing environmental land use conflicts within the Batatais SP municipality is illustrated in Figure 6. The percentage of municipal area occupied by conflicts is 65%, distributed as follows per conflict class: class 1—32.3%; class 2—16.4%, and class 3—16.4% (Table 6). The cross-classification of flood vulnerability and environmental land use conflict exposes large percentages of municipality area (78.78%) occupying medium (52.01%), high (23.6%), or very high (3.17%) vulnerability regions where land use is in moderate (class 1; 40.23%), high (class 2; 16.61%) or severe (class 3; 21.94%) conflict with the soils' natural use.

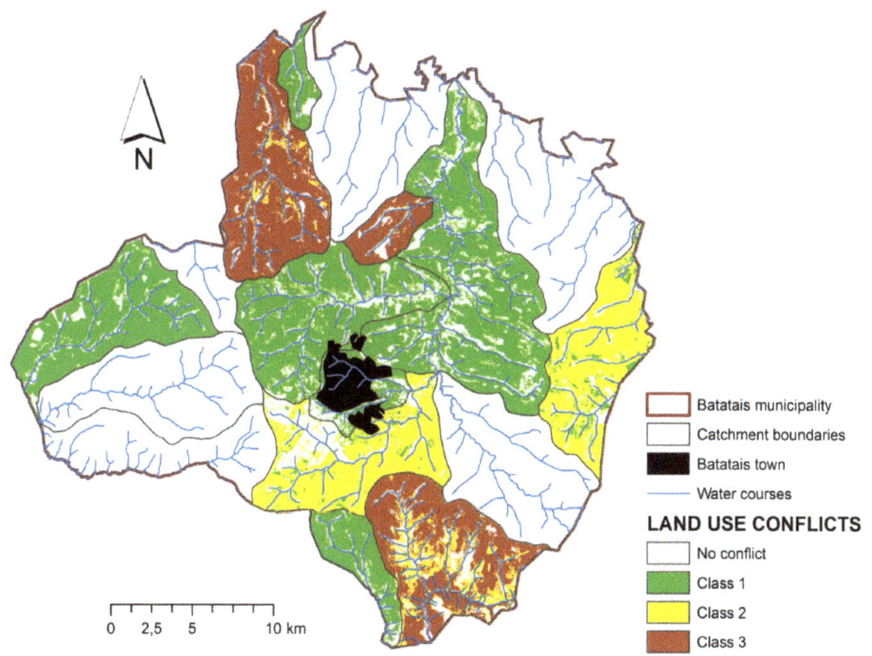

Figure 6. Map of land use conflicts within the Batatais municipality.

Table 6. Cross-classification of land use conflict and flood vulnerability in the municipality of Batatais SP. The grey shaded cells represent concern situations because they combine medium to very high flood vulnerability and land use conflict.

Land Use Conflict	Flood Vulnerability										Total	
	Very low		Low		Medium		High		Very High			
	km^2	%	km^2	%	km^2	%	km^2	%	km^2	%	km^2	%
Class 1	2.56	0.57	33.66	7.44	114.07	25.21	67.97	15.02	5.78	1.28	273.56	32.2
Class 2	2.79	0.62	31.27	6.91	60.23	13.31	14.92	3.3	4.93	1.09	139.37	16.4
Class 3	1.73	0.38	13.34	2.95	61.02	13.49	23.87	5.28	14.33	3.17	139.56	16.4

The illustration of land use conflict-flood vulnerability cross classification in the form of a Pareto diagram is depicted in Figure 7. In this case the Pareto 80/20 rule does not provide an unique solution, because the number of classes representing 20% of input (land use conflict–flood vulnerability classes) is different from the number of classes representing 80% of output (cross-classified area). The high frequency classes representing 20% of input comprise three classes, namely the areas of conflict classes 1 and 3 located in regions of medium flood vulnerability and the areas of conflict class 1 occupying regions of high flood vulnerability. The higher frequencies representing 80% of the cross classified area comprehend the previous three classes plus the areas of conflict class 2 occupying the regions of medium flood vulnerability and the areas of conflict class 1 and 2 located in regions of low flood vulnerability. From a decision-making standpoint, it was assumed that the optimal Pareto solution gathers four classes, namely all regions of medium flood vulnerability affected by land use conflicts (classes 1–3) and the regions of high flood vulnerability affected by class 1 conflict class. Under the Pareto principle, these would be the areas of Batatais SP, municipality requiring the implementation of mitigation measures in the first place.

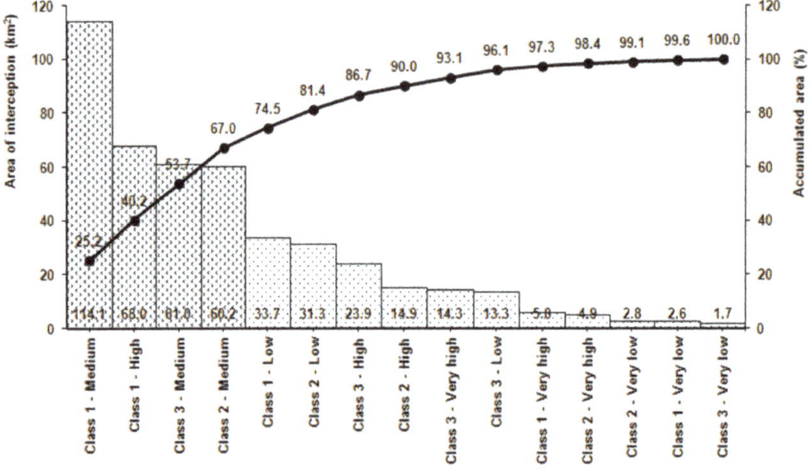

Figure 7. Pareto diagram illustrating the frequency of land use conflict versus flood vulnerability classes in the municipality of Batatais SP.

The cross classification of land use and flood vulnerability is quantified in Table 7 and illustrated in Figure 8 in the form of a Pareto diagram. In this case, the Pareto principle applies because the 80% of cross-classified area is associated with 20% of land use classes, namely agriculture located in regions of low to high flood vulnerability and forest occupation in regions of medium vulnerability.

Table 7. Cross-classification of land use and flood vulnerability in the municipality of Batatais SP. The grey shaded cells represent concern situations because they correspond to medium and high flood vulnerability over large agricultural areas.

Land Use	Flood Vulnerability										Total	
	Very Low		Low		Medium		High		Very High			
	km^2	%	km^2	%	km^2	%	km^2	%	km^2	%	km^2	%
Agriculture	3.4	0.4	112.47	13.22	315.6	37.09	201.2	23.64	26.97	3.17	659.64	77.5
Forest	13.63	1.6	31.39	3.69	44.09	5.18	25.93	3.05	4.29	0.5	119.33	14.0
Pasture	1.36	0.16	5.17	0.61	16.04	1.88	23.63	2.78	8.14	0.96	54.34	6.4
Urban	0	0	1.39	0.16	15.13	1.78	1.16	0.14	0	0	17.68	2.1

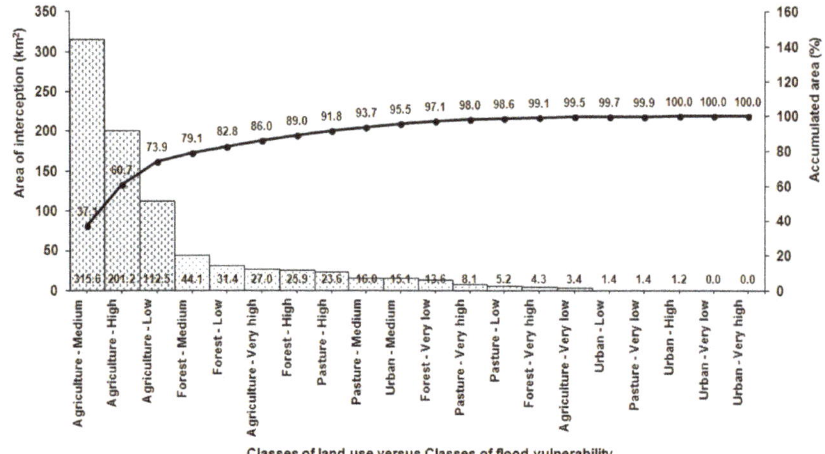

Figure 8. Pareto diagram illustrating the frequency of land use conflict versus flood vulnerability classes in the municipality of Batatais SP.

4. Discussion

The areas occupied by agriculture and simultaneously vulnerable to floods represent 516.80 km^2 of Batatais SP area, which means 60.73% of this municipality. From a decision-making stand point, these areas require intervention in the first place because they are agricultural areas located in regions of medium and high flood vulnerability. Although the scientific literature is scarce on actual or potential damage of floods in this region, the news periodically communicated by the social media (e.g., local television, newspapers) is enough reason to expose concerns about the potential harmful effects of recurrent floods for local agricultural systems and stream waters. Following the Pareto principle and common sense, these vulnerable areas require priority attention of public policies regarding decision making of preventive measures, public education and occupational rearrangement of the soil at catchment scale.

Floods submerge the fields used for agriculture, frequently causing production losses. The fields stay submersed for variable periods of time depending on factors such as topography, hydrologic barriers (e.g., dams), or return periods of the flood. The actual percentage of production loss will also depend on a number of factors, including crop variety, stage of plant development, flooded area, level of inundation and length of flooding period [32]. The present study does not provide numbers for all these factors but certainly indicate that large areas of Batatais SP municipality mostly used for sugar cane can be periodically submerged, especially where vulnerability is high. It is worth mentioning that the economic effects of river floods for sugar cane producers can be noteworthy. In a study in Pakistan [33], a reduction of sugar recovery from sugarcane was attributed to excessive water absorbed by a flooded sugarcane crop which reduced the sugar contents in the cane and increased in the cane weight. The overall consequence was more payments to the local growers with less sugar recovery. In the context of a changing climate, this economic damage can aggravate if extreme precipitation events become longer and more intense [34]. Besides the direct tangible losses, the flood consequences for farmers also comprise direct intangible (e.g., business interruption), and indirect (clean up and recovery, soil and water remediation) losses [35].

Land-use changes contribute to an increased frequency and intensity of floods by increasing surface runoff [36]. Flood hazard events often occur in areas with dense population or high levels of agricultural land reclamation [37,38]. This gives urbanization and expansion of agriculture decisive roles in increasing the intensity and frequency of floods. A common cause of flood change response in modern agriculture areas relate to soil compaction from ploughing and heavy machinery [39–41]. However, flood response to land use changes in the rural environment usually begins during land conversion. Land use changes such as shifts from forestry to agriculture, from pasture to arable land, from rain fed to irrigated agriculture or from agricultural use to urbanized areas act as drivers of flood frequency and intensity increases [42,43], because these changes reduce land capacity to attenuate runoff. Some of these changes can produce amplified flood responses in case they correspond to environmental land use conflicts. In these cases, the change is performed in disagreement with land capability (natural use), for example when agriculture expands to steep forested hillsides. Flood intensity under these settings is amplified because land reduces the runoff attenuation capacity offered by forest cover while the large slope gradients accelerate runoff increasing flood potential downstream. Given the large area occupied of class 3 conflicts (Figure 6), characterized by forest-agriculture conversions, the Batatais SP municipality is presumably very prone to amplified flood intensities.

Land use conflicts may accelerate runoff and amplify flood response, but floods can feedback and amplify soil, water and biodiversity damages in areas where the conflicts have developed. Land use conflicts trigger changes in the soil composition and structure, namely reduction of organic matter and the capacity of soil to form stable aggregates [31]. Consequently, a cascade of environmental impacts develop, such as amplified soil loss [14,15], deterioration of water quality [44,45], and biodiversity decline [46]. Under these settings, recurrent floods will tend to aggravate these damages. If no prevention or recovery measures are taken, the flood-conflict issue may turn into a negative spiral ultimately leading to land degradation [47]. The return of a degraded land to a neutral condition is a priority given the economic and social costs of land degradation [48]. The neutral condition implies no net loss of the land-based natural capital relative to a reference state. Actions to achieve land degradation neutrality include sustainable land management practices that avoid or reduce degradation, coupled with efforts to reverse degradation through restoration or rehabilitation of degraded land. This rationale is independent from the causes of land degradation and hence applies to flood-related causes. However, to achieve land degradation neutrality is important to have a strong engagement of planners, stakeholders and the society, and may need to involve law enforcement initiatives [49].

The first step towards flood prevention is to map flood vulnerability [5,6]. In the present study, a map of flood vulnerability was prepared for application in a rural environment, using flood control attributes already adopted by other work groups [50], and a similar multi-criteria analysis [51]. However, to our knowledge this was the first time a flood vulnerability map has been coupled with a land use conflict map to investigate the potential impact of these inadequate land use changes on floods, and vice versa. Following the production of vulnerability maps the next step should be focused on the implementation of measures to actively control the floods, as well as measures to improve soil conservation and rural households' flood preparedness.

In general, active flood control can be accomplished through construction of detention basins (structural measures), and recent models attempt to optimize their location in a catchment [52]. However, besides the construction costs of these dams concerns exist about installing large detention basins in catchments, because stream water stored in the dam reservoir can deteriorate extensively and rapidly [53]. To overcome this problematic situation, other recent models propose the construction of small, low-cost, and sustainable water detention basins, termed rainwater harvesting systems [54,55]. These models comprise tools to optimize the location of rain water harvesting systems in a catchment as to maximize their flood-control capacity and the complementary use for agro-forestry applications (irrigation, wildfire combat). In addition to flood control at the catchment scale, rain water harvesting systems can also be used to attenuate urban floods, which are important in various neighborhoods of Batatais town. In this case, water reservoirs are installed in rooftops to collect rainwater during storm events, thus reducing rainwater reaching the ground to become runoff. Furthermore, flood control in the urban environment the stored rainwater can be complementary used for sanitary or other purposes [56]. Therefore, the planning and installation of urban and rural rainwater harvesting systems is recommended as active flood-control measure in the studied area, especially in the context of a changing climate [57]. Other measures include non-structural flood management initiatives such as afforestation and restriction of urban development on floodplains. This set of measures has gained recognition in many countries as viable and cost-effective approaches to flood risk management [3,4,7].

For effective soil conservation, flood prone areas require the implementation of multiple complementary measures. Eventually, the most cost-effective measure is the reinforcement of manuring and composting of soil to raise the levels of organic matter [58] and produce stable aggregates that are resistant to detachment by rain drop action. Level plantation is another measure of soil erosion control, frequently used to reduce laminar erosion and increase soil water uptake. Level plantation can be successfully coupled with strip cropping [59] that involves the alternation of forages with strips of row crops, such as sugar cane. The control of soil erosion also comprises implementation of techniques that prevent soil compaction, such as no-tillage treatments [60], maintenance of crop residue to keep organic matter and nutrients in the topsoil, or more costly measures such as terraces in level or gradient since they reduce the ramp length reducing the surface drag of particles and nutrients [14].

The aforementioned measures of soil conservation act indirectly as stream water quality protective measures, because they tend to reduce sediment and nutrient loads transported in runoff. A direct protection of stream water quality from eroded soils and nutrients can be accomplished through creation of buffer strips along stream reaches [61]. Regardless of being a reliable measure to reduce deterioration of stream water quality, there is some debate around the buffer strip width capable to render these filters notable efficiency. For example, the new Forest Code in Brazil (Brazilian Law No. 12.651/12) has defined a minimum width of 30 m for efficient buffer strips, but some recent studies state that the reference should be raised to 50 m, at least [62].

All measures described above are likely to produce the proposed results (flood control and conservation of soil and water) if they can be aggregated as land consolidation–water management (LC-WM) plan capable to realize a multifunctional climate resilient rural area [63]. A decision support system will be required to adequately put this plan into practice [64,65]. Additionally, to be properly planned in the study area local watershed, committees, on which watershed management expectations stand, need to gain legitimacy to become mediators of sectorial and land use policies. To get such legitimacy, these committees need to acquire the status of State public entities offered by Brazilian Law No. 9433/97. Finally, to ensure effective implementation, a LC-WM plan needs to engage local communities, politicians and stakeholders in participatory design approaches [66] to evaluate the rural households' flood preparedness.

5. Conclusions

The Batatais SP municipality is a flood prone region characterized by environmental land use conflicts. Moderate floods can occur every 1–2 years, while severe floods have occurred every 10 years in this region. Natural flood vulnerability results from abundant precipitation over a territory with contrasting reliefs and altitude ranges. The most sensitive areas in this regard are the lower altitude lands that occupy 289 km^2 (34% of the municipal area). Environmental land use conflicts developed because a vast portion of land became occupied by sugar cane plantations, even where land capability (the natural use) advised occupation by native forest vegetation. The conflicts potentiate amplification of flood intensity and inundation levels, because they reduce runoff retention capacity of soils. The study revealed 60% occupation of Batatais SP municipality with areas in environmental land use conflicts. This very large coverage is preoccupying. The study also exposed the situation and alerted for the need to engage state water planners, municipality authorities, and local communities in a process of land consolidation–water management involving a diversity of specific flood control, as well as soil and water conservation measures. The implementation of such plan would constitute a notable contribution to the sustainable use of soil and water resources in the region, with unquestionable benefits for the environment and Batatais society.

Author Contributions: A.M.C. contributed data curation, investigation, conceptualization, methodology, and writing original draft; T.C.T.P. and F.C.R.N. contributed to formal analysis, supervision, project administration and funding acquisition; M.Z. and R.d.B.V.P. contributed to resources and formal analysis; R.C.A.C. contributed to software and visualization; F.A.L.P. and L.F.S.F. contributed to formal analysis, validation, writing-review, editing, and funding acquisition.

Funding: As regards the Portuguese authors, the research was funded by national funds (FCT-Portuguese Foundation for Science and Technology) under the strategic project of the Vila Real Chemistry Research Center (PEst-OE/QUI/UI0616) and the CITAB (Centre for the Research and Technology of Agro-Environmental and Biological Sciences) project UID/AGR/04033.

Acknowledgments: The Brazilian authors wish to thank the São Paulo State University (Unesp)/School of Agricultural and Veterinarian Sciences, Campus Jaboticabal, São Paulo, Brazil; the CAPES—Coordenação de Aperfeiçoamento de Pessoal de Nível Superior; and the CNPq—Conselho Nacional de Desenvolvimento Científico e Tecnológico IF—Instituto Florestal, São Paulo, Brazil. The authors also wish to thank the University of Trás-Montes and Alto Douro (UTAD), the Chemistry Research Center and the Center for the Research and Technology of Agro-Enviromental and Biological Sciences (CITAB), for technical support.

Conflicts of Interest: The authors declare no conflicts of interest.

References

1. Alfieri, L.; Cohen, S.; Galantowicz, J.; Schumann, G.J.; Trigg, M.A.; Zsoter, E.; Rudari, R. A global network for operational flood risk reduction. *Environ. Sci. Policy* **2018**, *84*, 149–158. [CrossRef]
2. Erdlenbruch, K.; Grelot, F. Economic Assessment of flood prevention projects. *Floods* **2017**, *2-Risk Management*, 321–335.
3. DAEE—Departamento de Águas e Energia Elétrica. *Caracterização dos Recursos Hídricos do Estado de São Paulo*; Departamento de Águas e Energia Elétrica: Centro, São Paulo, 1984.
4. Libardi, P.L. *Dinâmica da água no Solo*; Editora da Universidade de São Paulo: São Paulo, Brazil, 2005.
5. Samu, R.; Kentel, A.S. An Analysis of the Flood Management and Mitigation Measures in Zimbabwe for a Sustainable Future. *Int. J. Disaster Risk Reduct.* **2018**, *31*, 691–697. [CrossRef]
6. Solín, L'.M.; Sládeková, M.; Michaleje, L. Vulnerability assessment of households and its possible reflection in flood risk management: The case of the upper Myjava basin, Slovakia. *Int. J. Disaster Risk Reduct.* **2018**, *28*, 640–652.
7. Shaw, S.B.; Marrs, J.; Bhattarai, N.; Quackenbusch, L. Longitudinal study of the impacts of land cover change on hydrologic response in four mesoscale watersheds in New York State, USA. *J. Hydrol.* **2014**, *519*, 12–22. [CrossRef]
8. Ramos, M.C.; Martínez-Casasnovas, J.A. Nutrient losses from a vineyard soil in Northeastern Spain caused by an extraordinary rainfall event. *Catena* **2004**, *55*, 79–90. [CrossRef]
9. Marques, M.J.; García-Muñoz, S.; Muñoz-Organero, G.; Bienes, R. Soil conservation beneath grass cover in hillside vineyards under mediterranean climatic conditions (Madrid, Spain). *Land Degrad. Dev.* **2010**, *21*, 122–131. [CrossRef]
10. Bagagiolo, G.; Biddoccu, M.; Rabino, D.; Cavallo, E. Effects of rows arrangement, soil management, and rainfall characteristics on water and soil losses in Italian sloping vineyards. *Environ. Res.* **2008**, *166*, 690–704. [CrossRef] [PubMed]
11. Novara, A.; Gristina, L.; Saladino, S.S.; Santoro, A.; Cerdà, A. Soil erosion assessment on tillage and alternative soil managements in a Sicilian Vineyard. *Soil Tillage Res.* **2011**, *117*, 140–147. [CrossRef]

12. Blavet, D.; De Noni, G.; Le Bissonnais, Y.; Leonard, M.; Maillo, L.; Laurent, J.Y.; Asseline, J.; Leprun, J.C.; Arshad, M.A.; Roose, E. Effect of land use and management on the early stages of soil water erosion in French Mediterranean vineyards. *Soil Tillage Res.* **2009**, *106*, 124–136. [CrossRef]
13. Brown, G.; Raymond, C.M. Methods for identifying land use conflict potential using participatory mapping. *Landsc. Urban Plan.* **2013**, *122*, 196–208. [CrossRef]
14. Pacheco, F.A.L.; Varandas, S.G.P.; Sanches Fernandes, L.F.; Valle Junior, R.F. Soil losses in rural watersheds with environmental land use conflicts. *Sci. Total Environ.* **2014**, *485–486C*, 110–120. [CrossRef] [PubMed]
15. Valle Junior, R.F.; Varandas, S.G.P.; Sanches Fernandes, L.F.; Pacheco, F.A.L. Environmental land use conflicts: A threat to soil conservation. *Land Use Policy* **2014**, *41*, 172–185. [CrossRef]
16. Calzolari, C.; Ungaro, F. Predicting shallow water table depth at regional scale from rainfall and soil data. *J. Hydrol.* **2012**, *414*, 374–387. [CrossRef]
17. Chan, F.K.S.; Chuah, C.J.; Ziegler, A.D.; Dąbrowski, M.; Varis, O. Towards resilient flood risk management for Asian coastal cities: Lessons learned from Hong Kong and Singapore. *J. Clean. Prod.* **2018**, *187*, 576–589. [CrossRef]
18. Wadt, P.G.S.; Frade, E.F.; Marcolan, A.L. A relação pesquisa e extensão na Amazônia. *Boletim da Sociedade Brasileira de Ciência do Solo Viçosa* **2012**, *37*, 42–47.
19. Erdlenbruch, K.; Bonté, B. Simulating the dynamics of individual adaptation to floods. *Environ. Sci. Policy* **2018**, *84*, 134–148. [CrossRef]
20. Ehsanzadeh, E.; Spence, C.; van der Kamp, G.; Mcconkey, B. On the behaviour of dynamic contributing areas and flood frequency curves in North American Prairie watersheds. *J. Hydrol.* **2012**, *414*, 364–373. [CrossRef]
21. Rudari, R.; Gabellani, S.; Delogu, F. A simple model to map areas prone to surface water flooding. *Int. J. Disaster Risk Reduct.* **2014**, *10*, 428–441. [CrossRef]
22. Bodoque, J.M.; Díez-herrero, A.; Eguibar, M.A.; Benito, G.; Ruiz-Villanueva, V.; Ballesteros-Cánovas, J.A. Challenges in paleoflood hydrology applied to risk analysis in mountainous watersheds—A review. *J. Hydrol.* **2015**, *529*, 449–467. [CrossRef]
23. Valle Junior, R.F. Diagnóstico de Áreas de Risco de erosão e Conflito de uso dos Solos na Bacia do rio Uberaba. Ph.D. Thesis, Agronomy, State University of São Paulo, Faculty of Agrarian and Veterinary Sciences, Jaboticabal, Brazil, 2008.
24. IBGE—INstituto Brasileiro de Geografia E Estatística. *Levantamento Sistemático da Produção Agrícola;* INstituto Brasileiro de Geografia E Estatística: Rio de Janeiro, Brazil, 2012; Volume 25, pp. 1–88.
25. Zanata, M.; Pissarra, T.C.T. (Coord.). In*formações Básicas Para o Planejamento Ambiental;* Município de Batatais. Jaboticabal: Funep, Brazil, 2012; 70p.
26. Siqueira, H.E.; Pissarra, T.C.T.; Valle Junior, R.F.; Fernandes, L.F.S.; Pacheco, F.A.L. A multi criteria analog model for assessing the vulnerability of rural catchments to road spills of hazardous substances. *Environ. Impact Assess. Rev.* **2017**, *64*, 26–36. [CrossRef]
27. Machado, E.R.; Valle Júnior, R.F.; Sanches Fernandes, L.F.; Pacheco, F.A.L. The vulnerability of the environment to spills of dangerous substances on highways: A diagnosis based on Multi Criteria Modeling. *Transp. Res. Part D* **2018**, *62*, 748–759. [CrossRef]
28. FAO—Food and Agriculture Organization. *World Reference Base for Soil Resources. International Soil Classification System for Naming Soils and Creating Legends for Soil Maps;* World Soil Resources Reports No. 106; Food and Agriculture Organization: Rome, Italy, 2015.
29. Saaty, T.L. Decision making with the analytic hierarchy process. *Int. J. Serv. Sci.* **2008**, *1*, 83–98. [CrossRef]
30. Alonso, J.A.; Lamata, M.T. Consistency in the analytic hierarchy process: A new approach. *Int. J. Uncertain. Fuzziness Knowl.-Based Syst.* **2006**, *14*, 445–459. [CrossRef]
31. Valera, C.A.; Valle Junior, R.F.; Varandas, S.G.P.; Sanches Fernandes, L.F.; Pacheco, F.A.L. The role of environmental land use conflicts in soil fertility: A study on the Uberaba River basin, Brazil. *Sci. Total Environ.* **2016**, *562*, 463–473. [CrossRef] [PubMed]
32. Chau, V.N.; Holland, J.; Cassells, S.; Tuohy, M. Using GIS to map impacts upon agriculture from extreme floods in Vietnam. *Appl. Geogr.* **2013**, *41*, 65–74. [CrossRef]
33. Raza Zaidi, S.M.; Saeed, A.; Shahid, S.M.; Ali, N. Impact of flood on sugarcane industry of Pakistan. *Interdiscip. J. Contemp. Res. Bus.* **2013**, *4*, 92–105.
34. Zhang, Q.; Gu, X.; Singh, V.P.; Liu, L.; Kong, D. Flood-induced agricultural loss across China and impacts from climate indices. *Glob. Planet. Chang.* **2016**, *139*, 31–43. [CrossRef]

35. Nga, P.H.; Takara, K.; Van, N.C. Integrated approach to analyze the total flood risk for agriculture: The significance of intangible damages—A case study in Central Vietnam. *Int. J. Disaster Risk Reduct.* **2018**, *31*, 862–872. [CrossRef]
36. Wu, F.; Sun, Y.; Sun, Z.; Wu, S.; Wu, Q.Z.F.; Sun, Y.; Sun, Z.; Wu, S.; Zhang, Q. Assessing agricultural system vulnerability to floods: A hybrid approach using emergy and a landscape fragmentation index. *Ecol. Indic.* **2017**, in press. [CrossRef]
37. Adger, W.N.; Quinn, T.; Lorenzoni, I.; Murphy, C.; Sweeney, J. Changing social contracts in climate-change adaptation. *Nat. Clim. Chang.* **2013**, *3*, 330–333. [CrossRef]
38. Dankers, R.; Arnell, N.W.; Clark, D.B.; Falloon, P.D.; Fekete, B.M.; Gosling, S.N.; Heinke, J.; Kim, H.; Masaki, Y.; Satoh, Y.; et al. First look at changes in flood hazard in the Inter-Sectoral Impact Model Intercomparison Project ensemble. *Proc. Natl. Acad. Sci. USA* **2014**, *111*, 3257–3261. [CrossRef] [PubMed]
39. O'connell, P.; Ewen, J.; O'donnell, G.; Quinn, P. Is there a link between agricultural land-use management and flooding? *Hydrol. Earth Syst. Sci.* **2007**, *11*, 96–107. [CrossRef]
40. Pattison, I.; Lane, S.N. The link between land-use management and fluvial flood risk: A chaotic conception? *Prog. Phys. Geogr.* **2011**, *30*. [CrossRef]
41. Lee, Y.; Brody, S.D. Examining the impact of land use on flood losses in Seoul, Korea. *Land Use Policy* **2018**, *70*, 500–509. [CrossRef]
42. EEA (European Environment Agency). *Flood Risks and Environmental Vulnerability—Exploring the Synergies Between Floodplain Restoration, Water Policies and Thematic Policies*; EEA Rep. No. 1/2016; European Environment Agency: Copenhagen, Denmark; Publications Office of the European Union: Luxembourg, 2016; ISSN 1977-8449.
43. Akter, T.; Quevauviller, P.; Eisenreich, S.J.; Vaes, G. Impacts of climate and land use changes on flood risk management for the Schijn River, Belgium. *Environ. Sci. Policy* **2018**, *89*, 163–175. [CrossRef]
44. Pacheco, F.A.L.; Sanches Fernandes, L.F. Environmental land use conflicts in catchments: A major cause of amplified nitrate in river water. *Sci. Total Environ.* **2016**, *548–549*, 173–188. [CrossRef] [PubMed]
45. Valle Junior, R.F.; Varandas, S.G.P.; Sanches Fernandes, L.F.; Pacheco, F.A.L. Groundwater quality in rural watersheds with environmental land use conflicts. *Sci. Total Environ.* **2014**, *493*, 812–827. [PubMed]
46. Valle Junior, R.F.; Varandas, S.G.P.; Pacheco, F.A.L.; Pereira, V.R.; Santos, C.F.; Cortes, R.M.V.; Fernandes, L.F.S. Impacts of land use conflicts on riverine ecosystems. *Land Use Policy* **2015**, *43*, 48–62.
47. Pacheco, F.A.L.; Sanches Fernandes, L.F.; Valle Junior, R.F.; Pissarra, T.C.T.; Valera, C.A. Land degradation: Multiple environmental consequences and routes to neutrality. *Curr. Opin. Environ. Sci. Health* **2018**, *5*, 79–86. [CrossRef]
48. Sutton, P.C.; Anderson, S.J.; Costanza, R.; Kubiszewski, I. The ecological economics of land degradation: Impacts on ecosystem service values. *Ecol. Econ.* **2016**, *129*, 182–192. [CrossRef]
49. Valera, C.A.; Pissarra, T.C.T.; Martins Filho, M.V.; Valle Junior, R.F.; Sanches Fernandes, L.F.; Pacheco, F.A.L. A legal framework with scientific basis for applying the 'polluter pays principle' to soil conservation in rural watersheds in Brazil. *Land Use Policy* **2017**, *66*, 61–71. [CrossRef]
50. González-Arqueros, M.L.; Mendoza, M.E.; Bocco, G.; Castillo, B.S. Flood susceptibility in rural settlements in remote zones: The case of a mountainous basin in the Sierra-Costa region of Michoacán, Mexico. *J. Environ. Manag.* **2018**, *223*, 685–693. [CrossRef] [PubMed]
51. Mahmoud, S.H.; Gan, T.Y. Multi-criteria approach to develop flood susceptibility maps in arid regions of Middle East. *J. Clean. Prod.* **2018**, *196*, 216–229. [CrossRef]
52. Bellu, A.; Sanches Fernandes, L.F.; Cortes, R.M.V.; Pacheco, F.A.L. A framework model for the dimensioning and allocation of a detention basin system: The case of a flood-prone mountainous watershed. *J. Hydrol.* **2016**, *533*, 567–580. [CrossRef]
53. Santos, R.M.B.; Sanches Fernandes, L.F.; Cortes, R.M.V.; Varandas, S.G.P.; Jesus, J.J.B.; Pacheco, F.A.L. Integrative assessment of river damming impacts on aquatic fauna in a Portuguese reservoir. *Sci. Total Environ.* **2017**, *601–602*, 1108–1118. [CrossRef] [PubMed]
54. Terêncio, D.P.S.; Sanches Fernandes, L.F.; Cortes, R.M.V.; Pacheco, F.A.L. Improved framework model to allocate optimal rainwater harvesting sites in small watersheds for agro-forestry uses. *J. Hydrol.* **2017**, *550*, 318–330. [CrossRef]

55. Terêncio, D.P.S.; Sanches Fernandes, L.F.; Cortes, R.M.V.; Moura, J.P.; Pacheco, F.A.L. Rainwater harvesting in catchments for agro-forestry uses: A study focused on the balance between sustainability values and storage capacity. *Sci. Total Environ.* **2018**, *613–614*, 1079–1092. [CrossRef] [PubMed]
56. Sanches Fernandes, L.F.; Terêncio, D.P.S.; Pacheco, F.A.L. Rainwater harvesting systems for low demanding applications. *Sci. Total Environ.* **2015**, *529*, 91–100. [CrossRef] [PubMed]
57. Sanches Fernandes, L.F.; Pereira, M.G.; Morgado, S.G.; Macário, E.B. Influence of Climate Change on the Design of Retention Basins in Northeastern Portugal. *Water* **2018**, *10*, 743. [CrossRef]
58. Damodar Reddy, D.; Subba Rao, A.; Rupa, T.R. Effects of continuous use of cattle manure and fertilizer phosphorus on crop yields and soil organic phosphorus in a vertisol. *Bioresour. Technol.* **2000**, *75*, 113–118. [CrossRef]
59. Lenka, N.K.; Satapathy, K.K.; Lal, R.; Singh, R.K.; Singh, N.A.K.; Agrawal, P.K.; Choudhury, P.; Rathore, A. Weed strip management for minimizing soil erosion and enhancing productivity in the sloping lands of north-eastern India. *Soil Tillage Res.* **2017**, *170*, 104–113. [CrossRef]
60. Bogunovic, I.; Pereira, P.; Kisic, I.; Sajko, K.; Sraka, M. Tillage management impacts on soil compaction, erosion and crop yield in Stagnosols (Croatia). *Catena* **2018**, *160*, 376–384. [CrossRef]
61. Alemu, T.; Bahrndorff, S.; Alemayehu, E.; Ambelu, A. Agricultural sediment reduction using natural herbaceous buffer strips: A case study of the east African highland. *Water Environ. J.* **2017**, *31*, 522–527. [CrossRef]
62. Valera, C.A. Avaliação do Novo Código Florestal: As Áreas de Preservação Permanente—APPS, e a Conservação da Qualidade do solo e da água Superficial. Ph.D. Thesis, Universidade Estadual Paulista "Júlio De Mesquita Filho" (UNESP), Faculdade De Ciências Agrárias E Veterinárias Câmpus De Jaboticabal, Jaboticabal, Brazil, 2017.
63. Stańczuk-Gałwiaczek, M.; Sobolewska-Mikulska, K.; Henk Ritzema, H.; Van Loon-Steensma, J.M. Integration of water management and land consolidation in rural areas to adapt to climate change: Experiences from Poland and the Netherlands. *Land Use Policy* **2018**, *77*, 498–511. [CrossRef]
64. Sanches Fernandes, L.F.; Marques, M.J.; Oliveira, P.C.; Moura, J.P. Decision support systems in water resources in the demarcated region of Douro—Case study in Pinhão River Basin, Portugal. *Water Environ. J.* **2014**, *28*, 350–357.
65. Sanches Fernandes, L.F.; Santos, C.; Pereira, A.; Moura, J. Model of management and decision support systems in the distribution of water for consumption: Case study in North Portugal. *Eur. J. Environ. Civ. Eng.* **2011**, *15*, 411–426.
66. Monde Patrina Mabuku, M.P.; Senzanje, A.; Mudhara, M.; Jewitt, G.; Mulwafu, W. Rural households' flood preparedness and social determinants in Mwandi district of Zambia and Eastern Zambezi Region of Namibia. *Int. J. Disaster Risk Reduct.* **2018**, *28*, 284–297. [CrossRef]

Article

The Role of Landscape Configuration, Season, and Distance from Contaminant Sources on the Degradation of Stream Water Quality in Urban Catchments

António Carlos Pinheiro Fernandes [1], Luís Filipe Sanches Fernandes [1], Rui Manuel Vitor Cortes [1] and Fernando António Leal Pacheco [2,*]

1 Centre for the Research and Technology of Agro-Environment and Biological Sciences, University of Trás-os-Montes and Alto Douro, Ap. 1013, 5001-801 Vila Real, Portugal; acpf91@utad.pt (A.C.P.F); lfilipe@utad.pt (L.F.S.F.); rcortes@utad.pt (R.M.V.C.)

2 Chemistry Research Centre, University of Trás-os-Montes and Alto Douro, Ap. 1013, 5001-801 Vila Real, Portugal

* Correspondence: fpacheco@utad.pt

Received: 8 July 2019; Accepted: 25 September 2019; Published: 28 September 2019

Abstract: Water resources are threatened by many pollution sources. The harmful effects of pollution can be evaluated through biological indicators capable of tracing problems in life forms caused by the contaminants discharged into the streams. In the present study, the effects on stream water quality of landscape configuration, season, and distance from contaminant emissions of diffuse and point sources were accessed through the evaluation of a Portuguese macroinvertebrate index ($IPtI_N$) in 12 observation points distributed within the studied area (Ave River Basin, Portugal). Partial least-squares path models (PLS-PMs) were used to set up cause–effect relationships between this index, various metrics adapted to forest, agriculture, and artificial areas, and the aforementioned emissions, considering 13 distances from the contaminant sources ranging from 100 m to 56 km. The PLS-PM models were applied to summer and winter data to explore seasonality effects. The results of PLS-PM exposed significant scale and seasonal effects. The harmful effects of artificial areas were visible for distances larger than 10 km. The impact of agriculture was also distance related, but in summer this influence was more evident. The forested areas could hold onto contamination mainly in the winter periods. The impact of diffuse contaminant emissions was stronger during summer, when accessed on a short distance. The impact of effluent discharges was small, compared to the influence of landscape metrics, and had a limited statistical significance. Overall, the PLS-PM results evidenced significant cause–effect relationships between land use metrics and stream water quality at 10 km or larger scales, regardless of the season. This result is valid for the studied catchment, but transposition to other similar catchments needs to be carefully verified given the limited, though available, number of observation points.

Keywords: water quality; landscape metrics; PLS-SEM; scale; season; distance from pollution sources

1. Introduction

The growing population and demographic expansion threaten hydric resources, not only by inducing stressful water demands but also because of the continuous surge of pollution sources. The response to anthropogenic pressures relies on proper management that should always stand on environmental research. The risks to water quality are well known by experts, but continuous research should be applied since the world is in constant change [1,2]. Effluent discharges are an undeniable threat. The potential contamination by wastewaters is frequently reduced in urban and industrial areas where treatment stations are efficient, but in many regions proper treatment is not applied [3]. In those situations, surface waters are directly contaminated by bacteria [4], nutrient loads [5–7], heavy metals [8,9] and even microplastics [10]. The runoff transports herbicides and pesticides from agriculture [11] and high organic loads from livestock [7] to surface waters. The presence of forested areas or riparian vegetation can create a barrier that retains such flow of contaminants [12,13]. Wildfires are another threat to water quality [14,15], not only because they can destroy the aforementioned barriers [16], but also because soil erosion increases [17], and ash-derived contaminants are leached towards the streams [18]. Another factor that can affect water quality is land occupation planning. When land use is not conformed to land capability (natural use), land use conflict is generated [19], which amplifies soil losses [20] and accelerates other phenomena that cause water deterioration [21]. Land use type and configuration are other key aspects that have been studied by many authors in the context of water quality changes [22]. Relevant conclusions achieved in these studies are that landscapes retain nutrients [23]; a high edge density is an indicator of high anthropogenic activity [24]; as Shannon's diversity index (SHDI) increases, the water quality decreases [25]; and aggregated urban land uses are more suited to preserve surface water quality [24], among others.

The effects of landscape metrics on water quality are commonly accessed by Spearman or Pearson correlation coefficients and, when predictions are involved, through multiple linear regression analyses [26]. When these matters are studied, the authors are aware that the spatial resolution of land cover maps can affect the results [23,27]. But another critical aspect that is questioned by many authors is the spatial extent for statistical sampling [28]. This can vary from circular buffers, riparian extents, or catchments [29,30]. For proper management of river basins or even urban planning, it is essential to use an appropriate scale. By comparing different studies, some inconsistencies can be detected regarding the option for a suitable scale. Some authors infer that the use of entire watersheds provides better results [28,31–34], but other working groups argue that a riparian scale is more suitable [35–37]. These inconsistent results can be attributed to differences between study designs and study areas [38], but other factors such as stream order [28], season [39], and topology [34,39] can also play prominent roles.

Water quality research requires the use of statistical or process-based models [40]. The first type has the advantage to access the relationships between the pertinent variables, while in mechanistic models, the interactions are already determined by chemical, biological, and physical processes, which makes them preferable for prediction purposes [41].

An example of a statistical method is the multivariate method called partial least-squares path modeling (PLS-PM). The first steps of PLS-PM have been given in social sciences [42]. Nowadays, this method is being practiced in many studies in diverse research areas, namely the environment [43–46], geology [47], flood effects [48], and ecological conservation [49], among others. The authors have adopted this technique because it can exhibit cause–effect relations straightforwardly using a graphical interface. The present work continues a sequence of studies developed by this research team, who has been studying the quality of water in the Ave River Basin (Portugal) through multivariate statistics. In the first study, three PLS regression models were tested in a row to explain the pollution of surface waters and the resulting impacts on ecological integrity [50]. In a second study, an identical dataset was used in PLS-PM [51] to trace the difference of cause–effect relations between an anthropogenic (Ave River) and a rural (Sabor River) basin. Since the results were promising in both studies, in a third study they were used to predict the ecological status of the Ave River Basin in the near future [52]. The aim in the present work was to take another step forward and explore the influence of landscape metrics and contaminant emissions on ecological integrity, as well as the impact of season and scale on the results.

2. Materials and Methods

2.1. Study Area

The Ave River Basin is located in the northern region of Portugal (Figure 1A), occupying an area of approximately 1322 km^2. The main water course extends for 100 km, the most important tributary catchments are the Este (247 km^2) and Vizela (323 km^2) rivers. The altitude ranges from 0 m along the Atlantic coast to 1254 m at the Cabreira mountains, where the catchment headwaters are located. This river basin is surrounded to the west the by the Atlantic Ocean, to the south by the Leça River Basin, to the east by the Douro River Basin, and to the north by the Cávado River Basin. The group of three river basins, Ave, Leça and Cávado, belong to the same management unit, namely hydrographic region number 2 [53].

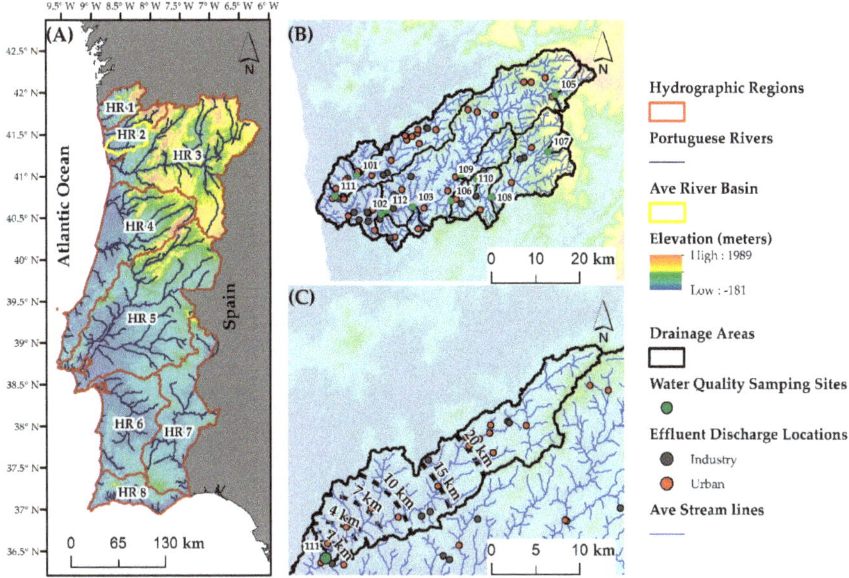

Figure 1. (**A**) Map of Portugal with the distribution of hydrographic regions. (**B**) Ave River Basin and sampling sites. (**C**) Drainage area of sampling site 111 and intersection of buffer limits with the drainage area.

In the second half of the 20th century, the Ave River Basin was heavily contaminated by untreated domestic and industrial effluents, being tagged as "Europe's Great Sewer". Following the construction of public wastewater treatment plants in the 90s, water quality increased, but some microbial contamination persisted related to improper functioning of some domestic plants. The heavy pollution of Ave comprised high concentrations of heavy metals in sediments and freshwater, especially in the Este, Selho, and Vizela rivers [9,54]. This condition improved after public investment in the wastewater treatment plants [55,56]. The abundant and persistent nutrient and metal contamination deteriorated the river's ecological status at the central area and lowlands of the Ave River Basin [57,58] where industrial areas have been settled on the river banks for years. Besides the effluents from domestic and industrial origins, contributions from agriculture and livestock production have also been reported as significant causes of water quality and ecological deterioration [59–61].

2.2. Workflow

The purpose of the present study was to show how the cause–effect relationships between ecological integrity and pollution sources/indicators changed with the season and distance from contaminant sources. Ecological integrity was assessed through the measurement of a macroinvertebrate index ($IPtI_N$; see Equation (1) below) in 12 sampling sites along the Ave River Basin (Figure 1B) during the winter and summer seasons of 2017. For these sites, the entire upstream drainage area was delineated and then sectioned at predefined distances from the sampling site (Figure 1C). Subsequent to catchment delineation and sectioning, land use and contaminant emission data were prepared for each section to be used in PLS-PM models. Two separate models were defined based on $IPtI_N$ values determined in winter and summer, respectively. The purpose was to explore how the effects of anthropogenic pressures could change as a function of season and scale. Figure 2 displays the adopted workflow, summarized in 6 steps.

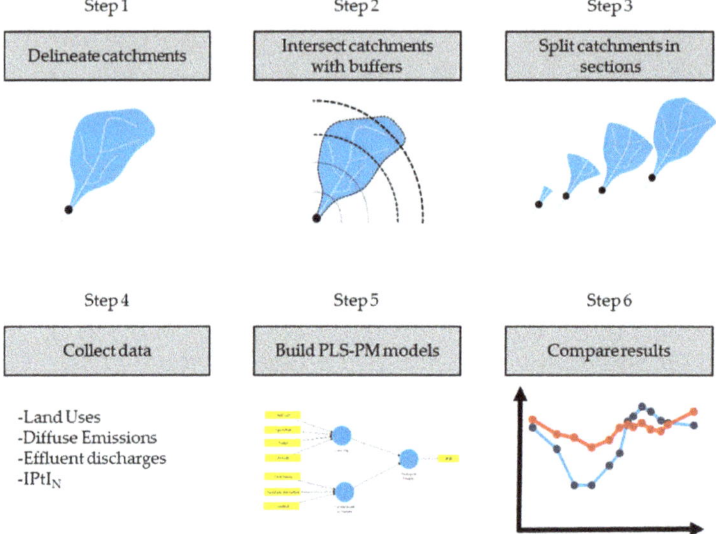

Figure 2. Workflow distributed in six steps.

For the delineation of drainage areas and drainage networks (Step 1), ArcMap [62] and its embedded module ArcHydro [63] were used. These computer packages are the key technical elements used in environmental studies with a strong spatial incidence [64–71]. As an input, a digital elevation model with a pixel resolution of 25 m was used. To design each drainage area, the tool "Batch Watershed Delineation" was used, followed by previous procedures [72]. For each sampling site, the "Buffer" ArcMap tool was used to create a circular area, while the created buffers extended for 100, 250, 500, 1000, 2000, 3000, 4000, 5000, 7000, 10,000, 15,000, 20,000, and 56,000 m (Step 2). Each drainage area intersected with the respective buffer in order to create drainage sections (Step 3). The collected spatial data (Table 1) were processed and applied in the PLS-PM models (Step 4). The land uses and discharge emissions were collected using the ArcMap "Intersect" tool. A total of 26 PLS-PM models were created, one for each season and distance combination. The algorithm was executed in SmartPLS software [73]. PLS-PM analysis was the chosen method because it could establish cause–effect relationships for the studied latent variables, which were termed "Land Use", "Contaminant Emissions", and "Ecological Integrity". Formative models were chosen because these have prediction capabilities and, at the same time, establish the wanted cause–effect relationships [74]. The 26 model outputs were measured with weights, path coefficients, and R-squared values. The outputs were compiled and compared using graphics designed in Excel [75] (Step 6).

Table 1. Data sources.

Data	Description	Source
Elevation model	Elevation model raster file, with a pixel size of 25 per 25 m.	http://www.eea.europa.eu
Effluent discharge points	Flow of discharged biological and chemical oxygen demands, nitrogen and phosphorous from urban effluents in surface water and underground water.	http://www.apambiente.pt
Diffuse discharge values	Nitrogen and phosphorous yields sourced from agriculture, forest, and livestock production areas.	http://www.apambiente.pt
Land use	Land uses map of Portuguese territory in 2015.	http://www.dgterritorio.pt
Macroinvertebrate index ($IPtI_N$)	Biodiversity of benthic macroinvertebrates	measured in field

2.3. Dataset

In this study, a group of variables was gathered (Table 1) that could be connected to the variation of $IPtI_N$. The actual data are provided as Supplementary Material. The elevation model was used for the delineation of drainage areas. The effluent discharge points values were provided by the Agência Portuguesa do Ambiente (APA; in English, Portuguese Environmental Agency) in the form of shapefiles. Each point contained was attached to information on the total discharge of nitrogen, phosphorous, and chemical and biological oxygen demands, expressed in released kilograms during the year of 2016. The discharge of nitrogen and phosphorous from livestock production, agriculture, and forestry were also provided by the APA in the form of shapefiles. Each polygon was a catchment containing the released kilograms of nutrients during the year of 2016. The most recent Portuguese land use map refers to 2015 (COS 2015) and is available at the Portuguese Territory Planning website. This map was obtained in the form of a shapefile, containing the land use or occupation from each zone. The land use types were assembled into 4 categories: agriculture, artificial surfaces, forest, and seminatural areas and water bodies. For the calculation of $IPtI_N$, samples were collected in situ during the summer and winter of 2017. After laboratory analyses, the index was calculated for the twelve locations.

For each drainage section, the discharging values of BOD (biochemical oxygen demand), COD (chemical oxygen demand), N, and P were summed and then divided by the drainage area, creating four variables representing the discharge of each type of nutrient and oxygen demand. For diffuse discharges, the total release discharge of N and P in the drainage sections was calculated and then divided by the respective area, resulting in 4 variables: the releases of N and P from livestock and forest/agriculture. For each section, landscape metrics were calculated using a python toolbox embedded in ArcMap [76]. A total of 17 metrics were calculated for all the drainage sections (Table 2).

Table 2. $IPtI_N$ values and classification for the twelve measurement sites for winter and summer of 2017. An identification color was linked to each class that shades the corresponding cells: red—Very Poor, orange—Poor, yellow—Moderate, green—Excellent. The shades were added to the table for illustration and prompt interpretation of the quality classes.

Site	Winter 2017		Summer 2017	
	Value	Classification	Value	Classification
101	0.415	Poor	0.979	Excellent
102	0.632	Moderate	0.398	Poor
103	0.389	Poor	0.364	Poor
104	0.113	Very Poor	0.498	Moderate
105	1.048	Excellent	1.017	Excellent
106	0.307	Poor	0.512	Moderate
107	1.050	Excellent	0.936	Excellent
108	0.493	Moderate	0.582	Moderate
109	0.138	Very Poor	0.346	Poor
110	0.522	Moderate	0.308	Poor
111	0.432	Moderate	0.397	Poor
112	0.219	Poor	0.275	Poor

The north invertebrate Portuguese index ($IPtI_N$) is widely used to evaluate the ecological status of stream waters in northern Portugal [77]. The $IPtI_N$ index reflects the abundance and diversity of benthic invertebrates that are sensitive to all forms of pollution [77–79].

For the measurement of this indicator, organism samples were collected from 12 surface water locations, illustrated in Figure 1C. For each site, the organisms were classified and counted, and then Equation (1) was used to calculate the $IPtI_N$ score. The equation is complex since it uses a variety of parameters, namely, the number of taxonomic groups present in the sample (N° taxa); the number of families that belong to Ephemeroptera, Plecoptera, and Trichoptera orders (EPT); Pioleu index or evenness [80,81]; biological monitoring working party index divided by the number of families included in this index (IASPT) [82]; and the sum of individuals belonging to Heptageniidae, Ephemeridae, Brachycentridae, Goeridae, Odontoceridae, Limnephilidae, Polycentropodidae, Athericidae, Dixidae, Dolichopodidae, Empididae, and Stratiomyidae families (Sel.ETD):

$$IPtI_N = N^{\circ}\ \text{Taxa} \times 0.25 + \text{EPT} \times 0.15 + \text{Evenness} \times 0.1 + (\text{IASPT} - 2) \times 0.3 + \text{Log}\,(\text{Sel.ETD} + 1) \times 0.2. \quad (1)$$

Among all possible variables that could be used in this study, only 8 were chosen for the PLS-PM models. The purpose was to reach low variance inflation factors (VIFs), and hence statistical significance, making a note that variables of the same domain can be strongly correlated in raising the VIFs. To represent the effluent discharges, the released annual flow divided by the drainage area was used, naming this variable as "Point Source". For diffuse contamination, the discharges of nitrogen from livestock, forestry, and agriculture were used, naming these variables as "Livestock" and "Forest and Agriculture", respectively. The chosen landscape metrics were Shannon's diversity, the edge density of forest and seminatural areas, the number of patches of artificial surfaces that were connected at a distance of 500 m, and the percentage of "agricultural areas" that were connected at a distance of 500 m. The land use variables were named as "Diversity", "Forest", "Artificial" and "Agriculture", respectively. The Portuguese index of macroinvertebrates was also used, named as "IPtIN" for the PLS-PM models, as an evaluator of biodiversity as well as water quality.

3. Results

3.1. Spatial Data

Figure 3 illustrates the spatial distribution of pressures in the Ave River Basin. Figure 3A depicts the land use map of 2015. It can be noted that 50% was occupied by the dominant land use, forest and seminatural areas. Agricultural areas occupied 30% of the river basin, 20% was artificial surfaces, and less than 0.5% was water bodies. The values of discharged nitrogen from livestock, forest, and agricultural areas in the river basin catchments can be seen in Figure 3B,D. The scattered effluent discharge sites are represented in Figure 3C. Among a total of 60 locations, 24 were discharge sites of industrial treatment plants, while 36 were from domestic sewage treatment plants.

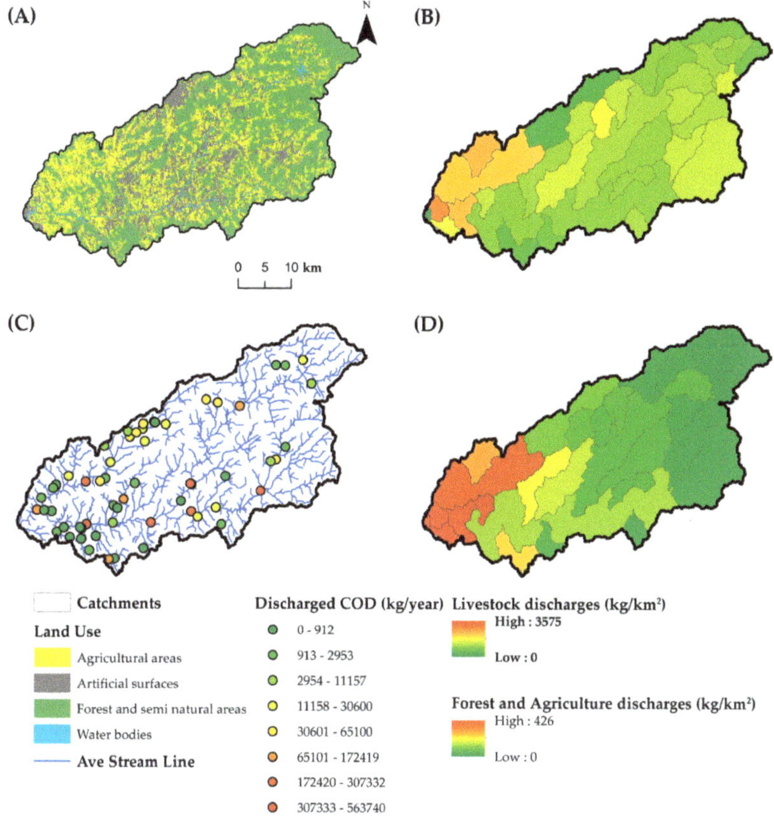

Figure 3. Spatial distribution of pressure data used in the partial least-squares path models (PLS-PMs). (**A**) Land Uses. (**B**) Forest and agriculture discharges of nitrogen. (**C**) Effluent discharge locations. (**D**) Livestock discharges of nitrogen.

The twelve river sites where the $IPtI_N$ was measured are represented in Figure 1, numbered from 101 to 112. Table 2 depicts the $IPtI_N$ values and respective classification. During the winter of 2017, the ecological status was classified as "Excellent" in two sites, "Moderate" in 4, "Poor" for another 4 sites, and "Very Poor" in 2. Overall, the values increased from winter to summer, as none of the sites was classified as "Very Poor" in summer, 6 were classified as "Poor", 3 as "Moderate", and 3 as "Excellent". In the locations 103, 105, 111, 112, 107, and 108, the classification changes were minimal. Besides, in site 111 the classification changed from "Moderate" to "Poor" because the $IPtI_N$ value was very close to the class threshold. In site 106 the classification changed from "Poor" to "Moderate", but in sites 109, 110, 103, and 104 the decrease in ecological status was startling because there was a significant decrease in $IPtI_N$ and subsequent classification of a less-rich class. For site 101 there was a significant increase, from "Poor" to the maximum class "Excellent". These changes of values were dependent upon seasonal effects but also on the pressures in surface waters.

3.2. Interpretation of a PLS-PM Example Model

The output models of SmartPLS were all similar to the one represented, as an example, in Figure 4. Each measured variable (MV) is represented as a yellow rectangle, and latent variables (LVs) as blue circles. Inside the LVs preceding other LVs, the R-squared value is portrayed. In the example model, only "Ecological Integrity" has an R-squared value, since this is the only variable that has a measured score (calculated by the sum of the product of MVs with the own weight) and a predicted score (calculated by the sum of the product between the LVs). In a PLS-PM model, weights and path coefficients are determined through an iterative process, termed the path algorithm [83], with the purpose to maximize the R-squared value.

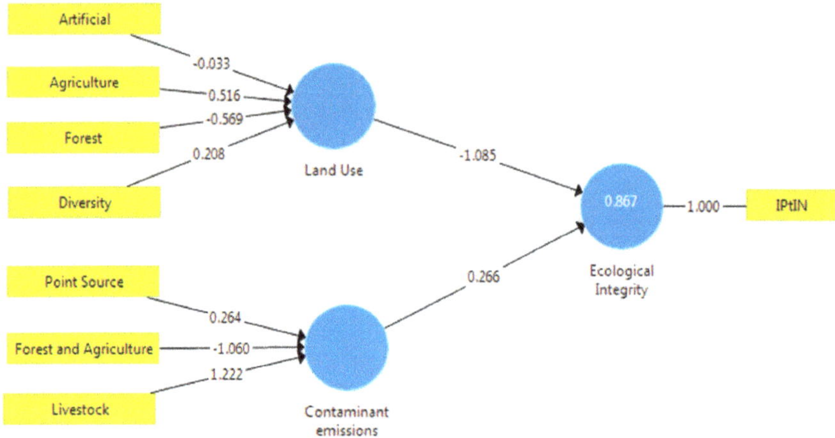

Figure 4. PLS-PM example model.

Several PLS-PM models were built and analyzed in this study. In order to exemplify how these models can be interpreted, a PLS-PM model is demonstrated in Figure 4, where the data were gathered for the drainage sections within a distance of 4 km and $IPtI_N$ values measured during the winter of 2017.

Each LV was formed by one or more MVs. For the present case study, an LV "Land Use" was created and composed of 4 MVs, namely "Diversity", "Forest", "Agriculture", and "Artificial". The LV "Contaminant Emissions" was composed of three MVs pertaining to different types of contaminant flows: "Point Source", "Livestock", and "Forest and Agriculture". "Ecological Integrity" was formed by a single MV, which is the $IPtI_N$. This LV accumulated the effects of the other LVs that were pressures in surface waters, which is why "Contaminant Emissions" and "Land Use" were connected to "Ecological Integrity". The PLS-PM model was divided into two sub-models, inner and outer. The equations of the measured scores of each LV were calculated according to Equations (2)–(4) for "Land Use", "Contaminant Emissions", and "Ecological Integrity", respectively, and are the equations that composed the outer models. The inner model was composed of relations between latent variables, which, in this case, is solely expressed by Equation (5).

$$\text{Land Use}_{\text{Measured Score}} = \text{Artificial} \times (-0.033) + \text{Agriculture} \times (0.516) + \text{Forest} \times (-0.569) + \text{Diversity} \times (0.208); \quad (2)$$

$$\text{Contaminant Emissions}_{\text{Measured Score}} = \text{Point Source} \times (-0.264) + \text{Forest and Agriculture} \times (-1.060) + \text{Livestock} \times (1.222); \quad (3)$$

$$\text{Ecological Integrity}_{\text{Measured Score}} = \text{IPtI}_N \times (1.000) \quad (4)$$

$$\text{Ecological Integrity}_{\text{Predicted Score}} = \text{Land Use}_{\text{Measured Score}} \times (-1.085) + \text{Contaminant Emissions}_{\text{Measured Score}} \times (0.266); \quad (5)$$

$$\begin{aligned}\text{Ecological Integrity}_{\text{Predicted Score}} &= \\ \text{Artificial} \times (0.035) + \text{Agriculture} \times (-0.560) &+ \text{Forest} \times (0.627) + \text{Diversity} \times (-0.226); \\ &+ \\ \text{Point Source} \times (0.070) + \text{Forest and Agriculture} &\times (-0.282) + \text{Livestock} \times (0.325).\end{aligned} \quad (6)$$

To interpret the example model, the weights and path coefficients should be analyzed simultaneously, which can be viewed in Equation (6), where the combination of Equations (2)–(5) is made. For example, the MV "Diversity" has a positive weight (0.208), so it increases the LV "Land Use", while the same applies to "Agriculture" (0.561). Conversely, "Forest" and "Artificial" have negative weights, namely −0.569 and −0.033, and therefore decrease "Land Use". But since the path coefficient of "Land Use" in "Ecological Integrity" is negative (−1.085), "Diversity" and "Agriculture" are variables that decrease "Ecological Integrity" because the product of the weight and path coefficient is negative: −0.226 and −0.560, respectively. On the other hand, "Forest" and "Artificial Areas" increase "Ecological Integrity", since the product between the path coefficient and weight is positive, respectively 0.617 and 0.036. Equation (6) expresses the total effect of each pressure in "Ecological Integrity". The results of this study were based on the analysis of the product between the path coefficients and weights (termed pcw) for the studied 26 models.

3.3. Results of All PLS-PM Models

As shown in Table 2, the $IPtI_N$ values were collected during two seasons, winter and summer. For this reason, the 26 PLS-PM models were divided into two groups, winter (2017) and summer (2017), and traced as two dot arrays colored as blue and red, respectively, in Figure 5. For each model in the respective group, the pressure values were used as input data, gathered from the 13 drainage sections and the $IPtI_N$ values collected in winter or summer. Figure 5 portrays the results of the PLS-PM models. Each graphic describes the pcw of a measured variable in all models (y axis). The x axis represents the logarithm of the buffer distance for the respective model. For the distances of 100, 250, 500, 1000, 2000, 3000, 4000, 5000, 7000, 10,000, 15,000, 20,000, and 56,000 meters, the $\log_{10}$ scores were 2, 2.4, 2.7, 3, 3.3, 3.5, 3.6, 3.7, 3.8, 4, 4.2, 4.3, and 4.7, respectively. The purpose of the plots was to illustrate the effects of the pressures in "Ecological Integrity" ("$IPtI_N$").

The effect of "Artificial" was independent of the season since the variations with distance were practically identical for both winter and summer. For distances shorter than 10 km, the effect was positive, but for longer distances the effect became negative. The strongest positive effects were detected for a distance of 100 m in summer (pcw = 0.386) and for 1000 m in winter (pcw = 0.310). The strongest negative effects were detected for the maximum distance (56 km) (i.e., for the entire drainage areas) either in winter (pcw= −0.247) or summer (pcw= −0.201).

For "Agriculture" it was seen that for both winter and summer periods, the effect was practically identical, but the summer line was below the winter line for a majority of buffer distances (only between 3 km and 5 km is the red line above the blue). The effect of agriculture was practically null for a distance of 100 m in summer (pcw = 0.008). For the same distance, it was positive during winter (pcw = 0.276) and practically null for a distance of 250 m (pcw = 0.01). For longer distances, the effect was negative, which indicated that agriculture decreased water quality. Peak values were found for distances of 4 km in winter (pcw = −0.560) and 10 km in summer (pcw = −0.648), but for distances larger than 10 km, the changes were minimal. The results lead to the conclusion that, for the Ave River basin, agriculture is a threat to water quality, while the impact seems to be stronger during the summer period.

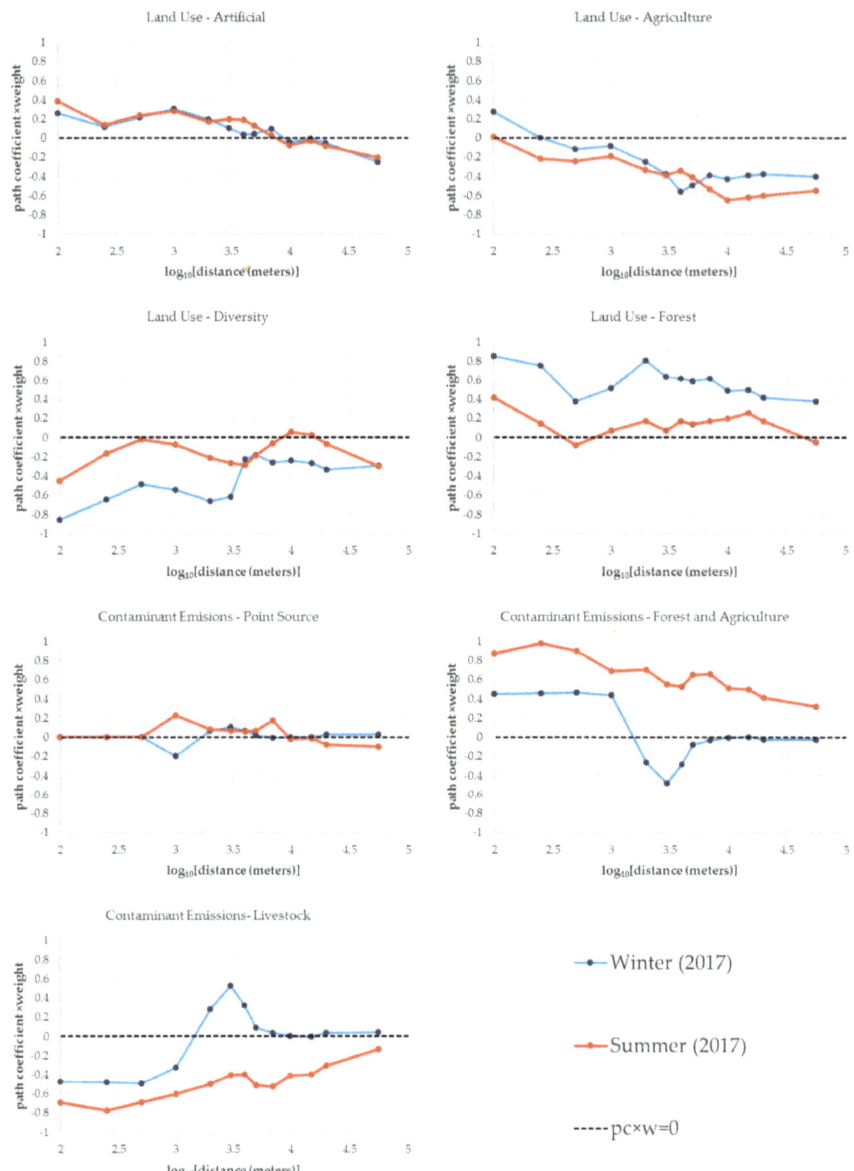

Figure 5. PLS-PM results.

Peak values of "Diversity" were detected for the minimum distance, 100 m, either in winter (pcw = −0.856) or in summer (pcw = −0.456). The effect was always negative for winter periods, and stronger in this season for almost all distances, except between 4 and 5 km. For the summer period, the effect was close to zero, but still negative; only distances of 500 m, 10, and 15 km were close to zero, since the pcw values were −0.024, 0.059, and 0.026, respectively. For the longest distance (56 km), the effect was practically the same for winter (pcw = −0.290) and summer (pcw = −0.298). The results provide evidence that the impact of "Diversity" is a threat in winter.

The effect of "Forest" was essentially positive. For the winter period, the effect was positive for all distances, and always higher than in summer, where the effect was negative but close to zero for the distances 500 m (pcw = −0.080) and 56 km (pcw = −0.049). Peak values occurred on the shortest scale, 100 m. In winter the pcw was 0.853, and in summer it was 0.419. As the buffer distances increased, the effects changed irregularly (drop and rise), but from an overall view, values were always between 0.853 and 0.374 during winter. It can be said that globally (for winter and summer), "Forest" favors water quality.

The variable that had less effects along all distances and seasons was "Point Source". For this variable and the models that comprehended distances between 100 to 500 m, the attributed weight was 0, since for short distances there were no discharge points. Even so, compared to all the other variables, it had less impact because the pcw values were contained in a short range that varied from −0.198 to 0.226. For winter, negative values were found in 1 km (pcw = −0.198) and 7 km (pcw = −0.006), while in summer the negative values were observed for distances longer than 7 km. This indicates that the effect of effluent discharges only decreased $IPtI_N$ values during the summer period when long distances were analyzed, but with minimal impact.

The discharges of nitrogen from diffuse pressures were represented by the variables "Forest and Agriculture" and "Livestock". When both graphics were compared, it was seen that there was an inverse relationship between these two variables for all models in both seasons. When the effect "Forest and Agriculture" increased, "Livestock" decreases. For the summer period, the effect of "Livestock" was always negative, while the effect of "Forest and Agriculture" was always positive. These effects were stronger over shorter distances, since maximum values for "Forest and Agriculture" and minimum values for "Livestock" appeared over the short distances. But as the distance increased, both effects approached zero. The variations of both variables were minimal for short distances (≤1 km), positive for "Forest and Agriculture", and negative for "Livestock". At the distances 2, 3, and 4 km, the effect became positive for "Livestock" and negative for "Forest and Agriculture", with a peak at 4 km. For distances longer than 4 km, the effect tended to zero.

The analysis of the pcw variations for the 7 measured variables for all the models is crucial to comprehend the cause–effect relationship changes as function of season and distance. But the analysis of the R-squared values (Figure 6) reveals the models' capacity to explain $IPtI_N$ variations.

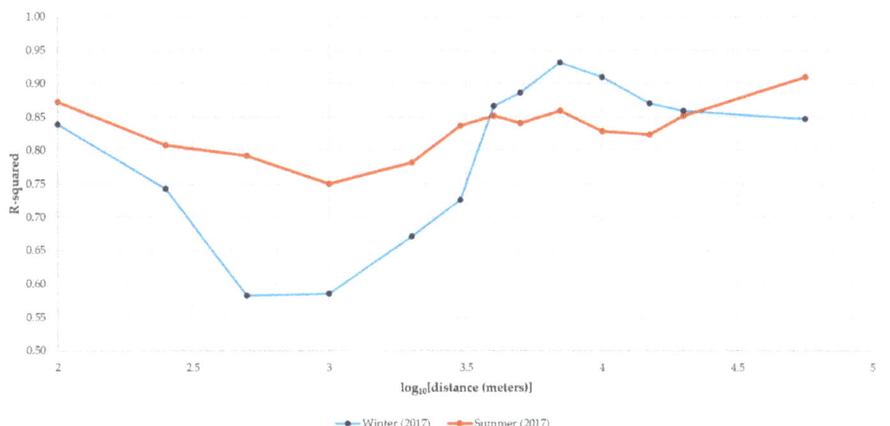

Figure 6. PLS-PM R-squared values.

The calculated R-squared values of summer varied less than the winter counterparts. The range of values varied from 0.75 (1 km) to 0.91 (56 km) in summer, while in winter they ranged from 0.58 (500 m) to 0.93 (7 km). For the winter period, for distances comprehended between 250 m and 3 km, the model's explicability was below 0.75, but the in winter models that comprehended distances between 4 to 20 km, the values were higher than in the corresponding summer models.

In order to assure that the models had no multicollinearity, it was ensured that all the VIF values were below 5 (please see Supplementary Material). The significance of weights and path coefficients was accessed through bootstrapping. By approaching the traditional threshold for statistical significance, p values larger than 0.05 were achieved for the weights (please see Supplementary Material). On the other hand, it was verified that the path coefficients of "Land Use" were significant, characterized by p values lower than 0.05 for long distances (Figure 7).

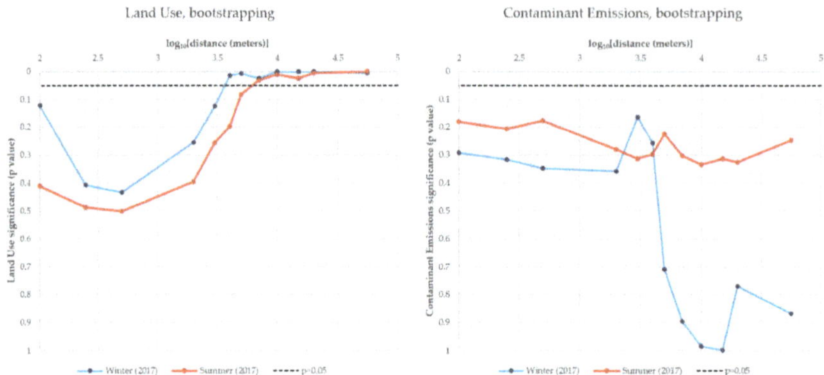

Figure 7. PLS-PM significance of path coefficients.

The significance of land uses seemed to follow a sigmoid pattern. For the summer period, statistical significance ($p < 0.05$) was achieved for distances larger than 5 km, but for the winter period, significance was achieved for distances larger than 3 km. When it comes to "Contaminant emissions" none of the models achieved statistical significance. For the summer period, p values increased with distance. For winter, p values seemed not to change for distances below 2 km, increased to 0.124 at 3 km, and then dropped consistently until a distance of 15 km. For 20 and 56 km, there was a notable loss of statistical significance.

4. Discussion

Before analyzing the results, some expectations regarding the effect of the variables are outlined. It was expected that all pressures had a negative impact on ecological integrity, while "Forest" was expected to have a positive effect. This was expected because in catchments with a high presence of forested areas, good water quality can be found [84]. It was also thought that the measured variables that belonged to the latent variable "Contaminant Emissions" would have a higher effect than "Land Use". This was because the discharges of COD from point sources and nitrogen from livestock, agriculture, and forestry represented the mass flow of contaminants that were transported to surface water, while land use metrics were only indicators of possible pollution. In terms of season and scale, no expectations were anticipated because authors already recognized in different studies that different conclusions can be achieved [34].

The positive effect of "Artificial" for distances below 10 km is hard to explain. By studying the impact of land uses on biological integrity, a positive response was found in urbanized areas [31] by accessing a riparian scale. Another author [85] compared the effects of land uses in the same Index, of Biotic Integrity (IBI) and noted that, in a riparian range, the impacts of urban activity were positively correlated to this index, but for the watershed scale, the outcome was negative, which is in concordance with the present study. This might happen because, at a short range, the impacts of urban presence may be hard to capture, compared to an extended scale. Possibly, urbanized areas only affect water quality when their predominance occurs over a long range.

The agricultural land uses are revealed as a threat to Ave River Basin. For distances larger than 200 m, the effect was negative and increased in scale for both winter and summer seasons. This can reflect that agriculture only affects water quality when the predominance is on a long scale, as there is a high accumulation of contaminants. Likewise, other studies revealed a negative impact over a long scale [24,34,39,86].

The edge density of forested areas "Forest" was the variable that had an expected impact on winter and summer (except at the distances of 56 km and 500 m, where the effects were null and negative, respectively). Many authors have concluded that the effect of forestry improves water quality. Positive impacts are found with biotic indexes [31,32], and negative correlations or effects with contaminant concentrations are found [23–25,28,39,86,87] independently of scale, season, and study area, or even accessed metric.

In light of the previously mentioned reasons, it was not expected that the effect of "Land use" would be greater than "Contaminant Emissions". It was noticed that in the winter period, the measured variables that belonged to "Land Use" were the ones with stronger effects (except in the model of the scale 500 m, where livestock had the strongest impact). But, in the summer period, the contaminant emissions from "Forest and Agriculture" and "Livestock" were the variables with the highest pcw. Only at a scale higher than 7 km did the effect of agricultural land use overlap contaminant emissions.

When it comes to significance, the path coefficient of "Land Use" achieved statistical significance in both seasons at a long scale, while "Contaminant emissions" did not. This leads to the conclusion that, in concordance with other authors, the effect of land use should be accessed on a long scale, also called a complete watershed. Both diffuse and point-source discharges have temporal changes [7,88–90]. To access the seasonal effects of contaminant emissions, it is important to trace the temporal changes in the contaminant flow, just like it was done in other studies [91]. In the present study, the point source flow of COD and diffuse discharge resulting from livestock production, forested, and agricultural activities was in the form of annual flow. In fact, it is quite hard to access the released contamination on shorter temporal scales for a whole river basin. But, if the data of point and diffuse sources from APA were monthly, or even seasonal, it is believed that the significance of "Contaminant Emissions" would be higher in the model and could possibly reveal higher effects than landscape metrics. In previous studies where the Ave River Basin water quality was assessed, it was noted that point-source pressures [51] and livestock production [52] were major threats to water quality. But, in those studies, landscape metrics were not used in such a detailed form, only the percentage of catchments occupied by agricultural and artificial areas were used. During summer, the contribution of underground water to river discharge increased due to the lack of rainfall [92]. In the presented models this is shown because contaminant emissions from agriculture, forest, and livestock effects were indeed larger in summer periods than in winter. During winter, the runoff effect was higher due to strong rainfalls, which explains why variables related to land use metrics had a stronger impact in winter rather than in summer.

Another important aspect is that when formative PLS-PM models are adopted, the number of explaining variables cannot be large because of the shortcomings of variance inflation. One technique to reduce variance inflation is by restraining the number of variables [93]. For this reason, the effects of other land use metrics were not accessed in this study, only the edge density of forested areas, Shannon's diversity index, and connectance metrics of agricultural and artificial surfaces. However, this study succeeded in demonstrating the seasonal and scale effects in the interaction between various pollution sources and ecological integrity. The low sample size (n = 12) was a limitation. As a consequence, the weights were not significant. Despite this, it was still possible to achieve significance for the path coefficient "Land Use" latent variable. So, it is recommended to use larger sample sizes in similar and future studies. Since the weights were not significant, the tested models may not be suited for prediction purposes. At any rate, the presented study indicates that, in terms of prediction in the Ave

River Basin, longer scales should be adopted, considering the consistent high R-squared values and significance of land use variables observed at these scales.

For proper river basin management, landscape metric variables should always be assessed, as it is quite easy to calculate them using computer packages such as FRAGSTATS [94] or the ArcGIS toolbox [76]. In terms of landscape management, when the consequences of anthropogenic pressures for water quality are to be assessed, the exercise should always be applied to various scales. This is recommended because this study concluded that the sense (positive or negative) of an impact can change with scale.

The most limiting factor in this study was probably the small number of sampling points used to assess the $IPtI_N$, just 12. However, these few samples allowed us to provisionally expose the significant role ($p < 0.05$) of land use metrics for a satisfactory ($R^2 \approx 0.85$) explanation of water quality ($IPtI_N$) in the long range (>4 km from the contaminant sources). This result is noteworthy. It can be (and was) argued that more samples would render the possibility to reveal the influence of other anthropogenic pressures, eventually hidden in this study by the sample's coarse resolution. Nevertheless, it would not be a surprise if the results obtained in this study were replicated with a finer resolution, as contaminant emissions are subject to larger inter annual variations than are land uses, and, hence, they could eventually be inefficient in the studied period. A larger sample would probably capture fine-resolution effects, for example, related to point-source contaminant emissions, but is not certain that would change the general outcomes and conclusions taken from this study. The main goals were achieved, which were to explore the influence of anthropogenic pressures on water quality as function of scale and season using a novel statistical method. The 26 PLS-PM models implemented in a predefined sequence were capable of identifying the most important variables and distances from contaminant sources that controlled water quality in the Ave River Basin in the studied period. The model results may not be directly used in management initiatives without prior verification using a larger sample, but they suggested how scale and season can affect the conclusions about cause–effect relationships involving anthropogenic pressures and water quality. In that context, the outcomes from this study provided interesting clues for managers of water quality at catchment scale, which are inherently an important scientific result.

5. Conclusions

This study has shown to be effective in demonstrating seasonal and scale impacts in the interplay between the effect of landscape metrics and contamination sources on water quality. This analysis plays an important role for decision makers to take into account that territory planning is intrinsically linked to water quality. As it was found in this study, the effect of metrics can be greater than contamination sources. From a statistical point of view, this study showed that the use of a long scale is preferable, since it obtained higher coefficients of determination in both winter and summer, but also high statistical significance for the latent variable "Land Use". Even so, it is advised that when water quality studies are carried out, effects should always be analyzed not at a long scale but also at a short scale. This is because each river basin is unique and reveals natural and anthropogenic interactions that are different in each river basin. So, in other locations, stronger effects can be found at shorter scales. In terms of water quality improvement, besides the constant monitoring and reduction of pollution sources, it is pointed out that the presence of forested areas improves the ecological integrity, not in total area but in terms of edge density. As far as agricultural areas are concerned for ecological integrity, strategic relocations can also be a key strategy in order to decrease connectance. While in urban areas, these hardly can be changed, given the enormous costs that such processes can entail.

Supplementary Materials: The following are available online at http://www.mdpi.com/2073-4441/11/10/2025/s1, The supplementary materials comprise an Excel file with base data and results of PLS-PM models.

Author Contributions: Conceptualization, A.C.P.F.; methodology, A.C.P.F. and F.A.L.P.; software, A.C.P.F.; validation, F.A.L.P.; formal analysis, L.F.S.F.; investigation, A.C.P.F.; resources, L.F.S.F. and R.M.V.C.; data curation, A.C.P.F., F.A.L.P., L.F.S.F. and R.M.V.C.; writing—original draft preparation, A.C.P.F.; writing—review and editing,

F.A.L.P.; visualization, A.C.P.F.; supervision, F.A.L.P. and L.F.S.F.; project administration, R.M.V.C.; funding acquisition, R.M.V.C.

Funding: This research was funded by the INTERACT project "Integrated Research in Environment, Agro-Chain and Technology", no. NORTE-01-0145-FEDER-000017, in the line of research entitled BEST "Bio-economy and Sustainability", and co-financed by the European Regional Development Fund (ERDF) through NORTE 2020 (the North Regional Operational Program 2014/2020). For authors at the CITAB research centre, this research was further financed by the FEDER/COMPETE/POCI—Operational Competitiveness and Internationalization Programme under Project POCI-01-0145-FEDER-006958, and by National Funds of FCT—Portuguese Foundation for Science and Technology under the project UID/AGR/04033/2019. For the author in the CQVR, this research was additionally supported by National Funds of FCT—Portuguese Foundation for Science and Technology under the project UID/QUI/00616/2019.

Conflicts of Interest: The authors declare no conflict of interest. The funders had no role in the design of the study; in the collection, analyses, or interpretation of data; in the writing of the manuscript or in the decision to publish the results.

References

1. Vorosmarty, C.J.; Green, P.; Salisbury, J.; Lammers, R.B. Global Water Resources: Vulnerability from Climate Change and Population Growth. *Science* **2000**, *289*, 284–288. [CrossRef] [PubMed]
2. Delpla, I.; Jung, A.-V.; Baures, E.; Clement, M.; Thomas, O. Impacts of climate change on surface water quality in relation to drinking water production. *Environ. Int.* **2009**, *35*, 1225–1233. [CrossRef] [PubMed]
3. Muga, H.E.; Mihelcic, J.R. Sustainability of wastewater treatment technologies. *J. Environ. Manag.* **2008**, *88*, 437–447. [CrossRef] [PubMed]
4. Wright, J.; Gundry, S.; Conroy, R. Household drinking water in developing countries: A systematic review of microbiological contamination between source and point-of-use. *Trop. Med. Int. Health* **2004**, *9*, 106–117. [CrossRef] [PubMed]
5. Hayet, C.; Saida, B.A.; Youssef, T.; Hédi, S. Study of biodegradability for municipal and industrial Tunisian wastewater by respirometric technique and batch reactor test. *Sustain. Environ. Res.* **2016**, *26*, 55–62. [CrossRef]
6. Serrano-Grijalva, L.; Sánchez-Carrillo, S.; Angeler, D.G.; Sánchez-Andrés, R.; Álvarez-Cobelas, M. Effects of shrimp-farm effluents on the food web structure in subtropical coastal lagoons. *J. Exp. Mar. Biol. Ecol.* **2011**, *402*, 65–74. [CrossRef]
7. Grizzetti, B.; Bouraoui, F.; De Marsily, G. Assessing nitrogen pressures on European surface water. *Glob. Biogeochem. Cycles* **2008**, *22*. [CrossRef]
8. Sheoran, A.S.; Sheoran, V. Heavy metal removal mechanism of acid mine drainage in wetlands: A critical review. *Miner. Eng.* **2006**, *19*, 105–116. [CrossRef]
9. Soares, H.M.V.; Boaventura, R.A.; Machado, A.A.S.; Esteves da Silva, J.C. Sediments as monitors of heavy metal contamination in the Ave river basin (Portugal): Multivariate analysis of data. *Environ. Pollut.* **1999**, *105*, 311–323. [CrossRef]
10. Murphy, F.; Ewins, C.; Carbonnier, F.; Quinn, B. Wastewater Treatment Works (WwTW) as a Source of Microplastics in the Aquatic Environment. *Environ. Sci. Technol.* **2016**, *50*, 5800–5808. [CrossRef]
11. Ayers, R.S.; Westcot, D.W. *Water Quality for Agriculture*; FAO United Nations: Rome, Italy, 1985.
12. Nilsson, C.; Svedmark, M. Basic Principles and Ecological Consequences of Changing Water Regimes: Riparian Plant Communities. *Environ. Manag.* **2002**, *30*, 468–480. [CrossRef]
13. Merritt, R.W.; Walker, E.D.; Small, P.L.C.; Wallace, J.R.; Johnson, P.D.R.; Benbow, M.E.; Boakye, D.A. Ecology and transmission of Buruli ulcer disease: A systematic review. *PLoS Negl. Trop. Dis.* **2010**, *4*, e911. [CrossRef] [PubMed]
14. Santos, R.M.B.; Sanches Fernandes, L.F.; Varandas, S.G.P.; Pereira, M.G.; Sousa, R.; Teixeira, A.; Lopes-Lima, M.; Cortes, R.M.V.; Pacheco, F.A.L. Impacts of climate change and land-use scenarios on *Margaritifera margaritifera*, an environmental indicator and endangered species. *Sci. Total Environ.* **2015**, *511*, 477–488. [CrossRef] [PubMed]
15. Townsend, S.A.; Douglas, M.M. The effect of a wildfire on stream water quality and catchment water yield in a tropical savanna excluded from fire for 10 years (Kakadu National Park, North Australia). *Water Res.* **2004**, *38*, 3051–3058. [CrossRef]

16. Arkle, R.S.; Pilliod, D.S. Prescribed fires as ecological surrogates for wildfires: A stream and riparian perspective. *For. Ecol. Manag.* **2010**, *259*, 893–903. [CrossRef]
17. Smith, H.G.; Sheridan, G.J.; Lane, P.N.J.; Nyman, P.; Haydon, S. Wildfire effects on water quality in forest catchments: A review with implications for water supply. *J. Hydrol.* **2011**, *396*, 170–192. [CrossRef]
18. Santos, R.M.B.; Sanches Fernandes, L.F.; Pereira, M.G.; Cortes, R.M.V.; Pacheco, F.A.L. A framework model for investigating the export of phosphorus to surface waters in forested watersheds: Implications to management. *Sci. Total Environ.* **2015**, *536*, 295–305. [CrossRef]
19. Júnior, R.; Galbiatti, J.A.; Pissarra, T.C.; Martins Filho, M. Diagnóstico do Conflito de Uso e Ocupação do Solo na Bacia do Rio Uberaba. *Glob. Sci. Technol.* **2013**, *6*, 40–52. [CrossRef]
20. Pacheco, F.A.L.; Varandas, S.G.P.; Sanches Fernandes, L.F.; Valle Junior, R.F. Soil losses in rural watersheds with environmental land use conflicts. *Sci. Total Environ.* **2014**, *485–486*, 110–120. [CrossRef]
21. Pacheco, F.A.L.; Sanches Fernandes, L.F. Environmental land use conflicts in catchments: A major cause of amplified nitrate in river water. *Sci. Total Environ.* **2016**, *548–549*, 173–188. [CrossRef]
22. McGarigal, K. Landscape Pattern Metrics. In *Wiley StatsRef: Statistics Reference Online*; John Wiley & Sons Ltd.: Hoboken, NJ, USA, 2014.
23. Uuemaa, E.; Roosaare, J.; Mander, Ü. Scale dependence of landscape metrics and their indicatory value for nutrient and organic matter losses from catchments. *Ecol. Indic.* **2005**, *5*, 350–369. [CrossRef]
24. Lee, S.-W.; Hwang, S.-J.; Lee, S.-B.; Hwang, H.-S.; Sung, H.-C. Landscape ecological approach to the relationships of land use patterns in watersheds to water quality characteristics. *Landsc. Urban Plan.* **2009**, *92*, 80–89. [CrossRef]
25. Łowicki, D. Prediction of flowing water pollution on the basis of landscape metrics as a tool supporting delimitation of Nitrate Vulnerable Zones. *Ecol. Indic.* **2012**, *23*, 27–33. [CrossRef]
26. Wu, M.Y.; Xue, L.; Jin, W.B.; Xiong, Q.X.; Ai, T.C.; Li, B.L. Modelling the Linkage Between Landscape Metrics and Water Quality Indices of Hydrological Units in Sihu Basin, Hubei Province, China: An Allometric Model. *Procedia Environ. Sci.* **2012**, *13*, 2131–2145. [CrossRef]
27. Wu, J.; Shen, W.; Sun, W.; Tueller, P.T. Empirical patterns of the effects of changing scale on landscape metrics. *Landsc. Ecol.* **2002**, *17*, 761–782. [CrossRef]
28. Ding, J.; Jiang, Y.; Liu, Q.; Hou, Z.; Liao, J.; Fu, L.; Peng, Q. Influences of the land use pattern on water quality in low-order streams of the Dongjiang River basin, China: A multi-scale analysis. *Sci. Total Environ.* **2016**, *551–552*, 205–216. [CrossRef] [PubMed]
29. Morley, S.A.; Karr, J.R. Assessing and restoring the health of urban streams in the Puget Sound Basin. *Conserv. Biol.* **2002**, *16*, 1498–1509. [CrossRef]
30. Allan, J.D.; Erickson, D.L.; Fay, J. The influence of catchment land use on stream integrity across multiple spatial scales. *Freshw. Biol.* **1997**, *37*, 149–161. [CrossRef]
31. Roth, N.E.; Allan, J.D.; Erickson, D.L. Landscape influences on stream biotic integrity assessed at multiple spatial scales. *Landsc. Ecol.* **1996**, *11*, 141–156. [CrossRef]
32. Potter, K.M.; Cubbage, F.W.; Schaberg, R.H. Multiple-scale landscape predictors of benthic macroinvertebrate community structure in North Carolina. *Landsc. Urban Plan.* **2005**, *71*, 77–90. [CrossRef]
33. Hunsaker, C.T.; Levine, D.A. Hierarchical Approaches to the Study of Water Quality in Rivers. *Bioscience* **1995**, *45*, 193–203. [CrossRef]
34. Zhang, J.; Li, S.; Dong, R.; Jiang, C.; Ni, M. Influences of land use metrics at multi-spatial scales on seasonal water quality: A case study of river systems in the Three Gorges Reservoir Area, China. *J. Clean. Prod.* **2019**, *206*, 76–85. [CrossRef]
35. Johnson, L.B.; Richards, C.; Host, G.E.; Arthur, J.W. Landscape influences on water chemistry in Midwestern stream ecosystems. *Freshw. Biol.* **1997**, *37*, 193–208. [CrossRef]
36. Dodds, W.K.; Oakes, R.M. Headwater influences on downstream water quality. *Environ. Manag.* **2008**, *41*, 367–377. [CrossRef] [PubMed]
37. Tran, C.P.; Bode, R.W.; Smith, A.J.; Kleppel, G.S. Land-use proximity as a basis for assessing stream water quality in New York State (USA). *Ecol. Indic.* **2010**, *10*, 727–733. [CrossRef]
38. Schiff, R.; Benoit, G. Effects of impervious cover at multiple spatial scales on coastal watershed streams. *J. Am. Water Resour. Assoc.* **2007**, *43*, 712–730. [CrossRef]
39. Yu, S.; Xu, Z.; Wu, W.; Zuo, D. Effect of land use types on stream water quality under seasonal variation and topographic characteristics in the Wei River basin, China. *Ecol. Indic.* **2016**, *60*, 202–212. [CrossRef]

40. Thakur, A.K. Model: Mechanistic vs Empirical. In *New Trends in Pharmacokinetics*; Springer: Boston, MA, USA, 1991; pp. 41–51.
41. Loucks, D.P.; van Beek, E. Water Quality Modelling and Prediction. In *Water Resources Systems Planning and Management*; UNESKO: Paris, France, 2005.
42. Astrachan, C.B.; Patel, V.K.; Wanzenried, G. A comparative study of CB-SEM and PLS-SEM for theory development in family firm research. *J. Fam. Bus. Strategy* **2014**, *5*, 116–128. [CrossRef]
43. Gorai, A.K.; Tuluri, F.; Tchounwou, P.B. Development of PLS–path model for understanding the role of precursors on ground level ozone concentration in Gulfport, Mississippi, USA. *Atmos. Pollut. Res.* **2015**, *6*, 389–397. [CrossRef]
44. Chenini, I.; Khmiri, S. Evaluation of ground water quality using multiple linear regression and structural equation modeling. *Int. J. Environ. Sci. Technol.* **2009**, *6*, 509–519. [CrossRef]
45. Levêque, J.G.; Burns, R.C. A Structural Equation Modeling approach to water quality perceptions. *J. Environ. Manag.* **2017**, *197*, 440–447. [CrossRef] [PubMed]
46. Nugroho, A.R.; Masduqi, A. Structural equation modelling as instrument for water pollution factor analysis. In Proceedings of the 4th International Conference Advances in Applied Science and Environmental Technology—ASET 2016, Bangkok, Thailand, 7–8 May 2016.
47. Martins, L.; Pereira, A.; Oliveira, A.; Fernandes, A.; Sanches Fernandes, L.F.; Pacheco, F.A.L. An assessment of groundwater contamination risk with radon based on clustering and structural models. *Water* **2019**, *11*, 1107. [CrossRef]
48. Salgado Terêncio, D.P.; Sanches Fernandes, L.F.; Vitor Cortes, R.M.; Moura, J.P.; Leal Pacheco, F.A. Can Land Cover Changes Mitigate Large Floods? A Reflection Based on Partial Least Squares-Path Modeling. *Water* **2019**, *11*, 684. [CrossRef]
49. Garcês, A.; Pires, I.; Pacheco, F.A.L.; Sanches Fernandes, L.F.; Soeiro, V.; Lóio, S.; Prada, J.; Cortes, R.; Queiroga, F.L. Preservation of wild bird species in northern Portugal—Effects of anthropogenic pressures in wild bird populations (2008–2017). *Sci. Total Environ.* **2019**, *650*, 2996–3006. [CrossRef] [PubMed]
50. Ferreira, A.R.L.; Sanches Fernandes, L.F.; Cortes, R.M.V.; Pacheco, F.A.L. Assessing anthropogenic impacts on riverine ecosystems using nested partial least squares regression. *Sci. Total Environ.* **2017**, *583*, 466–477. [CrossRef] [PubMed]
51. Sanches Fernandes, L.F.; Fernandes, A.C.P.; Ferreira, A.R.L.; Cortes, R.M.V.; Pacheco, F.A.L. A partial least squares—Path modeling analysis for the understanding of biodiversity loss in rural and urban watersheds in Portugal. *Sci. Total Environ.* **2018**, *626*, 1069–1085. [CrossRef] [PubMed]
52. Fernandes, A.C.P.; Sanches Fernandes, L.F.; Moura, J.P.; Cortes, R.M.V.; Pacheco, F.A.L. A structural equation model to predict macroinvertebrate-based ecological status in catchments influenced by anthropogenic pressures. *Sci. Total Environ.* **2019**, *681*, 242–257. [CrossRef]
53. APA—ARH-Norte. *Plano de Gestão da Região Hidrográfica do Cávado, Ave e Leça (RH2). Relatório Base 1º Ciclo Parte 1—Enquadramento e Aspectos Gerais*; Agência Portuguesa do Ambiente: Porto, Portugal, 2012.
54. Gonçalves, E.P.R.; Boaventura, R.A.R.; Mouvet, C. Sediments and aquatic mosses as pollution indicators for heavy metals in the Ave river basin (Portugal). *Sci. Total Environ.* **1992**, *114*, 7–24. [CrossRef]
55. Alves, C.; Boaventura, R.; Soares, H. Evaluation of Heavy Metals Pollution Loadings in the Sediments of the Ave River Basin (Portugal). *Soil Sediment. Contam.* **2009**, *18*, 603–618. [CrossRef]
56. Peixoto, F.P.; Carrola, J.; Coimbra, A.M.; Teixeira, P.; Coelho, L.; Conceição, I.; Oliveira, M.M.; Fontaínhas-Fernandes, A.; Fernandes, C. Oxidative stress responses and histological hepatic alterations in barbel, Barbus bocagei, from Vizela River, Portugal. *Rev. Int. Contam. Ambient.* **2013**, *29*, 29–38.
57. Pascoal, C.; Pinho, M.; Cássio, F.; Gomes, P. Assessing structural and functional ecosystem condition using leaf breakdown: Studies on a polluted river. *Freshw. Biol.* **2003**, *48*, 2033–2044. [CrossRef]
58. Pascoal, C.; Cássio, F.; Marvanová, L. Anthropogenic stress may affect aquatic hyphomycete diversity more than leaf decomposition in a low-order stream. *Arch. Hydrobiol.* **2005**, *162*, 481–496. [CrossRef]
59. Dunck, B.; Lima-Fernandes, E.; Cássio, F.; Cunha, A.; Rodrigues, L.; Pascoal, C. Responses of primary production, leaf litter decomposition and associated communities to stream eutrophication. *Environ. Pollut.* **2015**, *202*, 32–40. [CrossRef] [PubMed]

60. Pinto, D.; Fernandes, A.; Fernandes, R.; Mendes, I.; Pereira, S.; Vinha, A.; Herdeiro, T.; Santos, E.; Machado, M. Determination of heavy metals and other indicators in waters, soils and medicinal plants from Ave valley, in Portugal, and its correlation to urban and industrial pollution. *Sci. Microb. Pathog. Commun. Curr. Res. Technol. Adv.* **2011**, 303–309.
61. Ribeiro, C.M.R.; Maia, A.S.; Ribeiro, A.R.; Couto, C.; Almeida, A.A.; Santos, M.; Tiritan, M.E. Anthropogenic pressure in a Portuguese river: Endocrine-disrupting compounds, trace elements and nutrients. *J. Environ. Sci. Health Part. A* **2016**, *51*, 1043–1052. [CrossRef]
62. Environmental Systems Resource Institute. *ArcMap 10.2.*; ESRI: Redlands, CA, USA, 2014.
63. Maidment, D.R. *Arc Hydro: GIS for Water Resources*; ESRI Inc.: Redlands, CA, USA, 2002.
64. Santos, R.M.B.; Sanches Fernandes, L.F.; Cortes, R.M.V.; Varandas, S.G.P.; Jesus, J.J.B.; Pacheco, F.A.L. Integrative assessment of river damming impacts on aquatic fauna in a Portuguese reservoir. *Sci. Total Environ.* **2017**, *601*, 1108–1118. [CrossRef]
65. Fonseca, A.R.; Sanches Fernandes, L.F.; Fontainhas-Fernandes, A.; Monteiro, S.M.; Pacheco, F.A.L. The impact of freshwater metal concentrations on the severity of histopathological changes in fish gills: A statistical perspective. *Sci. Total Environ.* **2017**, *599*, 217–226. [CrossRef] [PubMed]
66. Pacheco, F.A.L. Regional groundwater flow in hard rocks. *Sci. Total Environ.* **2015**, *506*, 182–195. [CrossRef] [PubMed]
67. Pacheco, F.A.L.; Sousa Oliveira, A.; Van Der Weijden, A.J.; Van Der Weijden, C.H. Weathering, biomass production and groundwater chemistry in an area of dominant anthropogenic influence, the Chaves-Vila Pouca de Aguiar region, north of Portugal. *Water Air Soil Pollut.* **1999**, *115*, 481–512. [CrossRef]
68. Pacheco, F.A.L.; Landim, P.M.B. Two-Way Regionalized Classification of Multivariate Datasets and its Application to the Assessment of Hydrodynamic Dispersion. *Math. Geol.* **2005**, *37*, 393–417. [CrossRef]
69. Pacheco, F.A.L.; Landim, P.M.B.; Szocs, T. Anthropogenic impacts on mineral weathering: A statistical perspective. *Appl. Geochem.* **2013**, *36*, 34–48. [CrossRef]
70. Valle Junior, R.F.; Varandas, S.G.P.; Sanches Fernandes, L.F.; Pacheco, F.A.L. Multi Criteria Analysis for the monitoring of aquifer vulnerability: A scientific tool in environmental policy. *Environ. Sci. Policy* **2015**, *48*, 250–264. [CrossRef]
71. Valera, C.A.; Pissarra, T.C.T.; Martins Filho, M.V.; Valle Junior, R.F.; Sanches Fernandes, L.F.; Pacheco, F.A.L. A legal framework with scientific basis for applying the 'polluter pays principle' to soil conservation in rural watersheds in Brazil. *Land Use Policy* **2017**, *66*, 61–71. [CrossRef]
72. Environmental Systems Resource Institute. *ArcHydro Tools for ArcGIS 10—Tutorial*; ESRI: Redlands, CA, USA, 2012.
73. Ring, C.M.; Wende, S.; Will, A. Smart PLS. Hamburg, Germany. 2005. Available online: http//www.smartpls.de (accessed on 1 May 2019).
74. Hair, J.F., Jr.; Sarstedt, M.; Hopkins, L.; Kuppelwieser, V.G. Partial least squares structural equation modeling (PLS-SEM). *Eur. Bus. Rev.* **2014**, *26*, 106–121. [CrossRef]
75. Schindler, U.; Klinner, K.; Nestler, W. *Microsoft® Excel—Grundlagen der Makroprogrammierung*; Springer: Berlin/Heidelberg, Germany, 2013.
76. Adamczyk, J.; Tiede, D. ZonalMetrics—A Python toolbox for zonal landscape structure analysis. *Comput. Geosci.* **2017**, *99*, 91–99. [CrossRef]
77. APA Critérios de Classificação do Estado das Massas de Água—Rios e Albufeiras. Available online: https://www.apambiente.pt/dqa/criterios-classificacao.html (accessed on 22 June 2019).
78. UKTAG (UKTechnical Advisory Group on the WaterFramework Directive). Technical Report on Groundwater Hazardous Substances. 2016. Available online: https://www.wfduk.org (accessed on 22 June 2019).
79. European Commission. Monitoring under the Water Framework Directive. In *Common Implementation Strategy for the Water Framework Directive (2000/60/EC)*; European Commission: Brussels, Belgium, 2003.
80. Pielou, E.C. The measurement of diversity in different types of biological collections. *J. Theor. Biol.* **1966**, *13*, 131–144. [CrossRef]
81. Jost, L. The Relation between Evenness and Diversity. *Diversity* **2010**, *2*, 207–232. [CrossRef]
82. Devís-Devís, J.; Villamón-Herrera, M. A scientific field searching visibility. International Workshop on Physical Activity and Sport Sciences Quality Journals. *Proc. Prof. Inf.* **2008**, *17*, 242–246.
83. Ketchen, D.J. A Primer on Partial Least Squares Structural Equation Modeling. *Long Range Plan.* **2013**, *46*, 184–185. [CrossRef]

84. Neary, D.G.; Ice, G.G.; Jackson, C.R. Linkages between forest soils and water quality and quantity. *For. Ecol. Manag.* **2009**, *258*, 2269–2281. [CrossRef]
85. Snyder, C.D.; Young, J.A.; Villella, R.; Lemarié, D.P. Influences of upland and riparian land use patterns on stream biotic integrity. *Landsc. Ecol.* **2003**, *18*, 647–664. [CrossRef]
86. Sliva, L.; Dudley Williams, D. Buffer Zone versus Whole Catchment Approaches to Studying Land Use Impact on River Water Quality. *Water Res.* **2001**, *35*, 3462–3472. [CrossRef]
87. Bu, H.; Meng, W.; Zhang, Y.; Wan, J. Relationships between land use patterns and water quality in the Taizi River basin, China. *Ecol. Indic.* **2014**, *41*, 187–197. [CrossRef]
88. Grizzetti, B.; Bouraoui, F.; de Marsily, G.; Bidoglio, G. A statistical method for source apportionment of riverine nitrogen loads. *J. Hydrol.* **2005**, *304*, 302–315. [CrossRef]
89. Bowes, M.J.; Smith, J.T.; Jarvie, H.P.; Neal, C.; Barden, R. Changes in point and diffuse source phosphorus inputs to the River Frome (Dorset, UK) from 1966 to 2006. *Sci. Total Environ.* **2009**, *407*, 1954–1966. [CrossRef]
90. Heathwaite, A.L.; Dils, R.M.; Liu, S.; Carvalho, L.; Brazier, R.E.; Pope, L.; Hughes, M.; Phillips, G.; May, L. A tiered risk-based approach for predicting diffuse and point source phosphorus losses in agricultural areas. *Sci. Total Environ.* **2005**, *344*, 225–239. [CrossRef] [PubMed]
91. Spieles, D.J.; Mitsch, W.J. The effects of season and hydrologic and chemical loading on nitrate retention in constructed wetlands: A comparison of low- and high-nutrient riverine systems. *Ecol. Eng.* **1999**, *14*, 77–91. [CrossRef]
92. Jones, J.P.; Sudicky, E.A.; Brookfield, A.E.; Park, Y.-J. An assessment of the tracer-based approach to quantifying groundwater contributions to streamflow. *Water Resour. Res.* **2006**, *42*. [CrossRef]
93. Coltman, T.; Devinney, T.M.; Midgley, D.F.; Venaik, S. Formative versus reflective measurement models: Two applications of formative measurement. *J. Bus. Res.* **2008**, *61*, 1250–1262. [CrossRef]
94. McGarigal, K.; Marks, B.J. *FRAGSTATS: Spatial Pattern Analysis Program for Quantifying Landscape Structure*; U.S. Department Agriculture Forest Service/Pacific Northwest Research Station: Washington, DC, USA, 1995.

 *water* 

Article

The Buffer Capacity of Riparian Vegetation to Control Water Quality in Anthropogenic Catchments from a Legally Protected Area: A Critical View over the Brazilian New Forest Code

Carlos Alberto Valera [1,2,3], Teresa Cristina Tarlé Pissarra [2,3], Marcílio Vieira Martins Filho [2,3], Renato Farias do Valle Júnior [3,4], Caroline Fávaro Oliveira [3,4], João Paulo Moura [5] and Fernando António Leal Pacheco [3,6,*]

1 Coordenadoria Regional das Promotorias de Justiça do Meio Ambiente das Bacias dos Rios Paranaíba e Baixo Rio Grande, Rua Coronel Antônio Rios, 951, Uberaba MG 38061-150, Brazil; carlosvalera@mpmg.mp.br

2 Universidade Estadual Paulista, Faculdade de Ciências Agrárias e Veterinárias, Via de Acesso Prof. Paulo Donato Castellane, s/n, Jaboticabal SP 14884-900, Brazil; teresap1204@gmail.com (T.C.T.P.); marcilio.martins-filho@unesp.br (M.V.M.F.)

3 POLUS—Grupo de Política de Uso do Solo, Universidade Estadual Paulista (UNESP), Via de Acesso Prof. Paulo Donato Castellane, s/n, Jaboticabal SP 14884-900, Brazil; renato@iftm.edu.br (R.F.d.V.J.); caroline_favaro@hotmail.com (C.F.O.); lfilipe@utad.pt (L.F.S.F.)

4 Instituto Federal do Triângulo Mineiro, Campus Uberaba, Laboratório de Geoprocessamento, Uberaba MG 38064-790, Brazil

5 Centro de Investigação e Tecnologias Agroambientais e Biológicas, Universidade de Trás-os-Montes e Alto Douro, Ap. 1013, 5001-801 Vila Real, Portugal; jpmoura@utad.pt

6 Centro de Química de Vila Real, Universidade de Trás-os-Montes e Alto Douro, Ap. 1013, 5001-801 Vila Real, Portugal

* Correspondence: fpacheco@utad.pt; Tel.: +55-351-917519833

Received: 10 February 2019; Accepted: 12 March 2019; Published: 16 March 2019

Abstract: The riparian buffer width on watersheds has been modified over the last decades. The human settlements heavily used and have significantly altered those areas, for farming, urbanization, recreation and other functions. In order to protect freshwater ecosystems, riparian areas have recently assumed world recognition and considered valuable areas for the conservation of nature and biodiversity, protected by forest laws and policies as permanent preservation areas. The objective of this work was to compare parameters from riparian areas related to a natural watercourse less than 10 m wide, for specific purposes in Law No. 4761/65, now revoked and replaced by Law No. 12651/12, known as the New Forest Code. The effects of 15, 30 and 50 m wide riparian forest in water and soil of three headwater catchments used for sugar cane production were analyzed. The catchments are located in the Environmental Protection Area of Uberaba River Basin (state of Minas Gerais, Brazil), legally protected for conservation of water resources and native vegetation. A field survey was carried out in the catchments for verification of land uses, while periodical campaigns were conducted for monthly water sampling and seasonal soil sampling within the studied riparian buffers. The physico-chemical parameters of water were handled by ANOVA (Tukey's mean test) for recognition of differences among catchments, while thematic maps were elaborated in a geographic information system for illustration purposes. The results suggested that the 10, 30 or even 50 m wide riparian buffers are not able to fulfill the environmental function of preserving water resources, and therefore are incapable to ensure the well-being of human populations. Therefore, the limits imposed by the actual Brazilian Forest Code should be enlarged substantially.

Keywords: water pollution; riparian forest; environmental Law; anthropogenic catchment; watershed management; land use policy

1. Introduction

Riparian forests are woodlands in association with streams, rivers and lakes. The location of riparian forests adjacent to water courses ensures that they can exert a strong influence on the quality of freshwater and help to protect the whole ecosystem from anthropogenic activities taking place upwards in the watershed [1–4]. Besides protection, riparian forests provide multiple services such as habitat for aquatic species, soil biodiversity, sediment filtering, flood control, stream channel stability and aquifer recharge [5–13].

The water, soil and vegetation of riparian forests are state indicators of conservation and preservation of land and stream suitability [14]. The biotic community components act as integrator of ecological conditions [12,15,16] and form the transition between the aquatic environment and the anthropogenic pressure. From a different standpoint, the biotic community components express the different spatial and temporal scales of anthropogenic pressures, and therefore support the environmental assessment of watersheds [3,17,18]. For this reason, efforts should be made to understand the theory and metrics of soil attributes and water quality in riparian buffer ecosystems and their link to specific or aggregated types of anthropogenic disturbance [19].

Studies on the width of riparian forests are abundant and relevant [20–22], but only a few works had the main purpose to contribute, from scientific grounds, to the evaluation of environmental laws. In Brazil, riparian forests are called permanent preservation areas (PPA) under the terms of articles 4th, 5th and 6th of Law No. 12651/12 (the so-called New Forest Code), being defined as: *"protected area, covered or not by native vegetation, with the environmental function of preserving water resources, the landscape, the geological stability and the biodiversity, facilitating the gene flow of fauna and flora, protecting the soil and ensuring the well-being of human populations"* [23].

The technical concept of PPA was introduced in the first Brazilian Forest Code, published on the 23rd January 1934 (Federal Decree No. 23793/34), which has categorized the national forest into four types: protected forest, remaining forest, model forest and income forest. Among other roles, the protected forests were meant to preserve the water flow, minimize the erosion process and ensure public health conditions, and therefore fall into the current concept of permanent preservation area. It is worth to mention that the legal concept of PPA, already used in the revoked Federal Law No. 4771/65 and reproduced in the current Federal Law No. 12651/12, was not created by the legislator. Instead, the legislator has appropriated the existing technical and scientific knowledge for the normative definition of that ecosystem. On 1965, when the Forest Law was published, there was no Ministry of Environment, and the environmental terms were themselves incipient, resulting that Federal Law No. 4771/65 was created and managed by the Ministry of Agriculture, Livestock and Supply.

The 1965 and 2012 forest laws were mostly based on the concept of preservation. Other concepts and definitions equally relevant for the role of riparian forests as preservation, such as ecological function or ecosystem service, were not emphasized in these laws. The Ecological function is *"the operation by which the biotic and abiotic elements that are part of a given environment contribute, in their interaction, to the maintenance of the ecological balance and to the sustainability of the evolutionary processes"*. By fulfilling this function the PPA would provide ecosystem services through ecological and evolutionary processes, including gene flow, disturbance and nutrient cycling, besides the preservation issue. The ecosystem service concept and the practical assessment of ecosystem services [24] in watersheds [25] should be more explicitly applied to the PPAs of anthropogenic catchments.

The New Forest Code has also reduced the overall protection of riparian forests. The width of riparian buffers has not been altered in the new Law, but the location criteria used to measure it have changed. This has led to a smaller area of protected forests, besides the implications for the renting of such protected spaces as well as for transition rules (article No. 59 of Federal Law No. 12651/12). The reduction of riparian vegetation reduces the environmental protection of streams provided by these "green filters", and therefore the likelihood of ecological disasters is expected to increase compromising the sustainability of aquatic systems. A coherent forest code should look upon catchments as spaces where man and nature coexist and self-sustain. A different look inevitably opens the space for radically opposing goals based on the same concept [14]. Therefore, a scientifically based assessment of forest laws represents an environmental policy topic worthy of investigation.

This study aims to take that step forward, namely to compare riparian buffer widths as defined by the revoked (Law No. 4771/65; [26]) and current (Law No. 12651/12; [23]) forest laws for the marginal areas of streams less than 10 m wide, and verify their effects on water and soil resources. The specific goals are: (1) to study riparian buffer soils and water quality along watercourses of anthropogenic watersheds, namely watersheds used for sugar cane production. Watercourses in these catchments may be affected by a diversity of pollutants, including nitrogen and phosphorus from fertilizers or fine sediments from soil erosion. In this study, water quality was assessed by an index that involves the measurement of dissolved oxygen, turbidity, total dissolved solids, which means parameters that can be interpreted as proxies to those pollutants. The index is called the *IWQ*—Index for Water Quality and was proposed by the Environmental Company of São Paulo State—CETESB (https://cetesb.sp.gov.br) to be used in water quality assessments. The study was replicated in watercourses with 15, 30 and 50 m wide riparian forests; (2) to look upon the riparian buffer width defined by the New Forest Code and attempt to understand the underlying environmental function of conserving water resources; (3) to define metrics for the evaluation of water and soil resources within riparian buffers. Due to its regional and national importance, this research was carried out in the Uberaba River Basin, namely at the Municipal Environmental Protection Area.

2. Materials and Methods

2.1. Study Area

The study area comprises the Municipal Environmental Protection Area of Uberaba River Basin (EPA-URB), which is located in the Triângulo Mineiro Region, State of Minas Gerais, Brazil (Figure 1). The EPA-URB occupies an area of approximately 525 km^2 between the Meridian coordinates 188–220 km East and Parallel coordinates 7815–7840 km North of Universal Transverse Mercator coordinate system, 23K. The EPA-URB was acknowledged as a Sustainable Land Use Conservation Unit, which is a portion of Minas Gerais State territorial waters subject to a special regime of administration. The demarcation of the EPA-URB involved the recognition of important natural characteristics besides water resources, namely native vegetation (Cerrado Biome), worth of state protection by the Municipal Law No. 9892 of 28 December 2005.

According to Köppen's climate classification, the region is classified as Aw, tropical, and the climatic domain is classified as semi-humid with 4 to 5 dry months, with a relative humidity of 70–75%. The average annual temperature varies between 20 and 24 °C. The warmest months are October to February, with temperatures ranging between 21 and 25 °C. The month of July is the coldest month with temperatures ranging from 16 to 18 °C. The long-term (sixty two year record) mean annual precipitation in Uberaba municipality is 1584.2 mm. On a monthly basis, average rainfall varies between 42.8 and 541 mm (www.inmet.gov.br/).

The EPA-URB is located in the Central Brazil Plateau and northeast portion of Paraná Basin. Topography is characterized by undulated landscapes. Geology is dominated by a sedimentary sequence comprising two major geologic groups and associated formations: the Sao Bento Group and Serra Geral Formation; the Bauru Group and the Marilia and Uberaba formations. The São Bento Group is composed of basalts cropping out at lower altitudes. The Uberaba Formation is made up of Cenozoic sediments, with a predominance of sedimentary rocks with volcaniclastic contribution, and overlays the Serra Geral Formation along an erosive contact. The upper contact with the Marília

Formation is also considered abrupt, being marked by a silexite level and a conglomerate rich in quartz grains cemented by calcite [27].

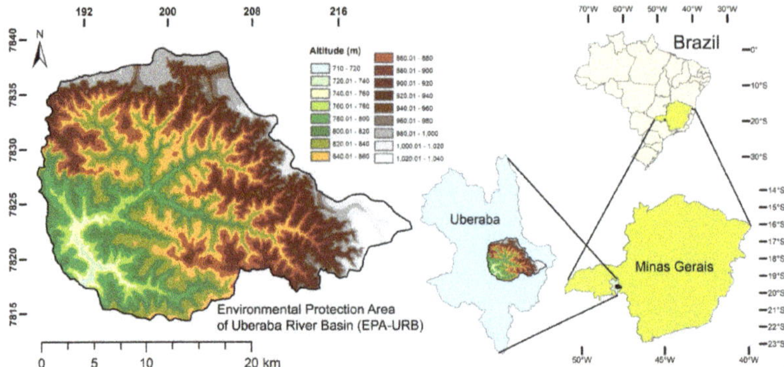

Figure 1. Location of the Environmental Protection Area of Uberaba River Basin (EPA-URB) in the Uberaba Municipality, State of Minas Gerais, and Brazil.

The main soil units are latosols (predominant) and argisols (small areas), according to the Brazilian system of soil classification (https://www.embrapa.br/solos/sibcs). These soil units correspond to ferralsols in the World Reference Base (http://www.fao.org/soils-portal) and oxisols in the classification scheme of Natural Resources Conservation Service (https://www.nrcs.usda.gov/wps/portal/nrcs/site/soils). The latosols are characterized by clayey texture whereas the argisols are characterized by sandy texture.

2.2. Experimental Sites (Sub-Catchments)

The experimental sites comprised three sub-basins selected within the EPA-URB, termed Mangabeira 1 (area: 373.09 ha), Mangabeira 2 (426.6 ha) and Lanhoso (1243.64 ha) (Figure 2). In all cases the catchments were mostly used for sugar cane plantations, which occupy 49.4, 39.5 and 34.3% of the area, respectively, and therefore could be considered anthropogenic basins. Besides this use, the catchments were substantially occupied by native forests (36.1, 30.9 and 53.1%). However, the riparian buffers marginal to the watercourses were characterized by quite different widths: on average, 15 m in Mangabeira 1, 30 m in Mangabeira 2 and 50 m in Lanhoso. The samples of soil and water were collected at the sub-basin outlet.

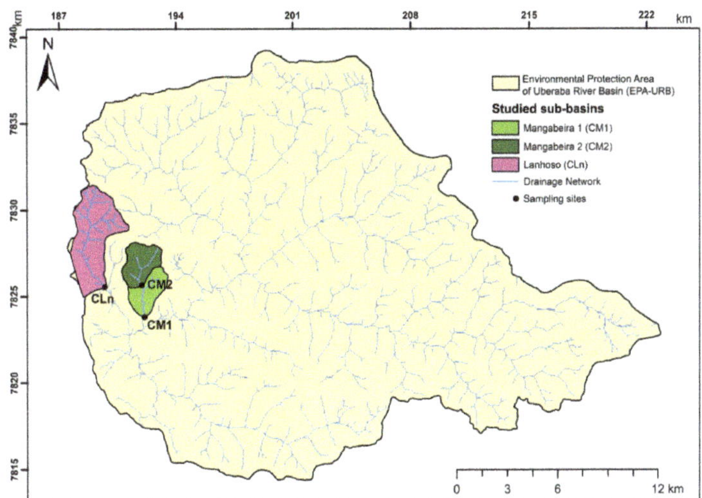

Figure 2. Experimental sites (sub-basins Mangabeira 1, Mangabeira 2 and Lanhoso).

Land uses in the three sub-basins are illustrated in Figure 3a–c and can be summarized as follows: *pasture*—natural or managed pastures (used for the grazing of domestic livestock), sometimes composed of grasses and forbs, in other cases including native vegetation; *sugar cane*—sugar cane plantations; *rural dwelling*—space occupied by the people who work on farms and related activities; *native forest*—area occupied by spontaneous native vegetation, sometimes deforested; *managed forest*—mainly eucalyptus stands; *water bodies*—lakes and reservoirs; *roads*—paved roads; *other land uses*—include orchards, and areas used for rain fed or irrigated corps.

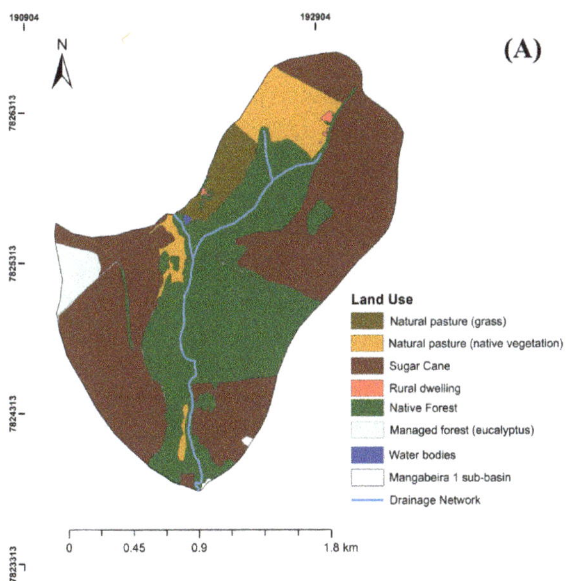

Figure 3. *Cont.*

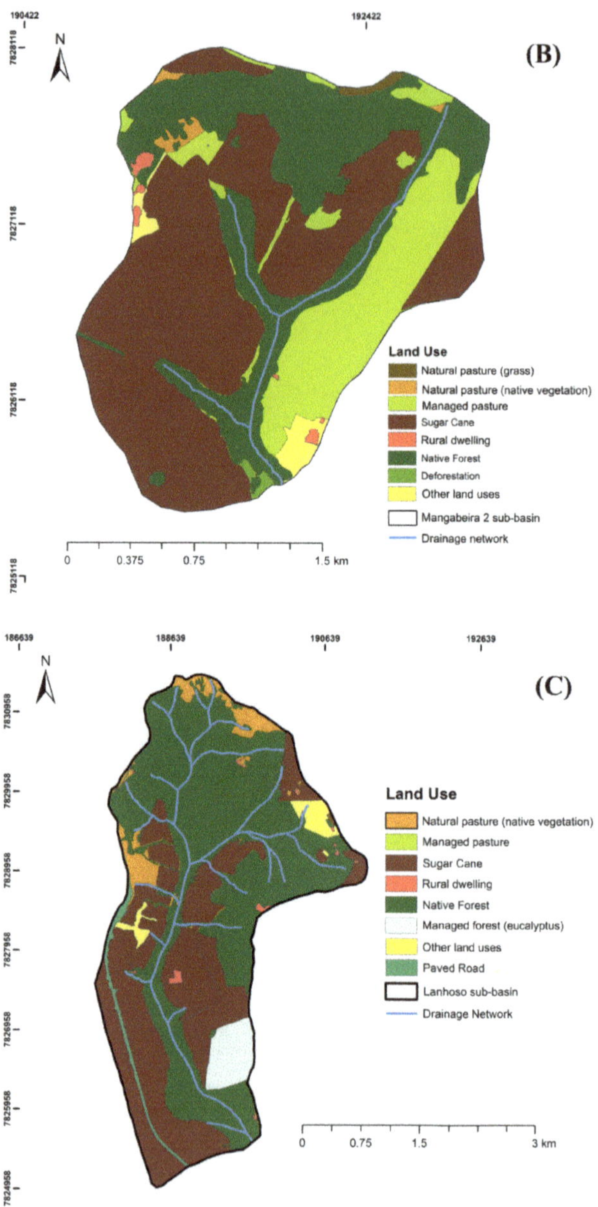

Figure 3. (**a**) Land use in Mangabeira 1 sub-basin. Buffer strip width: 15 m; (**b**) Land use in Mangabeira 2 sub-basin. Buffer strip width: 30 m.; (**c**) Land use in Lanhoso sub-basin. Buffer strip width: 50 m.

2.3. Sampling and Analysis

2.3.1. Soils

The sampling of soils was carried out in the areas occupied by riparian vegetation. Depending on the sub-basin, this represented a buffer extending 15, 30 or 50 m from the watercourse upwards. The sampling procedure followed the guidelines of São Paulo State Environmental Agency [28], and took place in April and November of 2015 at least 10 m away from the stream within the buffer zone. The campaigns involved the collection of undisturbed as well as disturbed samples. The undisturbed samples were collected at the 0–20 and 20–40 cm depth layers to determine soil density and other physical attributes. The number of sites per sub-basin was five, and therefore the total number of undisturbed samples was 60 spanning the two sampling seasons. The disturbed soils were collected within a 10 m square grid with 3 columns and 6 rows, in a total of 18 sites (repetitions). Considering the number of seasons (2), the number of sub-basins (3), and the number of sites per sub-basin (18), the amount of disturbed soil samples was 108.

Following collection, the soil samples were air dried, stripped and passed through a 2 mm mesh screen for chemical analyses. The analyses followed the methods of [29] and involved determination of: pH; Al^{3+}(cmol$_c$ dm^{-3}); Ca—exchangeable calcium (cmol$_c$ dm^{-3}); Mg—exchangeable magnesium (cmol$_c$ dm^{-3}); H+Al—potential acidity (cmol$_c$ dm^{-3}); SB—sum of bases (cmol$_c$ dm^{-3}); t = SB + Al^{3+}—Cation exchange capacity (cmol$_c$ dm^{-3}); T—Cation exchange capacity at pH 7.0, calculated as a function of (SB) + (H+Al), expressed as cmol$_c$ dm^{-3}; K—exchangeable potassium (mg dm^{-3}); P—available phosphorus (mg dm^{-3}); V = SB/CEC—Base Saturation (%); m—aluminum saturation (%); SOM—soil organic matter (dag kg^{-1}); OC—organic carbon (dag kg^{-1}); Sand (%); Silte (%); Clay (%).

2.3.2. Water

The stream water samples were collected immediately downstream from the soil sampling sites, in sectors of the stream that were adjacent to the riparian buffer. The sampling took place approximately 60 cm far from the stream margin, every month during 13 months (January 2016–January 2017). The annual rainfall in 2016 was 1214.4 mm. This value is smaller than the long-term average (1584.2 mm), meaning that 2016 was a dry year. Each month, the samples were collected between calendar days 15 and 20. The weather conditions in the sampling day as well as during the three antecedent days are summarized in Table 1. In the sampling day, rainfall was always <5 mm with the exception of February 2016 and January 2017 campaigns, when rainfall reached 5.9 and 10.9 mm, respectively. In the antecedent days, average rainfall was also small (3.5–5.8 mm), with few exceptions represented in bold in Table 1. The antecedent days with a substantial rainfall were 13 November 2016, and 16 January 2017, with precipitation >25 mm. Therefore, the average analytical results should reflect long-term effects of land use and buffer strip width on the quality of stream water rather than short term effects related with storm events. In the sampling site of a catchment, each campaign involved the measurement of water quality parameters in 10 samples (repetitions), according to CONAMA Resolution No. 357/2005. The parameters were measured using a Horiba U-50 Series multi-parameter probe, and comprised: T—water temperature (°C), pH, ORP—Oxidation Reduction Potential (mV), Ec—Electrical conductivity (μS cm^{-1}), Turbidity, measured in Nephelometric Turbidity Units (NTU), DO—Dissolved oxygen (mg L^{-1}), PDO—Percentage of Dissolved Oxygen (%), and TDS—total dissolved solids (mg L^{-1}).

Table 1. Weather conditions (rainfall) in the water sampling day and the three antecedent days (day-1 until day-3). Values larger than 10 mm day^{-1} are represented in boldface.

Water Sampling Date													
Year	2016												2017
Month	Jan	Feb	Mar	Apr	May	Jun	Jul	Aug	Sep	Oct	Nov	Dec	Jan
Day	19	16	15	19	17	21	19	16	20	18	15	20	17
Rainfall (mm)													
Day	2.0	5.9	4.0	2.4	0.5	0.2	0.0	0.8	0.1	3.7	1.9	2.3	**10.9**
Day-1	8.0	8.7	5.5	1.2	2.8	0.0	0.0	0.0	0.7	4.7	**13.9**	2.6	**27.0**
Day-2	3.4	2.9	2.3	0.9	0.1	0.0	0.0	0.0	0.1	4.9	**32.4**	5.0	**10.0**
Day-3	**16.6**	2.8	6.3	3.7	0.9	0.0	0.0	0.0	0.0	0.6	8.4	1.4	4.4

A subset of parameters was used to calculate the Index for Water Quality (*IWQ*) proposed by the Environmental Company of São Paulo State—CETESB (https://cetesb.sp.gov.br):

$$IWQ = \prod_{i=1}^{n} q_i^{w_i} \quad (1)$$

where $0 \leq IWQ \leq 100$, q_i is the quality of *i*th parameter obtained from standardization of the measured values into a 0–100 range, w_i is the weight of *i*th parameter, which varies in the $0 \leq w_i \leq 1$ interval as function of its importance to the overall quality, and n is the total number of parameters. According to CETESB, $n = 9$ and comprises water temperature, pH, dissolved oxygen, turbidity, total dissolved solids, biochemical oxygen demand, fecal coliforms, total nitrogen and total phosphorus. When data is lacking for some of these parameters, the index can still be calculated using a different set of weights as proposed by [30]. The calculus of *IWQ* in the present study was based on the first five parameters from the list ($n = 5$) and on the following weights: 0.10 (water temperature); 0.21 (pH); 0.17 (turbidity); 0.2 (dissolved oxygen); 0.17 (total dissolved solids). The standardization curves for these parameters, which transform the measured parameters into q scores (Equation (1)), are portrayed in Figure 4.

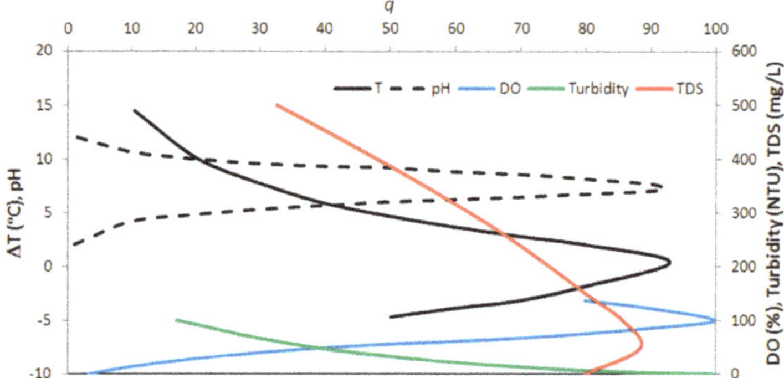

Figure 4. Standardization curves used to transform the water quality parameters into q scores (Equation (1)). Source: https://cetesb.sp.gov.br.

According to the *IWQ*, the quality of stream water is graded as follows: extremely poor ($IWQ \leq 19$), poor ($19 < IWQ \leq 36$), regular ($36 < IWQ \leq 51$), good ($51 < IWQ \leq 79$), excellent ($79 < IWQ \leq 100$). It is worth to note that the *IWQ* is rather sensitive to small changes in the bearing parameters, given the multiplicative formulation of Equation (1). As corollary of this conception, a good water quality ($IWQ > 51$) requires that all q values are high while an excellent quality ($IWQ > 79$) implies that all q scores are very high.

2.4. Thematic Maps and Statistical Treatment of Soil and Water Data

The thematic maps (e.g., Figures 1–3) were prepared in ArcMap software of ESRI [31], a common tool in spatial analysis of hydrologic and environmental data widely used in many recent studies [32,33]. The base information was compiled from various spatial databases, namely the maps published by the Brazilian Institute for Geography and Statistics (https://ww2.ibge.gov.br) on the 1:100,000 scale, and the digital terrain model obtained from the ASTER GDEN V2 satellite image with spatial resolution of 30 m. The statistical treatment of water data was based on the analysis of variance (ANOVA) and Tukey's mean test ($p < 0.05$). The data were processed in the *R* computer program (https://www.r-project.org/).

3. Results

The analytical results are depicted in Table 2 (soils) and Table 3 (water). The water quality index (*IWQ*) is depicted in Table 4.

Table 2. Analytical results for the soil samples. The symbols were defined in the text (Section 2.3.1).

Parameter	pH	Al^{3+}	Ca	Mg	H+Al	SB	t	T	K	P	V	m	SOM	OC	Sand	Silt	Clay
Unit				cmolc dm^{-3}					mg dm^{-3}		%		dag kg^{-1}			%	
								April									
CM1	5.7	0.4	2.7	1.1	5.7	4.1	4.6	9.8	93.6	5.4	40.1	17.4	3.0	1.8	44.5	24.9	30.6
CM2	5.6	0.5	1.0	0.3	4.6	1.4	1.9	6.0	41.5	9.8	25.9	32.0	4.9	2.9	63.7	23.8	12.5
Ln	5.8	0.6	2.2	0.9	4.5	3.2	3.8	7.7	54.6	2.8	43.4	17.3	3.1	1.8	60.5	23.4	16.1
								November									
CM1	5.7	0.4	2.7	1.1	5.7	4.1	4.6	9.8	93.6	5.4	40.1	17.4	3.0	1.8	44.5	24.9	30.6
CM2	5.6	0.5	1.0	0.3	4.6	1.4	1.9	6.0	41.5	9.8	25.9	32.0	4.9	2.9	63.7	23.8	12.5
CLn	5.8	0.6	2.2	0.9	4.5	3.2	3.8	7.7	54.6	2.8	43.4	17.3	3.1	1.8	60.5	23.4	16.1

Table 3. Analytical results for the water samples: average value, standard deviation, Tukey's mean test result (ANOVA; $p < 0.05$). The symbols were defined in the text (Section 2.3.2). Values with different label (lowercase letters *a*, *b*, *c* or *d*) are considered significantly different from each other by the Tukey's test, and therefore can be used to differentiate the sub-basins.

Sub-Basin	Mangabeira 1	Mangabeira 2	Lanhoso
	CM1 (15 m)	CM2 (30 m)	CLn (50 m)
T (°C)	19.04	20.08	19.67
	±2.6	±2.3	±2.6
	d	*bc*	*cd*
pH	7.10	7.00	7.43
	±0.5	±0.5	±0.4
	b	*b*	*a*
ORP (mV)	140.12	153.24	204.68
	±34.9	±43.2	±74.6
	cd	*bc*	*a*
Ec (μS/cm)	60	70	90
	±20	±20	±30
	bc	*b*	*a*
Turbidity (NTU)	6.25	3.75	2.03
	±6.1	±2.9	±2.0
	cd	*d*	*d*
DO (mg/L)	7.64	7.32	11.58
	±1.3	±1.6	±11.3
	b	*b*	*a*
PDO (%)	84.72	83.23	124.81
	±14.4	±18.7	±111.01
	b	*b*	*a*
TDS (mg/L)	40	50	60
	±10	±11	±19
	bc	*b*	*a*

Table 4. Water quality index (*IWQ*) and potentially related environmental variables.

Sub-Basin	Buffer Width (m)	Sugar Cane—*SC* (%)	Native Forest—*NF* (%)	*NF/SC*	*IWQ*	Water Quality
CM1	15	49.4	36.1	0.7	30.8	Poor
CM2	30	39.5	30.9	0.8	31.0	Poor
CLn	50	34.2	53.1	1.6	33.4	Poor

The relationship between soil parameters and the riparian buffer width (Table 2) was not detected for most parameters. However, the percentage of sand and the content of aluminum increased as function of width, while the silt content decreased. The results for sand and aluminum seem to expose the capacity of buffer strips to retain mineral aggregates, especially the more coarse grained. The results for silt may be apparent because sand, silt and clay in a texture analysis sum 100% and therefore when a fraction increases the other tend to decrease regardless of their abundance in the sample.

The analytical results for water (Table 3) indicate a statistical difference between the Mangabeira catchments (CM1 and CM2) and the Lanhoso catchment (CLn), in the case of pH, oxidation-reduction potential, conductivity, dissolved oxygen (in mg L^{-1} or %) and total dissolved solids, which means the majority of parameters. This is strong indication that the catchment with a wider riparian buffer is different from the catchments with a narrower buffer, as regards water quality. The results of the *IWQ* calculation showed reduced values in all basins. These results qualified the stream waters as poor.

When the *IWQ* parameter was plotted as a function of buffer width (Figure 5a) and a trend line was fitted to the scatter points, the fitting equation was parabolic:

$$IWQ = 0.003\ BW^2 - 0.1238BW + 31.971 \quad (2)$$

where *BW* means buffer width. A similar plot, but of *IWQ* as function of *BW* combined with land use/occupation (*NF*/*SC* = native forest/sugar cane ratio; Figure 5b), could be fitted to the following linear equation:

$$IWQ = 0.0391BW\frac{NF}{SC} + 30.26 \quad (3)$$

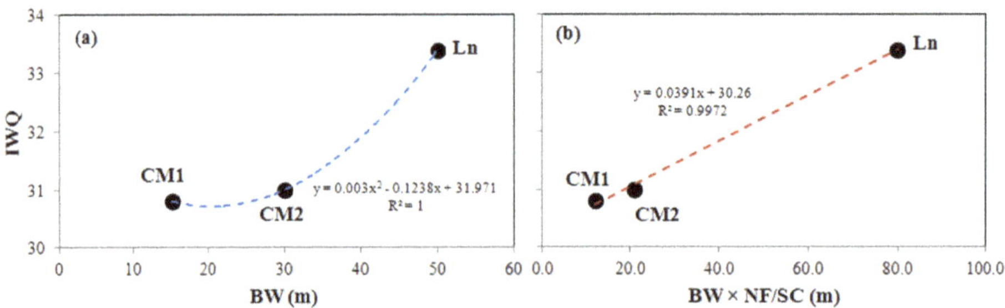

Figure 5. Plot of water quality index (*IWQ*) as function of: (**a**) riparian buffer width (*BW*); (**b**) buffer strip width combined with native forest/sugar cane ratio (*BW* × *NF/SC*). The points represent the values of *IWQ* versus *BW* or *IWQ versus BW* × *NF/SC* in the Mangabeira 1 (CM1), Mangabeira 2 (CM2) and Lanhoso (CLn) sub-basins.

Equation (2) attempts to describe the independent influence of buffer strip width on water quality, while Equation (3) analyzes this influence but coupled the potential interference of land use. In fact, the two effects are barely separable because an increase of buffer strip width tends to increase the NF/SC ratio.

The parabolic trend in Figure 5a may reflect the fact that water quality in CM1 is affected by water quality of CM2, besides the influence of local land use and buffer strip width. Because the *IWQ* in both catchments are similar, regardless the differences between buffer strip widths, the contribution from CM2 is probably large. Put another way, if water quality in CM1 reflected solely land use and buffer strip width, then a lower *IWQ* would be expected in this catchment. In the figure, the CM1 point would drop along the y-axis and the general trend would shift from the parabolic trend towards a (more likely) linear trend.

Using Equation (3) with constant *NF/SC* values allows relating water quality (*IWQ*) and buffer strip width for specific land occupations. Table 5 describes these relationships for three *NF/SC* ratios: 0.7 (the lowest ratio in the studied catchments), 1.6 (highest ratio) and 3.2 (twice the highest ratio, forecasting implementation of conservation practices through expansion of native forest). In the first case (*NF/SC* = 0.7), it is expected that a regular water quality is attained for *BW* = 205 m, while this value reduced to *BW* = 90 m for *NF/SC* = 1.6, and to *BW* = 45 m for *NF/SC* = 3.2. If the buffer strip width would duplicate in the Lanhoso catchment through implementation of conservation practices, the water quality would become good for *BW* = 155 m. Therefore, the 30 m threshold foreseen in Federal Law No. 12651/2012 may not satisfy the water quality requirements in the studied basins, while it is worth remembering that these catchments are located in a protected area for water resources. It should be admitted, however, that this study was based on a small number of catchments and that more general conclusions would require a more exhaustive analysis based on a larger sample. Besides, the interpretations so far hold for anthropogenic catchments used for sugar cane production and cannot be extended directly to other land uses. Finally, the weather reference for this study is a dry year.

Table 5. Riparian buffer width (*BW*), native forest/sugar cane ratio (*NF/SC*) and related water quality (*IWQ*), as predicted by Equation (3) for three constant values of *NF/SC*.

Water Quality	*IWQ*	Riparian Buffer Width—*BW* (m)		
		NF/SC = 0.7	*NF/SC* = 1.6	*NF/SC* = 3.2
Poor	19–36	15–205	15–90	15–45
Regular	36–51	205–300	90–300	45–155
Good	51–79	nd	nd	155–300
Excellent	79–100	nd	nd	nd

It is also worth recalling that water quality expose the aggregate effects of all natural processes and anthropogenic inputs [34,35] that can occur along the flow paths [36], namely chemical weathering [37–40], uptake/release from/to biota [41], leachates from fertilizers that also affect chemical weathering [42–44], discharge of domestic sewage [45,46], among others. The trend depicted in Figure 5b exposes the impact of sugar cane on *IWQ* combined with the buffering capacity of native forest. It is expected that in absence of native forest (*NF* = 0) the water quality index would drop to *IWQ* = 30.26 because *NF* = 0 implies $BW \times NF/SC = 0$. This rather low level of *IWQ* would represent the impact on water quality exclusively attributed to sugar cane production (IWQ_0), including the effects related to fertilizing and management (e.g., erosion control). According to Equation (3), if fertilizing and management conditions are kept unaltered in the future the regular water quality can only be achieved if the proportion of native forest over sugar cane plantations rises substantially. This may not be economically feasible, the reason why the route to follow is to improve management practices to raise the value of IWQ_0.

4. Discussion

The Environmental Protection Area of Uberaba River Basin (EPA-URB) and other similar conservation units exist to reconcile human occupation with the sustainable use of their natural resources, not to expel human populations. However, the activities and uses developed in these areas are subject to specific rules. This work exposed the need to manage properly the permanent preservation areas of three small catchments located in the EPA-URB to accomplish environmental sustainability.

The capacity of riparian buffers to retain particles and dissolved compounds from catchment uplands depend on the buffer width (50, 30 and 15 m) as well as on the types of land use or occupation and their management practices. This study exposed a stronger buffer capacity in the catchment where the riparian forests extend 50 m upwards from the stream margin, but even in this catchment the water quality is generally poor. A regular or good quality would require much wider strips and larger *NF/SC* ratios. In complement, better management practices could be implemented to prevent or at least reduce substantially the exports of sediment and nutrients towards the streams.

The Brazilian law defined the buffer width limits based on two scenarios: the Federal Law No. 4771 and the New Forest Code. In the first case, for watercourses up to 10 m wide the permanent preservation areas need to extend at least 30 m upwards from the stream margin considering the widest seasonal riverbank. In the second case, there are two rules: The transition rule takes into account the size of land property calculated as fiscal modules and creates a distance from the stream margin that goes from a minimum of 5 m to a maximum of 20 m, considering the regular river bank; the permanent rule defines 30 m as unique distance. By changing the reference from the largest riverbank (wet season) to the regular riverbank, the New Forest Code has decreased the legal riparian buffer width. A huge amount of scientific literature has reported the importance of riparian buffer width for water quality and ecological functions [3,11–14,18]. In the present study, it was suggested for a range of native forest/sugar cane occupational ratios (0.7–3) that the legal width should be at least 45 m, but preferably more, corroborating the studies of GAEMA [47]. Besides, efforts should be made to better understand the theory and the metrics of soil attributes and water quality in riparian forest ecosystems to develop ecological functions for these areas based on buffer width [19].

In general, the owner of a rural property has the legal right to use, enjoy, possess and dispose of it. However, this legal right is not applicable to permanent preservation areas included in the property. The permanent preservation area is a legal area. It is not an area for the socioeconomic use of a land owner. The permanent preservation areas are subject to a restriction of use imposed by the Brazilian Constitution, which aims to ensure a provisional ecosystem function, namely the provision of soil and water as resource. In this context, the management of permanent preservation areas is allowed solely when there is no local option of public and social interest, and the interventions are to be done with a low impact. The New Forest Code recommends a riparian buffer width that can keep fundamental ecosystem functions. This study suggested that this width should be increased to 45 m, at least. Besides the correct dimensioning of the riparian buffer width, a number of mitigation measures are ought of implementation to increase the IWQ_0 far above its current value (IWQ_0 = 30.26 = poor water quality). To become effective, the causes and paths of pollution should be assessed [48–51] and then the measures should be modeled in spatial decision support systems focused on water resources planning and management [52–56], evidenced and discussed by government agencies and public and private companies, and integrate public policies and environmental management plans [57]. From a broader standpoint, the specific widths along the drainage network should be reviewed, being defined as function of basin area, watercourse and the catchments' social and economic importance for public water supply.

The results illustrated in Figure 5b raised a striking question: What should be the area released to the agro system, in replacement of forest areas that have the aim of protecting the environment? The criticism we make to the New Forest Code in this case, is that the ratio of permanent preservation area over area used for agriculture or other anthropogenic activities, should be defined technically and not on the basis of political or socioeconomic convenience. This rationale is also valid for riparian buffer widths. The technical work of Kageyama, Cordeiro and Metzer [47] strongly suggested a minimum buffer strip width of 50 m to protect small streams from anthropogenic activities located upward of the catchment hillsides. In the present study, the data collected on water are in favor of a buffer strip width even larger than that threshold if good quality water is aimed at the studied catchments. If anthropogenic activities are practiced in this protected area, the erosion and transport processes, followed by the silting and eutrophication of stream water, will be accentuated, a situation that will be defined as environmental damage according to Art. No. 3 of the Federal Law No. 6938/81 and handled through the "polluter-pays" principle [58].

The present study is corroborated in the literature, and was founded on principles of Environmental Law, namely the principles of precaution and prevention. The New Forest Code should take this and other scientific studies as example, and always be interpreted as *pro-nature*: in favor of Environment. Thus, if field data suggest that environmental vulnerability occurs within 50 m from the stream margin at the widest riverbank, it is clear that one should opt for riparian buffer solutions that result in greater environmental protection.

5. Conclusions

The role of riparian vegetation and forest cover in the control of stream water quality in anthropogenic catchments was investigated in this study. The analysis involved three headwater catchments characterized by increasing buffer strip widths, namely 15, 30 and 50 m widths, as well as increasing native forest to sugar cane ratios (*NF/SC*). The studied basins are located in the Environmental Protection Area of Uberaba River Basin (EPA-URB; state of Minas Gerais, Brazil). The water quality analysis aimed to evaluate a recent forest law (Law No. 12651/12) in this very important water resources and native forest (Cerrado Biome) conservation unit. A linear trend was defined between a specific water quality index (*IWQ*) and the combined protective effects of buffer with (*BW*) and *NF/SC* ($BW \times NF/SC$). Presently, the quality of stream water in the three catchments is poor ($IWQ < 36\%$). The linear trend allows estimating a regular water quality ($36 \leq IWQ \leq 51\%$) if buffer widths were larger than 45 m, but only if the coverage by native forest increased substantially (e.g., duplicated) in the studied basins. Under the current land use ($0.7 \leq NF/SC \leq 1.6$) the regular water quality would be reached for buffer strip widths in the 90–205 m range. While keeping the current *BW* an *NF/SC* values, water quality could be improved if conservation practices were implemented in the sugar cane fields to reduce the export of sediments and nutrients towards the aquatic media. Overall, it was suggested in this study that the 30 m buffer strip width proposed in the New Forest Code, is barely capable of protecting water quality in the EPA-URB.

Author Contributions: Conceptualization, C.A.V. and T.C.T.P.; methodology, C.A.V. and T.C.T.P.; software, R.F.d.V.J. and J.P.M.; validation, C.A.V., L.F.S.F. and F.A.L.P.; formal analysis, C.A.V., F.A.L.P., T.C.T.P. and M.V.M.F.; investigation, C.A.V.; resources, T.C.T.P.; data curation, C.A.V., C.F.O. and T.C.T.P.; writing—original draft preparation, C.A.V.; writing—review and editing, F.A.L.P.; visualization, R.F.d.V.J.; supervision, T.C.T.P. and M.V.M.F.; project administration, T.C.T.P. and M.V.M.F.; funding acquisition, T.C.T.P., M.V.M.F. and R.F.d.V.J.

Funding: The present study was carried out within the framework of the Post Graduation Research Programme of Coordenação de Aperfeiçoamento de Pessoal de Nível Superior (CAPES); Conselho Nacional de Desenvolvimento Científico e Tecnológico (CNPq); Agência do Ministério da Ciência, Tecnologia, Inovações e Comunicações (MCTIC); and Land Use Policy Brazilian Group (POLUS). The author is affiliated with IFTM Renato Farias do Valle Júnior wishes to acknowledge the funding through the CNPq research scholarship Proc. 307921/2018-2. The authors integrated in the CITAB research center were financed by National Funds of the FCT–Portuguese Foundation for Science and Technology POCI-01-0145-FEDER-006958, under the project UID/AGR/04033/2019. The author integrated in the CQVR was funded by National Funds of the FCT–Portuguese Foundation for Science and Technology POCI-01-0145-FEDER-006958, under the project UID/QUI/00616/2019.

Acknowledgments: Hygor Evangelista Siqueira, Mauro Ferreira Machado and Renata Cristina Araújo Costa are acknowledged for fruitful discussions, sharing of information and mapping of the area.

Conflicts of Interest: The authors declare no conflict of interest. The funders had no role in the design of the study; in the collection, analyses, or interpretation of data; in the writing of the manuscript, or in the decision to publish the results.

References

1. Valle Junior, R.F.; Varandas, S.G.P.; Pacheco, F.A.L.; Pereira, V.R.; Santos, C.F.; Cortes, R.M.V.; Fernandes, L.F.S. Impacts of land use conflicts on riverine ecosystems. *Land Use Policy* **2015**, *43*, 48–62. [CrossRef]
2. Santos, R.M.B.; Sanches Fernandes, L.F.; Varandas, S.G.P.; Pereira, M.G.; Sousa, R.; Teixeira, A.; Lopes-Lima, M.; Cortes, R.M.V.; Pacheco, F.A.L. Impacts of climate change and land-use scenarios on Margaritifera margaritifera, an environmental indicator and endangered species. *Sci. Total Environ.* **2015**, *511*, 477–488. [CrossRef] [PubMed]
3. Angelstam, P.; Lazdinis, M. Tall herb sites as a guide for planning, maintenance and engineering of riparian continuous forest cover. *Ecol. Eng.* **2017**, *103*, 470–477. [CrossRef]
4. Li, K.; Chi, G.; Wang, L.; Xie, Y.; Wang, X.; Fan, Z. Identifying the critical riparian buffer zone with the strongest linkage between landscape characteristics and surface water quality. *Ecol. Indic.* **2018**, *93*, 741–752. [CrossRef]
5. Choi, J.Y. Establishment and management of riparian buffer zones in Han River basin, Korea. *WIT Trans. Ecol. Environ.* **1970**, *48*, 1–7.
6. Tockner, K.; Ward, J.V. Biodiversity along riparian corridors. *Large Rivers* **1999**, *115*, 293–310. [CrossRef]
7. Poiani, K.A.; Richter, B.D.; Anderson, M.G. Biodiversity conservation at multiple scales: Functional sites, landscapes, networks. *Bioscience* **2000**, *50*, 133–146. [CrossRef]
8. Arscott, D.B.; Tockner, K.; van der Nat, D.; Ward, J.V. Aquatic habitat dynamics along a braided alpine river ecosystem (Tagliamento River, Northeast Italy). *Ecosystems* **2002**, *5*, 802–814. [CrossRef]
9. Meleason, M.A.; Quinn, J.M. Influence of riparian buffer width on air temperature at Whangapoua Forest, Coromandel Peninsula, New Zealand. *For. Ecol. Manag.* **2004**, *191*, 365–371. [CrossRef]
10. Dwire, K.A.; Lowrance, R.R. Riparian ecosystems and buffers-multiscale struture, function, and management Introduction. *J. Am. Water Resour. Assoc.* **2006**, *42*, 1–4. [CrossRef]
11. Anderson, P.D.; Poage, N.J. The Density Management and Riparian Buffer Study: A large-scale silviculture experiment informing riparian management in the Pacific Northwest, USA. *For. Ecol. Manag.* **2014**, *316*, 90–99. [CrossRef]
12. Phoebus, I.; Segelbacher, G.; Stenhouse, G.B. Do large carnivores use riparian zones? Ecological implications for forest management. *For. Ecol. Manag.* **2017**, *402*, 157–165. [CrossRef]
13. Nelson, J.L.; Hunt, L.G.; Lewis, M.T.; Hamby, K.A.; Hooks, C.R.; Dively, G.P. Arthropod communities in warm and cool grass riparian buffers and their influence on natural enemies in adjacent crops. *Agric. Ecosyst. Environ.* **2018**, *257*, 81–91. [CrossRef]
14. Rosot, M.A.; Maran, J.C.; da Luz, N.B.; Garrastazú, M.C.; de Oliveira, Y.M.; Franciscon, L.; Clerici, N.; Vogt, P.; de Freitas, J.V. Riparian forest corridors: A prioritization analysis to the Landscape Sample Units of the Brazilian National Forest Inventory. *Ecol. Indic.* **2018**, *93*, 501–511. [CrossRef]
15. Shirley, S.M.; Smith, J.N. Bird community structure across riparian buffer strips of varying width in a coastal temperate forest. *Biol. Conserv.* **2005**, *125*, 475–489. [CrossRef]
16. Awade, M.; Metzger, J.P. Using gap-crossing capacity to evaluate functional connectivity of two Atlantic rainforest birds and their response to fragmentation. *Austral Ecol.* **2008**, *33*, 863–871. [CrossRef]
17. Fierro, P.; Bertrán, C.; Tapia, J.; Hauenstein, E.; Peña-Cortés, F.; Vergara, C.; Cerna, C.; Vargas-Chacoff, L. Effects of local land-use on riparian vegetation, water quality, and the functional organization of macroinvertebrate assemblages. *Sci. Total Environ.* **2017**, *609*, 724–734. [CrossRef] [PubMed]
18. De Mello, K.; Valente, R.A.; Randhir, T.O.; Dos Santos, A.C.A.; Vettorazzi, C.A. Effects of land use and land cover on water quality of low-order streams in Southeastern Brazil: Watershed versus riparian zone. *Catena* **2018**, *167*, 130–138. [CrossRef]
19. Hénault-Ethier, L.; Larocque, M.; Perron, R.; Wiseman, N.; Labrecque, M. Hydrological heterogeneity in agricultural riparian buffer strips. *J. Hydrol.* **2017**, *546*, 276–288. [CrossRef]
20. Clinton, B.D. Stream water responses to timber harvest: Riparian buffer width effectiveness. *For. Ecol. Manag.* **2011**, *261*, 979–988. [CrossRef]
21. Yang, S.; Bai, J.; Zhao, C.; Lou, H.; Zhang, C.; Guan, Y.; Zhang, Y.; Wang, Z.; Yu, X. The assessment of the changes of biomass and riparian buffer width in the terminal reservoir under the impact of the South-to-North water diversion project in China. *Ecol. Indic.* **2018**, *85*, 932–943. [CrossRef]

22. Zhang, J.; Li, S.; Dong, R.; Jiang, C.; Ni, M. Influences of land use metrics at multi-spatial scales on seasonal water quality: A case study of river systems in the Three Gorges Reservoir Area, China. *J. Clean. Prod.* **2019**, *206*, 76–85. [CrossRef]
23. Brasil, 2012. Lei No. 12.651, de 25 de maio de 2012. Dispõe sobre a proteção da vegetação nativa; altera as Leis Nos. 6.938, de 31 de agosto de 1981, 9.393, de 19 de dezembro de 1996, e 11.428, de 22 de dezembro de 2006; revoga as Leis Nos. 4.771, de 15 de setembro de 1965, e 7.754, de 14 de abril de 1989, e a Medida Provisória No. 2.166-67, de 24 de agosto de 2001; e dá outras providências. Diário Oficial da República Federativa do Brasil, Poder Executivo, Brasília, DF, 28 maio 2012. Available online: www.planalto.gov.br (accessed on 26 September 2018).
24. Chan, K.M.; Anderson, E.; Chapman, M.; Jespersen, K.; Olmsted, P. Payments for ecosystem services: Rife with problems and potential—For transformation towards sustainability. *Ecol. Econ.* **2017**, *140*, 110–122. [CrossRef]
25. Villeneuve, B.; Piffady, J.; Valette, L.; Souchon, Y.; Usseglio-Polatera, P. Direct and indirect effects of multiple stressors on stream invertebrates across watershed, reach and site scales: A structural equation modelling better informing on hydromorphological impacts. *Sci. Total Environ.* **2018**, *612*, 660–671. [CrossRef] [PubMed]
26. Brasil, 1965. Lei No. 4.771, de 15 de setembro de 1965. Institui o novo Código Florestal. Diário Oficial da República Federativa do Brasil, Poder Executivo, Brasília, DF, 16 set. 1965. Seção 1. p. 9529. Available online: www.planalto.gov.br (accessed on 26 September 2018).
27. Quintão, D.A.; Caxito, F.D.; Karfunkel, J.; Vieira, F.R.; Seer, H.J.; Moraes, L.C.; Ribeiro, L.C.; Pedrosa-Soares, A.C. Geochemistry and sedimentary provenance of the Upper Cretaceous Uberaba Formation (Southeastern Triângulo Mineiro, MG, Brazil). *Braz. J. Geol.* **2017**, *47*, 159–182. [CrossRef]
28. Brasil, 2014 Conama. Conselho Nacional Do Meio Ambiente. Resolução No. 344, de 25 de março de 2004. Estabelece as diretrizes gerais e procedimentos mínimos para avaliação do material a ser dragado em águas jurisdicionais brasileiras, e dá outras providências. Available online: http://www.suape.pe.gov.br (accessed on 26 September 2018).
29. EMBRAPA (Empresa Brasileira De Pesquisa Agropecuária). *Manual de métodos de análises de solo*, 2nd ed.; Centro Nacional de Pesquisa de Solos: Rio de Janeiro, Brazil, 1997; 212p.
30. De Ribeiro, I.V.A.S.; Bouchonneau, N.; Da Silva, A.C.; Fernandes, R.M.C.; De Pinheiro, L.S. Cálculo do índice de qualidade de água (IQA), com estudo de caso nos rios Cocó e Maranguapinho, Ceará. In *Simpósio Brasileiro de Recursos Hídricos*; Associação Brasileira de Recursos Hídricos: Campo Grande, Brazil, 2009; p. 18.
31. ESRI (Environmental Systems Research Institute). *ArcGIS Professional GIS for the Desktop*; Versão 10; ESRI: Lisbon, Portugal, 2012.
32. Pacheco, F.A.L. Regional groundwater flow in hard rocks. *Sci. Total Environ.* **2015**, *506–507*, 182–195. [CrossRef] [PubMed]
33. Fonseca, A.R.; Sanches Fernandes, L.F.; Monteiro, S.M.; Fontainhas-Fernandes, A.; Pacheco, F.A.L. From catchment to fish: Impact of anthropogenic pressures on gill histopathology. *Sci. Total Environ.* **2016**, *550*, 972–986. [CrossRef]
34. Pacheco, F.A.L. Application of correspondence analysis in the assessment of groundwater chemistry. *Math. Geol.* **1998**, *30*, 129–161. [CrossRef]
35. Pacheco, F.A.L.; Landim, P.M.B. Two-way regionalized classification of multivariate data sets and its application to the assessment of hydrodynamic dispersion. *Math. Geol.* **2005**, *37*, 393–417. [CrossRef]
36. Pacheco, F.A.L.; Van Der Weijden, C.H. Weathering of plagioclase across variable flow and solute transport regimes. *J. Hydrol.* **2012**, *420–421*, 46–58. [CrossRef]
37. Pacheco, F.A.L.; Van Der Weijden, C.H. Integrating Topography, Hydrology and Rock Structure in Weathering Rate Models of Spring Watersheds. *J. Hydrol.* **2012**, *428–429*, 32–50. [CrossRef]
38. Pacheco, F.A.L.; Van Der Weijden, C.H. Role of hydraulic diffusivity in the decrease of weathering rates over time. *J. Hydrol.* **2014**, *512*, 87–106. [CrossRef]
39. Pacheco, F.A.L.; Van Der Weijden, C.H. Modeling rock weathering in small watersheds. *J. Hydrol.* **2014**, *513C*, 13–27. [CrossRef]
40. Pacheco, F.A.L.; Alencoão, A.M.P. Role of fractures in weathering of solid rocks: Narrowing the gap between experimental and natural weathering rates. *J. Hydrol.* **2006**, *316*, 248–265. [CrossRef]
41. Pacheco, F.A.L.; Sousa Oliveira, A.; Van Der Weijden, A.J.; Van Der Weijden, C.H. Weathering, biomass production and groundwater chemistry in an area of dominant anthropogenic influence, the Chaves-Vila Pouca de Aguiar region, north of Portugal. *Water Air Soil Pollut.* **1999**, *115*, 481–512. [CrossRef]

42. Pacheco, F.A.L.; Van Der Weijden, C.H. Mineral weathering rates calculated from spring water data: A case study in an area with intensive agriculture, the Morais massif, NE Portugal. *Appl. Geochem.* **2002**, *17*, 583–603. [CrossRef]
43. Pacheco, F.A.L.; Szocs, T. "Dedolomitization Reactions" driven by anthropogenic activity on loessy Sediments, SW Hungary. *Appl. Geochem.* **2006**, *21*, 614–631. [CrossRef]
44. Pacheco, F.A.L.; Landim, P.M.B.; Szocs, T. Anthropogenic impacts on mineral weathering: A statistical perspective. *Appl. Geochem.* **2013**, *36*, 34–48. [CrossRef]
45. Ferreira, A.R.L.; Sanches Fernandes, L.F.; Cortes, R.M.V.; Pacheco, F.A.L. Assessing anthropogenic impacts on riverine ecosystems using nested partial least squares regression. *Sci. Total Environ.* **2017**, *583*, 466–477. [CrossRef] [PubMed]
46. Sanches Fernandes, L.F.; Fernandes, A.C.P.; Ferreira, A.R.L.; Cortes, R.M.V.; Pacheco, F.A.L. A partial least squares—Path modeling analysis for the understanding of biodiversity loss in rural and urban watersheds in Portugal. *Sci. Total Environ.* **2018**, *626*, 1069–1085. [CrossRef]
47. GAEMA—Grupo de Atuação Especial de Defesa do Meio Ambiente. *Função ecológica é a operação pela qual os elementos bióticos e abióticos que compõem determinado meio contribuem, em sua interação, para a manutenção do equilíbrio ecológico e para a sustentabilidade dos processos evolutivos*; Technical Report; Ministério Público do Estado de São Paulo: São Paulo, Brasil, 2012.
48. Pacheco, F.A.L.; Sanches Fernandes, L.F. Environmental land use conflicts in catchments: A major cause of amplified nitrate in river water. *Sci. Total Environ.* **2016**, *548–549*, 173–188. [CrossRef] [PubMed]
49. Valle Junior, R.F.; Varandas, S.G.P.; Sanches Fernandes, L.F.; Pacheco, F.A.L. Groundwater quality in rural watersheds with environmental land use conflicts. *Sci. Total Environ.* **2014**, *493*, 812–827. [CrossRef] [PubMed]
50. Pacheco, F.A.L.; Santos, R.M.B.; Sanches Fernandes, L.F.; Pereira, M.G.; Cortes, R.M.V. Controls and forecasts of nitrate fluxes in forested watersheds: A view over mainland Portugal. *Sci. Total Environ.* **2015**, *537*, 421–440. [CrossRef]
51. Santos, R.M.B.; Sanches Fernandes, L.F.; Pereira, M.G.; Cortes, R.M.V.; Pacheco, F.A.L. A framework model for investigating the export of phosphorus to surface waters in forested watersheds: Implications to management. *Sci. Total Environ.* **2015**, *536*, 295–305. [CrossRef]
52. Sanches Fernandes, L.F.; Marques, M.J.; Oliveira, P.C.; Moura, J.P. Decision support systems inwater resources in the demarcated region of Douro—Case study in Pinhão River Basin, Portugal. *Water Environ. J.* **2014**, *28*, 350–357.
53. Sanches Fernandes, L.F.; Santos, C.; Pereira, A.; Moura, J. Model of management and decision support systems in the distribution of water for consumption: Case study in North Portugal. *Eur. J. Environ. Civ. Eng.* **2011**, *15*, 411–426. [CrossRef]
54. Santos, R.M.B.; Sanches Fernandes, L.F.; Pereira, M.G.; Cortes, R.M.V.; Pacheco, F.A.L. Water resources planning for a river basin with recurrent wildfires. *Sci. Total Environ.* **2015**, *526*, 1–13. [CrossRef]
55. Terêncio, D.P.S.; Sanches Fernandes, L.F.; Cortes, R.M.V.; Pacheco, F.A.L. Improved framework model to allocate optimal rainwater harvesting sites in small watersheds for agro-forestry uses. *J. Hydrol.* **2017**, *550*, 318–330. [CrossRef]
56. Terêncio, D.P.S.; Sanches Fernandes, L.F.; Cortes, R.M.V.; Moura, J.P.; Pacheco, F.A.L. Rainwater harvesting in catchments for agro-forestry uses: A study focused on the balance between sustainability values and storage capacity. *Sci. Total Environ.* **2018**, *613–614*, 1079–1092. [CrossRef]
57. Valle Junior, R.F.; Varandas, S.G.P.; Sanches Fernandes, L.F.; Pacheco, F.A.L. Multi Criteria Analysis for the monitoring of aquifer vulnerability: A scientific tool in environmental policy. *Environ. Sci. Policy* **2015**, *48*, 250–264. [CrossRef]
58. Valera, C.A.; Pissarra, T.C.T.; Martins Filho, M.V.; Valle Junior, R.F.; Sanches Fernandes, L.F.; Pacheco, F.A.L. A legal framework with scientific basis for applying the 'polluter pays principle' to soil conservation in rural watersheds in Brazil. *Land Use Policy* **2017**, *66*, 61–71. [CrossRef]

Article

A Regression Model of Stream Water Quality Based on Interactions between Landscape Composition and Riparian Buffer Width in Small Catchments

Teresa Cristina Tarlé Pissarra [1,6], Carlos Alberto Valera [1,2,6], Renata Cristina Araújo Costa [1,6], Hygor Evangelista Siqueira [1,6], Marcílio Vieira Martins Filho [1,6], Renato Farias do Valle Júnior [3,6], Luís Filipe Sanches Fernandes [4,6] and Fernando António Leal Pacheco [5,6,*]

1. Faculdade de Ciências Agrárias e Veterinárias, Universidade Estadual Paulista (UNESP), Via de Acesso Prof. Paulo Donato Castellane, s/n, Jaboticabal SP 14884-900, Brazil
2. Coordenadoria Regional das Promotorias de Justiça do Meio Ambiente das Bacias dos Rios Paranaíba e Baixo Rio Grande, Rua Coronel Antônio Rios, 951, Uberaba MG 38061-150, Brazil
3. Laboratório de Geoprocessamento, Instituto Federal do Triângulo Mineiro, Campus Uberaba, Uberaba MG 38064-790, Brazil
4. Centro de Investigação e Tecnologias Agroambientais e Biológicas, Universidade de Trás-os-Montes e Alto Douro, Ap. 1013, 5001-801 Vila Real, Portugal
5. Centro de Química de Vila Real, Universidade de Trás-os-Montes e Alto Douro, Ap. 1013, 5001-801 Vila Real, Portugal
6. POLUS—Grupo de Política de Uso do Solo, Universidade Estadual Paulista (UNESP), Via de Acesso Prof. Paulo Donato Castellane, s/n, Jaboticabal SP 14884–900, Brazil

* Correspondence: fpacheco@utad.pt; Tel.: +351-917519833

Received: 22 June 2019; Accepted: 20 August 2019; Published: 23 August 2019

Abstract: Riparian vegetation represents a protective barrier between human activities installed in catchments and capable of generating and exporting large amounts of contaminants, and stream water that is expected to keep quality overtime. This study explored the combined effect of landscape composition and buffer strip width (L) on stream water quality. The landscape composition was assessed by the forest (F) to agriculture (A) ratio (F/A), and the water quality by an index (IWQ) expressed as a function of physico-chemical parameters. The combined effect (F/A × L) was quantified by a multiple regression model with an interaction term. The study was carried out in eight catchments of Uberaba River Basin Environmental Protection Area, located in the state of Minas Gerais, Brazil, and characterized by very different F/A and L values. The results related to improved water quality (larger IWQ values) with increasing values of F/A and L, which were not surprising given the abundant similar reports widespread in the scientific literature. But the effect of F/A × L on IWQ was enlightening. The interaction between F/A and L reduced the range of L values required to sustain IWQ at a fair level by some 40%, which is remarkable. The interaction was related to the spatial distribution of infiltration capacity within the studied catchments. The high F/A catchments should comprise a larger number of infiltration patches, allowing a dominance of subsurface flow widespread within the soil layer, a condition that improves the probability of soil water to cross and interact with a buffer strip before reaching the stream. Conversely, the low F/A catchments are prone to the generation of an overland flow network, because the absence of permanent vegetation substantially reduces the number of infiltration patches. The overland flow network channelizes runoff and conveys the surface water into specific confluence points within the stream, reducing or even hampering an interaction with a buffer strip. Notwithstanding the interaction, the calculated L ranges (45–175 m) are much larger than the maximum width imposed by the Brazilian Forest Code (30 m), a result that deserves reflection.

Keywords: water pollution; riparian buffer width; landscape composition; regression model; interaction term; Brazilian Forest Code

1. Introduction

Riparian buffers represent undeniable benefits to stream water quality in catchments affected by agricultural nonpoint source pollution [1–4]. The benefits occur because riparian buffer strips favor physical processes, such as infiltration, sediment deposition or adsorption, as well as biochemical mechanisms, such as nutrient uptake or denitrification [5–7]. The role of riparian vegetation in the retention of nutrients and sediments has been reviewed in various studies [8–12]. A key parameter of riparian buffers is the width. Several studies refer minimum thresholds for riparian buffer width [13], while in some countries, such as Brazil or the United States, this width has been legally imposed or recommended [14,15]. However, other studies consider fixed-width recommendations problematic because riparian ecological responses are highly variable [8], and hence, optimal buffer widths are expected to vary from site to site [13]. On the other hand, width is just one among other major factors influencing stream water quality. Other key factors are landscape composition and patterns.

Following an early work by Kuehne, dated from the 1960s [16], numerous studies investigated the impacts of landscape composition and patterns on stream and lake water quality [17], even reporting that landscape characteristics are critical to water quality [18,19]. The reports evolved from cases where the relationship between the composition of landscape and the variation in water quality indicators was explored [20], to cases where the focus was moved to the spatial arrangement of the landscape (patterns) [21,22]. In the earlier studies, good water quality was generally associated with undeveloped watersheds dominated by forest land use, while poor water quality was linked to human development activities, such as agriculture [23]. In the more recent studies, a variety of landscape metrics was used to explain the correlation between landscape patterns and stream water quality, including patch density, largest patch index, landscape shape index or contagion [24–27].

The assessment of stream water quality based on the correlation with buffer strip widths or on the relationship with landscape composition usually leads to distinct results, and sometimes the standpoints are antagonistic. Various studies have shown that landscape composition within the river basin are decisive to identify the impacts of human activities on water quality [28,29], while other studies stated that patterns at the riparian buffer zones are more powerful to explain those impacts [30,31]. Eventually, the observed discrepancies were related to the dataset structure. Riparian variables are expected to become better explanatory variables when the land use is fairly homogenous and/or one land use category is widely dominant, as occurred in the [30,31] studies. In other cases, the landscape composition is almost always the first explanatory variable. However, discrepancies can also be interpreted as a scale problem [32]. The combined roles of the whole basin and buffer strip scales have been discussed in recent studies [33–36]. It has been reported that water quality varies along the direction of flow, due to human activities and the changing size of the buffer zone, and the self-purification ability of the water is influenced by the landscape composition in the river basin [37].

Despite the abundance of scientific literature on the relationship between stream water quality, buffer strip widths and landscape composition, a specific issue has not been tackled so far: The potential interaction among buffer strip widths and landscape composition. The studies mentioned in the previous paragraph link stream water quality variations to changes in buffer strip width and(or) landscape composition, acting independently as main effects, but fail to address potential joint effects. However, interactions among main effects are widely discussed in the scientific literature about regression models [38–40], and can play a role in the context of water quality assessments and their correlation with multi-scale factors. For example, stream water quality could be improved with narrower buffer strips if an enhanced self-purification of runoff was accomplished within a rural catchment by a large proportion of forest areas relative to agricultural areas. The negative consequences of overlooking interaction effects have been discussed in some forums (https://statisticsbyjim.com/regression/interaction-effects/). When interaction effects are statistically significant, the main effects cannot be interpreted without considering the interactions, under the sentence of severe misinterpretation of results and prognosis.

The general purpose of this study was to explore the relationship between water quality variation, landscape composition, buffer strip widths, and potential interactions between composition and widths. This general goal comprised the following specific objectives: (1) Investigate potential interactions between landscape composition and buffer strip widths using a linear regression model with and without an interaction term; (2) determine the range of buffer strip widths that ensure a regular water quality, as a function of a landscape composition index (ratio native forest/agriculture), with and without considering the interaction effects; (3) interpret the results from a management standpoint. The research was carried out in the Uberaba River Basin (state of Minas Gerais, Brazil), namely within the Municipal Environmental Protection Area (EPA-MURB) located at the headwaters. The study area comprises eight sub-basins of EPA-MURB. Watercourses in these catchments may be affected by a diversity of non-point (diffuse) pollutants, including nitrogen and phosphorus from fertilizers or fine sediments from soil erosion, mostly derived from sugar cane plantations. In this study, water quality was assessed by an index that involves the measurement of dissolved oxygen, turbidity, total dissolved solids, which means parameters that can be interpreted as proxies to those pollutants. The index is called IWQ—Index for Water Quality and was proposed by the Environmental Company of São Paulo State—CETESB (https://cetesb.sp.gov.br) to be used in water quality assessments. The regression models were first applied to the IWQ and then to its formation variables, with the purpose to identify the most influencing ones. The studied watercourses are characterized by approximately 15, 30 and 50 m wide riparian forests. The reason for selecting these buffer widths relates to the rules imposed in the Brazilian Forest Code (Law No. 12651/12). There are two rules: The transition rule takes into account the size of land property calculated as fiscal modules and creates a distance from the stream margin that goes from a minimum of 5 m to a maximum of 20 m, considering the regular river bank; the permanent rule defines 30 m as unique distance. The rationale was, therefore, to span buffer widths that enclose these limits, considering the real buffer widths observed in the field.

2. Materials and Methods

2.1. Study Area

The study area encompasses the Municipal Environmental Protection Area of Uberaba River Basin (EPA-MURB), which covers 528.1 km^2 of Triângulo Mineiro Region, State of Minas Gerais, Brazil (Figure 1). The EPA-MURB is located in the Brazilian Central Plateau, and the Northeast portion of Paraná River Basin, between the altitudes 710–1040 m and within the planimetric coordinates 188–220 km East and 7815–7840 km North expressed in the Universal Transverse Mercator system, zone 23K. The topography is characterized by undulated landscapes, sometimes incised by steep valleys.

Geology is dominated by two groups and associated formations (Figure 2a). The São Bento Group and associated Serra Geral Formation is composed of Lower Cretaceous grey to black basaltic lava flows (15–70 m thick), cropping out at lower altitudes. The Bauru Group, which overlays the São Bento Group along with an erosive contact, comprises the Uberaba Formation overlaid by the Marília Formation, being both composed of Cretaceous sandstones and conglomerates. The contact between the two sequences is abrupt, being marked by a silexite level and a conglomerate rich in quartz grains cemented with calcite [41]. The latosols dominate the landscape while the argisols occupy just small areas (Figure 2b) (https://www.embrapa.br/solos/sibcs). The latosols are characterized by clayey texture, whereas the argisols are characterized by sandy texture. Soil erosion may be intense because the land is prepared for seeding in the Autumn season, a period characterized by erosive rainfall events [42].

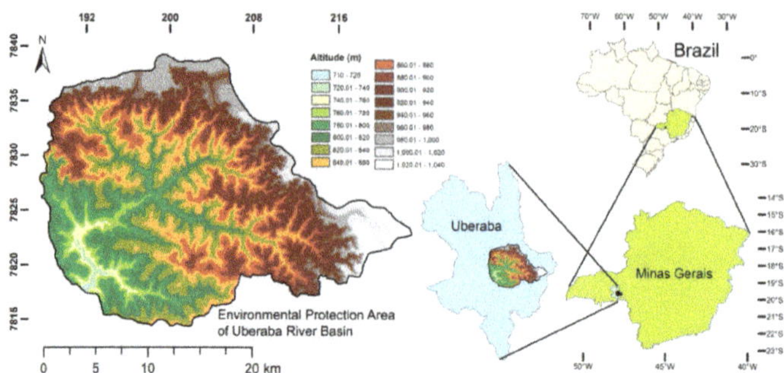

Figure 1. Location of the Environmental Protection Area of Uberaba River Basin in the Uberaba Municipality, State of Minas Gerais, and Brazil. Adapted from Valera et al. [15].

Climate is tropical (Aw in the Köppen's classification scheme) and the climatic domain is semi-humid with 4 to 5 dry months and relative humidity of 70–75%. The temperature ranges between 20 and 24 °C, on the annual average. The period from October to February observes the warmest temperatures that vary between 21 and 25 °C. The coldest month arrives in July, when temperatures drop to 16 to 18 °C. Based on a sixty-two years record, mean annual precipitation of 1584.2 mm is estimated for Uberaba municipality. The amount of rainfall varies considerably during the year, with average values ranging between 42.8 and 541 mm (www.inmet.gov.br/).

Agriculture and livestock production are important economic activities in the EPA-MURB. These areas form a mosaic where they alternate with spots of native vegetation (Cerrado), as illustrated in Figure 3. The landscape has changed significantly in the past half-century. In the 1960s, the Cerrado was the dominating land cover, reaching 40% of the EPA-MURB. In the following decades, expansion of managed pastures and (to a smaller extent) areas used for short-cycle agriculture (mostly corn and rice) caused a significant reduction in the share of native vegetation in the region. More recently, large areas have been converted to sugar cane plantations related to the production of energy from ethanol [43].

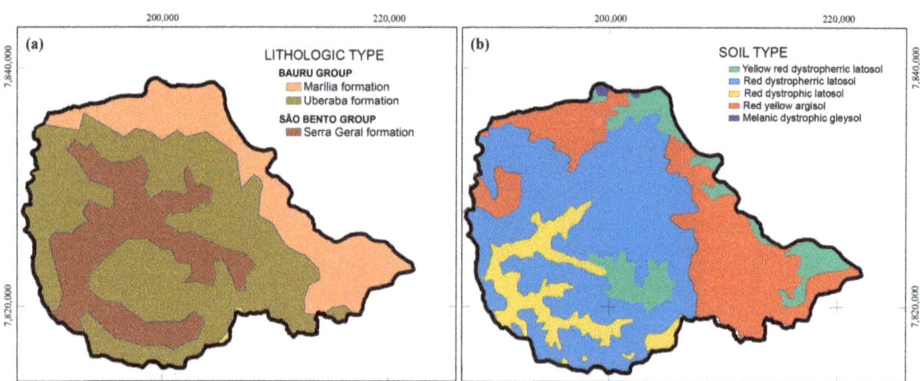

Figure 2. Geology (**a**) and soils (**b**) in the Environmental Protection Area of Uberaba River Basin. The soil map was adapted from Siqueira et al. [44].

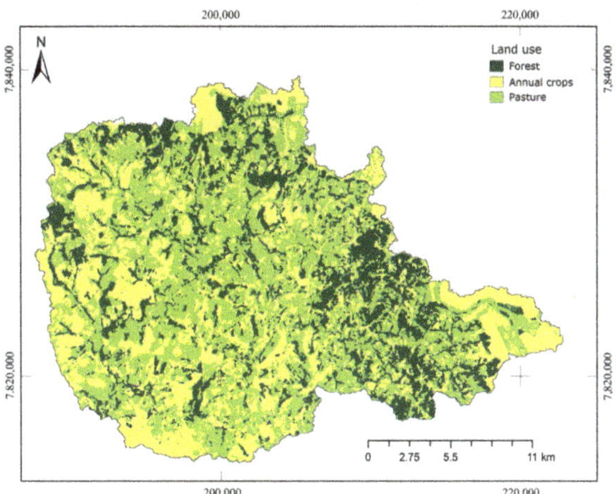

Figure 3. Land cover in the Environmental Protection Area of Uberaba River Basin. Adapted from Siqueira et al. [44].

On 20 January 1999, the EPA-MURB was legally protected through the State Law No. 13183, for an area of 463 km^2, being recognized as Sustainable Land Use Conservation Unit because of its important groundwater resources and remnants of Cerrado Biome. On 28 December 2005, the protection has been considered at Municipal level by the Law No. 9892 for an area of 528.1 km^2. The management plan for the EPA-MURB was published by the Municipal Secretary of the Environment (http://www.uberaba.mg.gov.br/), which divided the area into five zones: (1) Urban consolidation zone; (2) tourism and/or leisure development zone; (3) agricultural area; (4) conservation area of natural resources and (5) recovery zone. Recently (2017), Complementary Law No. 561 has regulated the urban perimeter within the EPA-MURB.

2.2. Studied Sub-Basins

The studied sites comprised eight small to medium sub-basins selected within the EPA-MURB (Figure 4), with areas ranging from 136.3 hectares (Borá 1 sub-basin) and 3764 hectares (Lajeado sub-basin). The distribution of main land uses or covers is summarized in Table 1. In all cases the catchments were mostly used for agriculture, namely sugar cane plantations, as well as natural or managed (used for the grazing of domestic livestock) pastures. The use for agriculture ("A" column in Table 1) represents 56.7% (Borá 2 sub-basin) to 88.8% (Borá 2 sub-basin) of the catchment areas, and therefore, they can be considered basins with significant anthropogenic influence. Besides the use for agriculture, the eight catchments are substantially covered with native (Cerrado) and managed (eucalyptus) forests ("F" column), with proportions to the agriculture use ("F/A" column) ranging from 0.1 to 0.7. The other uses or covers ("Other" column) comprise the rural dwellings, water bodies (lakes and reservoirs), roads, orchards and areas used for rainfed or irrigated corps. The riparian buffers marginal to the watercourses ("L" column) were characterized by quite different widths: On average and approximately, 15 m in the Alegria, Lageado and Mangabeira 1 sub-basins; 30 m in the Borá 1, Mangabeira 2 and Uberaba sub-basins; and 50 m in the Borá 2 and Lanhoso sub-basins. The water samples were collected at the sub-basins' outlets.

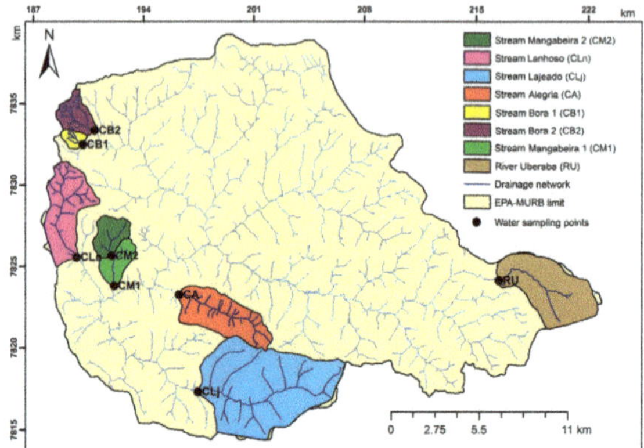

Figure 4. Distribution of studied sub-basins within the Environmental Protection Area of Uberaba River Basin.

Table 1. Land use and cover within the studied sub-basins. Symbols: L—approximate average width of riparian buffer marginal to the water course, observed in satellite images; F—forest cover (native or managed); A—use for agriculture (sugar cane plantations; natural and managed pastures); Other—other uses or covers.

Sub-Basin	Area (ha)	L (m)	F (%)	A (%)	Other (%)	F/A	L × F/A
Alegria	1263.4	15	21.3	75.4	3.3	0.3	4.2
Lajeado	3764.8	15	30.5	67.9	1.6	0.4	6.7
Mangabeira 1	373.1	15	36.1	63.7	0.3	0.6	8.5
Borá 1	136.3	30	9.3	88.8	2.0	0.1	3.1
Mangabeira 2	426.6	30	31.1	68.0	0.9	0.5	13.7
Uberaba	1737.0	30	3.4	81.5	15.0	0.0	1.3
Borá 2	409.4	50	40.5	56.7	2.8	0.7	35.7
Lanhoso	1243.6	50	37.5	62.1	0.3	0.6	30.2

2.3. Water Sampling and Physico-Chemical Analyses

The stream water samples were collected in the streams, 60 cm far from the stream margin, in sectors that were adjacent to the riparian buffer. The sampling campaigns were conducted on a monthly basis, from January 2016 to January 2017 (13 months). The sampling took place from calendar days 15 and 20, every month. The year of 2016 was dry because the annual rainfall (1214.4 mm) was smaller than the local long-term average precipitation (1584.2 mm). Table 2 reviews the prevailing weather conditions in the sampling days, as well as during the three antecedent days. The sampling days were characterized by low rainfall (<5 mm), with the exception of February 2016 and January 2017 campaigns. In these two campaigns, rainfall during the sampling day has reached 5.9 and 10.9 mm, respectively. Average rainfall in the three antecedent days was also small (3.5–5.8 mm), with the few exceptions represented in Table 2 in boldface. The exceptional days were 13 November 2016, and 16 January 2017, with precipitation >25 mm. Taken altogether, the average analytical results should be related to long-term effects of landscape composition and buffer strip width on the quality of stream water, rather than short term effects associated with storm events.

Table 2. Weather conditions (rainfall) in the water sampling day and the three antecedent days (day-1 until day-3). Values larger than 10 mm day-1 are represented in boldface. Source: Valera et al. [15].

Year	Water Sampling Date 2016												2017
Month	January	February	March	April	May	June	July	August	September	October	November	December	January
Day	19	16	15	19	17	21	19	16	20	18	15	20	17
	Rainfall (mm)												
Day	2.0	5.9	4.0	2.4	0.5	0.2	0.0	0.8	0.1	3.7	1.9	2.3	10.9
Day-1	8.0	8.7	5.5	1.2	2.8	0.0	0.0	0.0	0.7	4.7	13.9	2.6	27.0
Day-2	3.4	2.9	2.3	0.9	0.1	0.0	0.0	0.0	0.1	4.9	32.4	5.0	10.0
Day-3	16.6	2.8	6.3	3.7	0.9	0.0	0.0	0.0	0.0	0.6	8.4	1.4	4.4

The measurement of water quality parameters in every campaign involved 10 repetitions, as recommended in the CONAMA's Resolution No. 357/2005. The parameters were measured using a Horiba U-50 Series multi-parameter probe, and comprised: T—water temperature (°C), pH, ORP—Oxidation Reduction Potential (mV), Ec—Electrical conductivity (μS cm^{-1}), Turbidity, measured in Nephelometric Turbidity Units (NTU), DO—Dissolved oxygen (mg L^{-1}), PDO—Percentage of Dissolved Oxygen (%), and TDS—total dissolved solids (mg L^{-1}). The analytical results are provided as Supplementary Material.

A subset of parameters was used to calculate the Index for Water Quality (IWQ) proposed by the Environmental Company of São Paulo State—CETESB (https://cetesb.sp.gov.br):

$$IWQ = \prod_{i=1}^{n} {q_i}^{w_i}, \quad (1)$$

where $0 \leq IWQ \leq 100$, q_i is the quality of ith parameter obtained from standardization of the measured values into a 0–100 range, w_i is the weight of ith parameter, which varies in the $0 \leq w_i \leq 1$ interval as a function of its importance to the overall quality, and n is the total number of parameters. This *IWQ* evaluates water quality on the basis of nine parameters ($n = 9$), including water temperature, pH, dissolved oxygen, turbidity, total dissolved solids, biochemical oxygen demand, fecal coliforms, total nitrogen and total phosphorus. Data may be lacking for some parameters, but the index can still be calculated using the available data and adjusting the weights to different values. In the present study, the calculus of *IWQ* was based on five parameters ($n = 5$), namely the first five from the CETESB list, while the weights were set to—0.10 (water temperature); 0.21 (pH); 0.17 (turbidity); 0.2 (dissolved oxygen); 0.17 (total dissolved solids), as proposed in [45]. The transformation of measured parameters into q scores (Equation (1)) is based on standardization curves, which are portrayed in Figure 5 for the selected parameters. The average of each parameter in the studied catchments, as well as the corresponding *IWQ* values, are depicted in Table 3.

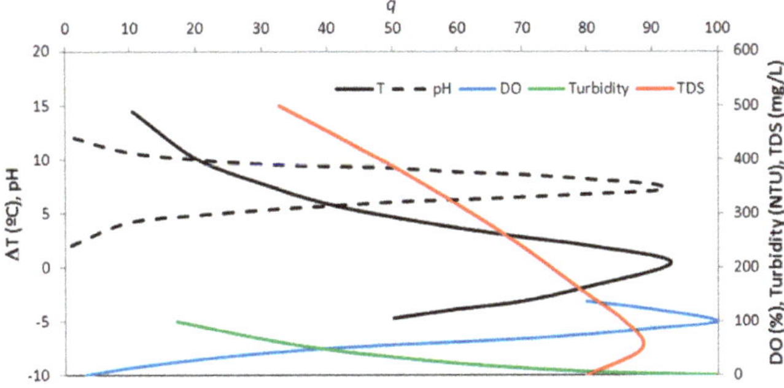

Figure 5. Standardization curves used to transform the water quality parameters into q scores (Equation (1)). Source: https://cetesb.sp.gov.br.

Table 3. Average water quality based on the five parameters used to calculate the IWQ (Index for Water Quality) index (Equation (1)). The full inventory of values, obtained within the monitoring period (January 2016–January 2017), is provided as Supplementary Material.

Sub-Basin	Temperature (°C)	pH	Turbidity (NTU)	DO (%)	TDS (mg/L)	IWQ
Alegria	20.7	7.3	35.9	80.0	0.04	28.3
Lajeado	21.1	7.4	15.8	83.8	0.03	30.0
Mangabeira 1	19.0	7.1	6.2	84.7	0.04	30.8
Borá 1	21.3	7.5	21.9	83.3	0.02	29.4
Mangabeira 2	20.1	7.0	3.8	83.2	0.05	31.0
Uberaba	21.1	6.2	4.3	80.8	0.03	28.7
Borá 2	20.3	7.5	27.3	86.7	0.05	34.0
Lanhoso	19.7	7.4	2.0	124.8	0.06	33.4

According to https://cetesb.sp.gov.br, the quality of stream water is graded as follows: Extremely poor (IWQ ≤ 19), poor (19 < IWQ ≤ 36), regular (36 < IWQ ≤ 51), good (51 < IWQ ≤ 79), excellent (79 < IWQ ≤ 100). It is worth to note that the IWQ index is rather sensitive to small changes in the bearing parameters, given the multiplicative formulation of Equation (1). As corollary of this conception, a good water quality (IWQ > 51) requires that all q values are high, while an excellent quality (IWQ > 79) implies that all q scores are very high.

2.4. Multiple Linear Regression with Interactions

A typical multiple linear regression model involving a dependent variable (Y) and two independent variables (X_1 and X_2) is written as:

$$Y = \beta_1 X_1 + \beta_2 X_2, \tag{2}$$

where β_1 and β_2 are regression coefficients representing the main effects of X_1 and X_2 on the predicted values of Y. Sometimes, besides the main effects, the dependence of Y on X_1 and X_2 is further constrained by interaction effects. An interaction exists when the effect of one independent variable changes, depending on the value of the other independent variable. In those cases, Equation (2) is recast as [46]:

$$Y = \beta_1 X_1 + \beta_2 X_2 + \beta_3 X_1 X_2, \tag{3}$$

where β_3 is the joint effect of X_1 and X_2 on Y and the product X_1X_2 is the interaction between X_1 and X_2 producing that effect. This product is also called a two-way interaction term, because it is the interaction between two independent variables.

The presence of interactions in multiple regression can be identified through statistical tests, namely through assessing the statistical significance of the interaction term, and comparing the coefficient of determination with and without the interaction term. If the interaction term is statistically significant, the interaction term is probably important. And if the coefficient of determination is also much bigger with the interaction term, it is definitely important. If neither of these outcomes is observed, the interaction term can be removed from the regression equation. As alerted in various forums (e.g., https://statisticsbyjim.com/regression/interaction-effects/), when interaction effects are present, it means that interpretation of main effects without considering joint effects may be incomplete or misleading.

In the present study, the independent variables used to model water quality with multiple regression were the riparian buffer strip width and the ratio between forest and agriculture, represented by the variables L and F/A in Table 1. In the first run, the dependent variable was the water quality index represented as IWQ in Table 3. In a second run, the regression analysis was replicated for water temperature, pH, dissolved oxygen, turbidity and total dissolved solids, which are the formation parameters of IWQ, to evaluate their specific roles in the studied area. The dataset comprehended the values of all these variables in the eight catchments. These data were processed for estimation of main effects and joint effects in the *STATISTICA* computer program (http://www.statsoft.com).

2.5. Thematic Maps

The thematic maps (e.g., Figures 1–3) were prepared in ArcMap software of ESRI [47], a common tool in spatial analysis of hydrologic and environmental data widely used in many studies [42,44,48–66]. The base information was compiled from various spatial databases, namely the maps published by the Brazilian Institute for Geography and Statistics (https://ww2.ibge.gov.br) on the 1: 100,000 scale, and the digital terrain model obtained from the ASTER GDEN V2 satellite image with a spatial resolution of 30 m.

3. Results

The scatter plots representing the IWQ index as a function of variables L (buffer strip width) and F/A (ratio forest over agriculture), as well as of interaction term L× F/A, are displayed in Figure 6a, 6b and 6c, respectively. The corresponding coefficients of determination are R^2 = 0.61, R^2 = 0.72 and R^2 = 0.93, meaning that the variance explained by the models raises from the main effects (L, F/A) to the interaction effect (L × F/A). The scatter plot of Figure 6a may be influenced by the small number of buffer strip widths (just three different values). This may limit the use of buffer strip width as a continuous variable in a regression model. The results of multiple regression support the conclusions taken from observation of Figure 6. In Table 4, it is evidenced that all multiple regression terms are significant at $p \leq 0.05$ and that the adjusted coefficient of determination is slightly higher (R^2 = 0.99) when the interaction term is incorporated in the model, relative to the no interaction case (R^2 = 0.97).

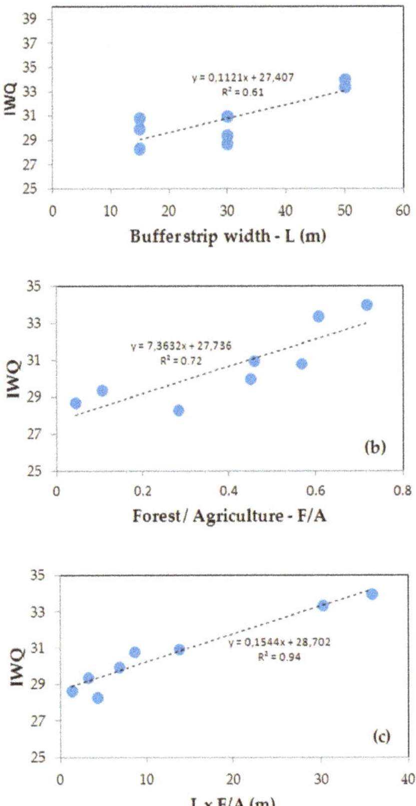

Figure 6. IWQ scatter plots. (**a**) This variable is projected as a function of buffer strip width (L), (**b**,**c**) as a function of ratio forest over agriculture (F/A) and interaction term L × F/A, respectively. The data used to draw the plots are depicted in Table 1 (L, F/A and L × F/A) and 3 (IWQ).

Table 4. Results of multiple regression model. In this run, the IWQ was used as a dependent variable.

Term	β	StdError (β)	B	StdError (B)	*p*-Level
		Without interaction (Adjusted $R^2 = 0.97$)			
Intercept			26.14	0.32	0.0000
L	0.54	0.07	0.08	0.01	0.0005
F/A	0.65	0.07	5.64	0.58	0.0002
		With interaction (Adjusted $R^2 = 0.99$)			
Intercept			23.80	0.94	0.00
L	1.10	0.22	0.16	0.03	0.01
F/A	1.19	0.22	10.31	1.87	0.01
L × F/A	−0.91	0.36	−0.15	0.06	0.05

In keeping with the results of multiple regression (Table 4), the relationship between water quality (IWQ), buffer strip width (L) and landscape composition (F/A) can be expressed by the following Equations:

$$IWQ = 26.14 + 0.08 \times L + 5.64 \times \frac{F}{A}, \quad (4a)$$

$$IWQ = 23.80 + 0.16 \times L + 10.31 \times \frac{F}{A} - 0.15 \times L \times \frac{F}{A}, \quad (4b)$$

where Equation (4a) represents the relationship without considering the interaction between L and F/A and Equation (4b) the case where this interaction is accounted for. The graphical representation of Equation (4a,b) are illustrated in Figure 7a,b. Figure 7a portrays a couple of parallel lines describing the evolution of IWQ as a function of L, for the extreme values of F/A (0 and 1). The lines are parallel because the model predicts no interaction [67]. The buffer strip widths required to ensure a regular water quality in the studied sub-basins ($36 < IWQ \leq 51$) range from $L = 130$ m to $L = 310$ m, when $F/A = 0$, and from L = 60 m to $L = 250$ m when $F/A = 1$. These results change considerably when the interaction term is incorporated in the regression analysis, as demonstrated in Figure 7b. Now, the non-parallel lines indicate much smaller L ranges to attain the regular water quality, namely $L = 75$–175 m for $F/A = 0$, and $L = 45$–155 m for $F/A = 1$. The consequences for planning of adopting one or the other model are evident, either for the planning of landscape composition or buffer strip widths. The interaction model may be more realistic because of its larger R^2 and interaction term significance.

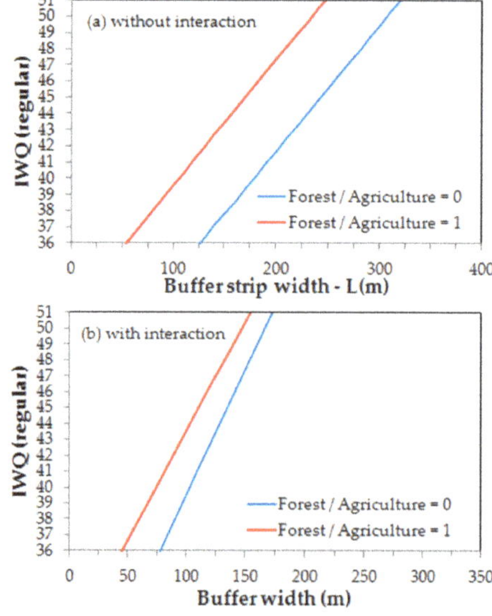

Figure 7. Interaction multiple regression plots: (**a**) No interaction model; (**b**) interaction model.

The results of multiple regression applied to the IWQ parameters are depicted in Table 5. Only the regressions with interaction term were considered in this second run. The results suggest a dominance of turbidity in the control of IWQ in the studied sub-basins. For this parameter, the coefficient of determination ($R^2 = 0.6$) is satisfactory, and all regression coefficients are significant at p-level ≤ 0.05. The results for total dissolved solids are characterized by a moderate $R^2 = 0.5$, but the regression coefficients are not significant. The results for the other parameters are characterized by a low $R^2 = 0.1$ and non-significant regression coefficients.

Table 5. Results of multiple regression model. In this run, the formation parameters of IWQ were used as dependent variables, namely water temperature, pH, turbidity, dissolved oxygen and total dissolved solids. The significant values (p-level ≤ 0.05) are represented in boldface.

Term	b	StdError (b)	B	StdError (B)	p-Level
		Water temperature (Adjusted $R^2 = 0.10$)			
Intercept		23.89	2.98	8.01	0.00
L	−1.69	−0.09	0.10	−0.92	0.41
F/A	−2.10	−6.94	5.93	−1.17	0.31
L × F/A	2.58	0.16	0.18	0.87	0.43
		pH (Adjusted $R^2 = 0.10$)			
Intercept		8.18	1.84	4.46	0.01
L	−1.56	−0.05	0.06	−0.75	0.49
F/A	−1.03	−1.85	3.65	−0.51	0.64
L × F/A	2.58	0.09	0.11	0.77	0.48
		Turbidity (Adjusted $R^2 = 0.6$)			
Intercept		130.65	31.49	4.15	0.01
L	−4.50	−3.93	1.07	−3.68	0.02
F/A	−4.29	−225.90	62.62	−3.61	0.02
L × F/A	7.19	6.99	1.91	3.66	0.02
		Dissolved Oxygen (Adjusted $R^2 = 0.1$)			
Intercept		42.48	56.08	0.76	0.49
L	1.34	1.37	1.90	0.72	0.51
F/A	0.99	60.97	111.50	0.55	0.61
L × F/A	−1.28	−1.46	3.40	−0.43	0.69
		Total dissolved solids (Adjusted $R^2 = 0.45$)			
Intercept		0.02	0.04	0.55	0.61
L	0.10	0.00	0.00	0.07	0.95
F/A	0.67	0.04	0.08	0.48	0.65
L × F/A	0.13	0.00	0.00	0.05	0.96

4. Discussion

The results of multiple regression, with and without interaction effects (Equation (4a,b)), indicate the improvement of water quality as a function of increasing forest to agriculture ratios (landscape composition) and buffer strip widths. The non-scaled regression coefficients (β; Table 4) point to a 45% contribution of L and 55% of F/A to IWQ values in the studied catchments, when the non-interaction model is used, and an equal contribution (50%) from both variables when the interaction model is adopted. These results are not surprising because many studies so far reported the benefits of forest cover and buffer strip widths to stream water quality, as quoted in the Introduction section [8–12].

The striking result refers to the interaction between F/A and L (Equation (4b)), because it describes a substantial reduction of L values required to sustain a regular water quality (36 < IWQ ≤ 51) in the streams, relative to the non-interaction model. The reductions related to the interaction reach 44% when the lines representing F/A = 0 are compared for L values across Figure 7a,b, and 38% in the case of F/A = 1. This is remarkable and requires a thorough interpretation. It seems that the positive effects on water quality resulting from the independent actions of F/A and L are amplified by a combined action (F/A × L) that ought to identify and justify.

The runoff that reaches a riparian buffer is mostly generated upstream within the catchment. It is, therefore, acceptable to link the combined action to hydrologic processes taken place away from the buffer strips, and hence, related to F/A. These processes must, however, describe a hydrologic connection between forested areas (high F/A) and buffer strips, because a combined action inherently assumes an interplay between the two land cover parameters. A potential strong candidate is the infiltration capacity of soils, and, more importantly, its spatial distribution [68]. This is a crucial control on hydrological processes at the interface between the ground surface and soil, including the distribution of flowing water by overland flow and shallow underground flow. The continuity of overland flow or alternation with subsurface flow depends on the intensity of rainfall and the locations of relatively high infiltration patches isolated on hillslopes [69–71]. In this context, it is expected that high F/A catchments comprise a larger number of infiltration patches, and that subsurface flow dominates in these catchments. It is also expected that subsurface flow is widespread within the soil layer, a condition that improves the probability of soil water to cross a buffer strip before reaching the stream. The higher the F/A ratios, the larger will be the chance of soil water to interact with the buffer strip. This scenario could explain a combined action of F/A and L on IWQ. Contrarily to high F/A catchments, low F/A catchments (e.g., dominated by agriculture) are prone to the generation of an overland flow network, because the absence of permanent vegetation substantially reduces the number of infiltration patches. The overland flow network channelizes runoff and conveys the surface water into specific confluence points within the stream, reducing or even hampering an interaction with a buffer strip [72]. Taken altogether, the F/A and L parameters represent the buffering capacity of vegetated areas distributed away and along the stream banks, respectively, while the F/A × L parameter represents the additional capacity promoted by a hydraulic connection between the two areas.

Besides the infiltration capacity issue, it is worth to explore the potential influence of rainfall intensity in the interaction effect, considered relevant in areas where annual climate fluctuations are stronger, including the tropical regions [73,74]. Turbidity is extremely sensitive to rainfall intensity and played a dominant role in the regression analyses of individual parameters (Table 5). The turbidity records of all studied catchments (see Supplementary Materials) are represented graphically in Figure 8, along with the corresponding daily precipitation record. It is evident the response of turbidity to larger values of daily precipitation, namely in the Borá 1 (January 2016), Lageado (April 2016) and Alegria (January 2017) catchments. A close inspection of Table 1 reveals that these catchments are among those with a lower L × F/A value. There are, however, striking exceptions: The Uberaba River catchment has the lowest L × F/A value (1.3 m), but barely responds to precipitation events; the Borá 2 catchment turbidity peak in September 2016 is not justified by a precipitation event. It is worth recalling, however, that the control of turbidity in streams is not only determined by the studied parameters, namely F/A, L, infiltration capacity and flow convergence related to L × F/A, and rainfall intensity. Topography is also a key parameter [75], which was not addressed in this study because the focus here was put on the interaction between landscape composition and buffer strip widths. We believe topography would help to explain the Uberaba River exception. This is a headwater catchment located in a relatively flat area (compare Figure 4 with Figure 1). These areas are well acquainted with retain sediments because they generate lower overland flow velocity while maximizing infiltration and particle deposition, in opposition to steep slope areas [76].

The regression results based on individual parameters (Table 5) exposed a significant relationship between catchment variables (landscape composition, buffer strip width), including their interactions, and water turbidity, but did not reveal identical influences of those variables or their interactions on other parameters measured in water. It should be remembered, however, that water quality parameters may respond differently to catchment variables depending on the spatial scale or antecedent rainfall conditions, as noted in Uriarte [31]. We, therefore, clarify that our results are valid at the studied spatial scales (Figure 4) and antecedent rainfall conditions (Table 2). Transposition to other settings needs verification.

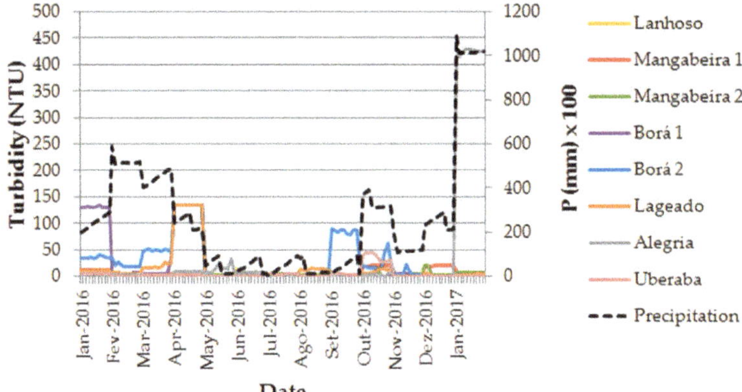

Figure 8. Values of daily precipitation and stream water turbidity observed in the studied catchments during the sampling campaigns. The discrete values are provided as Supplementary Material.

The recognition of an interaction effect between landscape composition and buffer strip widths capable of amplifying stream water quality improvements is certainly beneficial for water resources planning and management. The dominant role of turbidity in the regression analyses of individual terms suggests that water quality deterioration in the studied sub-basins is mostly related to soil erosion and sediment transport rather than with leaching of dissolved fertilizers from the arable land into the stream. Therefore, conservation measures related to the control of soil erosion should be prioritized in this protected area. Eventually, the most cost-effective measure is the reinforcement of manuring and composting of soil to raise the levels of organic matter [77] and produce stable aggregates that are resistant to detachment by rain drop action. The level plantation is another measure of soil erosion control, frequently used to reduce laminar erosion and increase soil water uptake. The level plantation can be successfully coupled with strip cropping [78] that involves the alternation of forages with strips of row crops, such as sugar cane. The control of soil erosion also comprises implementation of techniques that prevent soil compaction, such as no-tillage treatments [79], maintenance of crop residue to keep organic matter and nutrients in the topsoil, or more costly measures, such as terraces in level or gradient, since they reduce the ramp length reducing the surface drag of particles and nutrients [80]. To become effective, implementation of conservation measures should be monitored within spatial decision support systems focused on water resources planning and management [81], and integrate public policies and environmental management plans.

The regression results showed that, even considering the interaction effect regular water quality in the studied catchments is only attained when buffer strip widths are within the ranges $75 \leq L \leq 175$ m (F/A = 0) or $45 \leq L \leq 155$ m (F/A = 1) (Figure 7). These values are much larger than legal values in force in Brazil. So far, the Brazilian law has defined the buffer width limits based on two scenarios: The Old Forest Code (Federal Law No. 4771/1965) and the New Forest Code (Federal Law No. 12651/2012). In the first case, for watercourses up to 10 m wide the permanent preservation areas needed to extend at least 30 m upwards from the stream margin considering the widest seasonal riverbank. In the second case, there are two rules: The transition rule takes into account the size of land property calculated as fiscal modules and creates a distance from the stream margin that goes from a minimum of 5 m to a maximum of 20 m, considering the regular river bank; the permanent rule defines 30 m as unique distance. This study reinforces the suggestion of Valera et al. [15], who alerted that a 30 m buffer strip width, as proposed in the New Forest Code, is barely capable of protecting water quality in the EPA-MURB. The discussion on buffer strips, their geometry and composition, optimal widths, cost-benefit analysis for implementation [82–85], among other topics, is still a matter of debate. The discussion on interaction effects is expected to become another topic on this so challenging analysis.

5. Conclusions

The results of a multiple regression model involving an interaction term revealed the combined positive influence of landscape composition and buffer strip widths (L) on stream water quality, in eight catchments of Uberaba River Basin Environmental Protection Area (Minas Gerais State, Brazil). The landscape composition was characterized by the forest to agriculture ratio (F/A), and high F/A catchments were viewed as basins where forested areas located away from the stream are in hydraulic connection with riparian vegetation distributed along the stream banks. This hydraulic connection presumably amplifies the buffering capacity of forested areas and riparian buffers acting independently. To our best knowledge, this is a new finding in the study of buffer strips and their relationship with stream water quality. In practice, the combined effect reduced the width of buffer strips required to keep water quality at a fair level. Without the interaction, the calculated L range was 60–310 m. It decreased to 45–175 m when the interaction term was accounted for in the regression model. The reduction was expressive, but not enough to drop below the maximum legal value imposed by the Brazilian Forest Code (30 m). Eventually, this legal framework could be adapted to our scientific findings. The problem of setting thresholds to buffer strip widths is not exclusive from Brazil, and therefore, our results could serve as alert to other national realities.

Supplementary Materials: The following are available online at http://www.mdpi.com/2073-4441/11/9/1757/s1.

Author Contributions: Conceptualization, C.A.V. and T.C.T.P.; methodology, C.A.V. and T.C.T.P.; software, R.F.V.J. and H.E.S.; validation, C.V., T.C.T.P., L.F.S.F. and F.A.L.P.; formal analysis, C.A.V., F.A.L.P., T.C.T.P. and M.V.M.F.; investigation, T.C.T.P. and C.A.V.; resources, T.C.T.P.; data curation, C.A.V. and T.C.T.P.; writing—original draft preparation, T.C.T.P.; writing—review and editing, F.A.L.P.; visualization, R.C.A.C.; supervision, T.C.T.P. and M.V.M.F.; project administration, T.C.T.P. and M.V.M.F.; funding acquisition, T.C.T.P., M.V.M.F. and R.F.V.J.

Funding: The present study was carried out within the framework of the Post Graduation Research Programme of Coordenação de Aperfeiçoamento de Pessoal de Nível Superior (CAPES); Conselho Nacional de Desenvolvimento Científico e Tecnológico (CNPq); Agência do Ministério da Ciência, Tecnologia, Inovações e Comunicações (MCTIC); and Land Use Policy Brazilian Group (POLUS). The author affiliated to IFTM Renato Farias do Valle Júnior wishes to acknowledge the funding through the CNPq research scholarship Proc. 307921/2018-2. For the author integrated into the CITAB research center, it was further financed by the FEDER/COMPETE/POCI—Operational Competitiveness and Internationalization Program, under Project POCI-01-0145-FEDER-006958, and by National Funds of FCT—Portuguese Foundation for Science and Technology, under the project UID/AGR/04033/2019. For the author integrated into the CQVR, the research was additionally supported by National Funds of FCT–Portuguese Foundation for Science and Technology, under the project UID/QUI/00616/2019.

Acknowledgments: Mauro Ferreira Machado is acknowledged for fruitful discussions and sharing of information.

Conflicts of Interest: The authors declare no conflict of interest. The funders had no role in the design of the study; in the collection, analyses, or interpretation of data; in the writing of the manuscript, or in the decision to publish the results.

References

1. Cooper, J.R.; Gilliam, J.W. Phosphorus Redistribution from Cultivated Fields into Riparian Areas1. *Soil Sci. Soc. Am. J.* **1987**, *51*, 1600–1604. [CrossRef]
2. Lowrance, R.; Todd, R.; Fail, J.; Hendrickson, O.; Leonard, R.; Asmussen, L. Riparian Forests as Nutrient Filters in Agricultural Watersheds. *BioScience* **1984**, *34*, 374–377. [CrossRef]
3. Peterjohn, W.T.; Correll, D.L. Nutrient Dynamics in an Agricultural Watershed: Observations on the Role of a Riparian Forest. *Ecology* **1984**, *65*, 1466–1475. [CrossRef]
4. Lowrance, R.; Sheridan, J.M. Surface Runoff Water Quality in a Managed Three Zone Riparian Buffer. *J. Environ. Qual.* **2005**, *34*, 1851–1859. [CrossRef] [PubMed]
5. Schoonover, J.E.; Williard, K.W.J.; Zaczek, J.J.; Mangun, J.C.; Carver, A.D. Agricultural Sediment Reduction by Giant Cane and Forest Riparian Buffers. *Water Air Soil Pollut.* **2006**, *169*, 303–315. [CrossRef]
6. Daniels, R.B.; Gilliam, J.W. Sediment and Chemical Load Reduction by Grass and Riparian Filters. *Soil Sci. Soc. Am. J.* **1996**, *60*, 246–251. [CrossRef]

7. Lowrance, R. Groundwater Nitrate and Denitrification in a Coastal Plain Riparian Forest. *J. Environ. Qual.* **1992**, *21*, 401–405. [CrossRef]
8. Hansen, B.D.; Reich, P.; Cavagnaro, T.R.; Lake, P.S. Challenges in applying scientific evidence to width recommendations for riparian management in agricultural Australia. *Ecol. Manag. Restor.* **2015**, *16*, 50–57. [CrossRef]
9. Hill, A.R. Nitrate Removal in Stream Riparian Zones. *J. Environ. Qual.* **1996**, *25*, 743–755. [CrossRef]
10. Hoffmann, C.C.; Kjaergaard, C.; Uusi-Kämppä, J.; Hansen, H.C.B.; Kronvang, B. Phosphorus Retention in Riparian Buffers: Review of Their Efficiency. *J. Environ. Qual.* **2009**, *38*, 1942–1955. [CrossRef]
11. Liu, X.; Zhang, X.; Zhang, M. Major Factors Influencing the Efficacy of Vegetated Buffers on Sediment Trapping: A Review and Analysis. *J. Environ. Qual.* **2008**, *37*, 1667–1674. [CrossRef]
12. Roberts, W.M.; Stutter, M.I.; Haygarth, P.M. Phosphorus Retention and Remobilization in Vegetated Buffer Strips: A Review. *J. Environ. Qual.* **2012**, *41*, 389–399. [CrossRef] [PubMed]
13. Sweeney, B.W.; Newbold, J.D. Streamside Forest Buffer Width Needed to Protect Stream Water Quality, Habitat, and Organisms: A Literature Review. *JAWRA J. Am. Water Resour. Assoc.* **2014**, *50*, 560–584. [CrossRef]
14. Mayer, P.M.; Reynolds, S.K.; Canfiel, T.J. *Riparian Buffer Width, Vegetative Cover, and Nitrogen Removal Effectiveness: A Review of Current Science and Regulations*; US Environmental Protection Agency: Washington, DC, USA, 2005.
15. Valera, C.; Pissarra, T.; Filho, M.; Valle Júnior, R.; Oliveira, C.; Moura, J.; Sanches Fernandes, L.; Pacheco, F. The Buffer Capacity of Riparian Vegetation to Control Water Quality in Anthropogenic Catchments from a Legally Protected Area: A Critical View over the Brazilian New Forest Code. *Water* **2019**, *11*, 549. [CrossRef]
16. Kuehne, R.A. A Classification of Streams, Illustrated by Fish Distribution in an Eastern Kentucky Creek. *Ecology* **1962**, *43*, 608–614. [CrossRef]
17. Uriarte, M.; Yackulic, C.B.; Lim, Y.; Arce-Nazario, J.A. Influence of land use on water quality in a tropical landscape: A multi-scale analysis. *Landsc. Ecol.* **2011**, *26*, 1151–1164. [CrossRef]
18. Beckert, K.A.; Fisher, T.R.; O'Neil, J.M.; Jesien, R.V. Characterization and Comparison of Stream Nutrients, Land Use, and Loading Patterns in Maryland Coastal Bay Watersheds. *Water Air Soil Pollut.* **2011**, *221*, 255–273. [CrossRef]
19. Turner, M.G. Landscape Ecology: The Effect of Pattern on Process. *Annu. Rev. Ecol. Syst.* **1989**, *20*, 171–197. [CrossRef]
20. Townsend, C.R.; Doledec, S.; Norris, R.; Peacock, K.; Arbuckle, C. The influence of scale and geography on relationships between stream community composition and landscape variables: Description and prediction. *Freshw. Biol.* **2003**, *48*, 768–785. [CrossRef]
21. Stryjecki, R.; Zawal, A.; Stępień, E.; Buczyńska, E.; Buczyński, P.; Czachorowski, S.; Szenejko, M.; Śmietana, P. Water mites (Acari, Hydrachnidia) of water bodies of the Krąpiel River valley: Interactions in the spatial arrangement of a river valley. *Limnology* **2016**, *17*, 247–261. [CrossRef]
22. Zhou, T.; Wu, J.; Peng, S. Assessing the effects of landscape pattern on river water quality at multiple scales: A case study of the Dongjiang River watershed, China. *Ecol. Indic.* **2012**, *23*, 166–175. [CrossRef]
23. Tu, J. Spatially varying relationships between land use and water quality across an urbanization gradient explored by geographically weighted regression. *Appl. Geogr.* **2011**, *31*, 376–392. [CrossRef]
24. Rajaei, F.; Sari, A.E.; Salmanmahiny, A.; Delavar, M.; Bavani, A.R.M.; Srinivasan, R. Surface drainage nitrate loading estimate from agriculture fields and its relationship with landscape metrics in Tajan watershed. *Paddy Water Environ.* **2017**, *15*, 541–552. [CrossRef]
25. Teixeira, Z.; Marques, J.C. Relating landscape to stream nitrate-N levels in a coastal eastern-Atlantic watershed (Portugal). *Ecol. Indic.* **2016**, *61*, 693–706. [CrossRef]
26. Awoke, A.; Beyene, A.; Kloos, H.; Goethals, P.L.M.; Triest, L. River Water Pollution Status and Water Policy Scenario in Ethiopia: Raising Awareness for Better Implementation in Developing Countries. *Environ. Manag.* **2016**, *58*, 694–706. [CrossRef] [PubMed]
27. Campagnaro, T.; Frate, L.; Carranza, M.L.; Sitzia, T. Multi-scale analysis of alpine landscapes with different intensities of abandonment reveals similar spatial pattern changes: Implications for habitat conservation. *Ecol. Indic.* **2017**, *74*, 147–159. [CrossRef]

28. Ding, J.; Jiang, Y.; Liu, Q.; Hou, Z.; Liao, J.; Fu, L.; Peng, Q. Influences of the land use pattern on water quality in low-order streams of the Dongjiang River basin, China: A multi-scale analysis. *Sci. Total Environ.* **2016**, *551–552*, 205–216. [CrossRef]
29. Meneses, B.M.; Reis, R.; Vale, M.J.; Saraiva, R. Land use and land cover changes in Zêzere watershed (Portugal)—Water quality implications. *Sci. Total Environ.* **2015**, *527–528*, 439–447. [CrossRef]
30. McMillan, S.K.; Tuttle, A.K.; Jennings, G.D.; Gardner, A. Influence of Restoration Age and Riparian Vegetation on Reach-Scale Nutrient Retention in Restored Urban Streams. *JAWRA J. Am. Water Resour. Assoc.* **2014**, *50*, 626–638. [CrossRef]
31. Goldstein, R.M.; Carlisle, D.M.; Meador, M.R.; Short, T.M. Can Basin Land Use Effects on Physical Characteristics of Streams Be Determined at Broad Geographic Scales? *Environ. Monit. Assess.* **2007**, *130*, 495–510. [CrossRef]
32. Feld, C.K. Response of three lotic assemblages to riparian and catchment-scale land use: Implications for designing catchment monitoring programmes. *Freshw. Biol.* **2013**, *58*, 715–729. [CrossRef]
33. Shen, Z.; Hou, X.; Li, W.; Aini, G. Relating landscape characteristics to non-point source pollution in a typical urbanized watershed in the municipality of Beijing. *Landsc. Urban Plan.* **2014**, *123*, 96–107. [CrossRef]
34. Vikman, A.; Sarkkola, S.; Koivusalo, H.; Sallantaus, T.; Laine, J.; Silvan, N.; Nousiainen, H.; Nieminen, M. Nitrogen retention by peatland buffer areas at six forested catchments in southern and central Finland. *Hydrobiologia* **2010**, *641*, 171–183. [CrossRef]
35. Li, K.; Chi, G.; Wang, L.; Xie, Y.; Wang, X.; Fan, Z. Identifying the critical riparian buffer zone with the strongest linkage between landscape characteristics and surface water quality. *Ecol. Indic.* **2018**, *93*, 741–752. [CrossRef]
36. Shen, Z.; Hou, X.; Li, W.; Aini, G.; Chen, L.; Gong, Y. Impact of landscape pattern at multiple spatial scales on water quality: A case study in a typical urbanised watershed in China. *Ecol. Indic.* **2015**, *48*, 417–427. [CrossRef]
37. Toothaker, L.E.; Aiken, L.S.; West, S.G. Multiple Regression: Testing and Interpreting Interactions. *J. Oper. Res. Soc.* **1994**, *45*, 119.
38. Burks, J.J.; Randolph, D.W.; Seida, J.A. Modeling and interpreting regressions with interactions. *J. Account. Lit.* **2019**, *42*, 61–79. [CrossRef]
39. Kam, C.D.; Franzese, R.J. Modeling and interpreting interactive hypotheses in regression analysis. *Choice Rev. Online* **2008**, *45*, 45–3841.
40. Quintão, D.A.; Caxito, F.D.A.; Karfunkel, J.; Vieira, F.R.; Seer, H.J.; Moraes, L.C.; Ribeiro, L.C.B.; Pedrosa-Soares, A.C. Geochemistry and sedimentary provenance of the Upper Cretaceous Uberaba Formation (Southeastern Triângulo Mineiro, MG, Brazil). *Braz. J. Geol.* **2017**, *47*, 159–182. [CrossRef]
41. Valera, C.A.; Pissarra, T.C.T.; Martins Filho, M.V.; Valle Junior, R.F.; Sanches Fernandes, L.F.; Pacheco, F.A.L. A legal framework with scientific basis for applying the 'polluter pays principle' to soil conservation in rural watersheds in Brazil. *Land Use Policy* **2017**, *66*, 61–71. [CrossRef]
42. Valera, C.A.; Valle Junior, R.F.; Varandas, S.G.P.; Sanches Fernandes, L.F.; Pacheco, F.A.L. The role of environmental land use conflicts in soil fertility: A study on the Uberaba River basin, Brazil. *Sci. Total Environ.* **2016**, *562*, 463–473. [CrossRef] [PubMed]
43. Siqueira, H.E.; Pissarra, T.C.T.; do Valle Junior, R.F.; Fernandes, L.F.S.; Pacheco, F.A.L. A multi criteria analog model for assessing the vulnerability of rural catchments to road spills of hazardous substances. *Environ. Impact Assess. Rev.* **2017**, *64*, 26–36. [CrossRef]
44. De Ribeiro, I.V.A.S.; Bouchonneau, N.; Da Silva, A.C.; Fernandes, R.M.C.; De Pinheiro, L.S. Cálculo do índice de qualidade de água (IQA), com estudo de caso nos rios Cocó e Maranguapinho, Ceará. In Proceedings of the Simpósio Brasileiro de Recursos Hídricos, Associação Brasileira de Recursos Hídricos, Campo Grande, Brazil, 22–26 November 2009; p. 18.
45. *ESRI ArcGIS Professional GIS for the Desktop*. 2010. Versão 10. Available online: http://desktop.arcgis.com (accessed on 23 August 2019).
46. Fonseca, A.R.; Sanches Fernandes, L.F.; Fontainhas-Fernandes, A.; Monteiro, S.M.; Pacheco, F.A.L. The impact of freshwater metal concentrations on the severity of histopathological changes in fish gills: A statistical perspective. *Sci. Total Environ.* **2017**, *599–600*, 217–226. [CrossRef] [PubMed]

47. Fonseca, A.R.; Sanches Fernandes, L.F.; Fontainhas-Fernandes, A.; Monteiro, S.M.; Pacheco, F.A.L. From catchment to fish: Impact of anthropogenic pressures on gill histopathology. *Sci. Total Environ.* **2016**, *550*, 972–986. [CrossRef] [PubMed]
48. Valle Junior, R.F.; Varandas, S.G.P.; Sanches Fernandes, L.F.; Pacheco, F.A.L. Multi Criteria Analysis for the monitoring of aquifer vulnerability: A scientific tool in environmental policy. *Environ. Sci. Policy* **2015**, *48*, 250–264. [CrossRef]
49. Ferreira, A.R.L.; Sanches Fernandes, L.F.; Cortes, R.M.V.; Pacheco, F.A.L. Assessing anthropogenic impacts on riverine ecosystems using nested partial least squares regression. *Sci. Total Environ.* **2017**, *583*, 466–477. [CrossRef]
50. Santos, R.M.B.; Sanches Fernandes, L.F.; Cortes, R.M.V.; Varandas, S.G.P.; Jesus, J.J.B.; Pacheco, F.A.L. Integrative assessment of river damming impacts on aquatic fauna in a Portuguese reservoir. *Sci. Total Environ.* **2017**, *601–602*, 1108–1118. [CrossRef] [PubMed]
51. Sanches Fernandes, L.F.; Fernandes, A.C.P.; Ferreira, A.R.L.; Cortes, R.M.V.; Pacheco, F.A.L. A partial least squares—Path modeling analysis for the understanding of biodiversity loss in rural and urban watersheds in Portugal. *Sci. Total Environ.* **2018**, *626*, 1069–1085. [CrossRef]
52. Álvarez, X.; Valero, E.; Santos, R.M.B.; Varandas, S.G.P.; Sanches Fernandes, L.F.; Pacheco, F.A.L. Anthropogenic nutrients and eutrophication in multiple land use watersheds: Best management practices and policies for the protection of water resources. *Land Use Policy* **2017**, *69*, 1–11. [CrossRef]
53. Pacheco, F.A.L. Regional groundwater flow in hard rocks. *Sci. Total Environ.* **2015**, *506–507*, 182–195. [CrossRef]
54. Fernandes, L.F.S.; Marques, M.J.; Oliveira, P.C.; Moura, J.P. Decision support systems in water resources in the demarcated region of Douro-case study in Pinhão river basin, Portugal. *Water Environ. J.* **2014**, *28*, 350–357. [CrossRef]
55. Fernandes, L.F.S.; dos Santos, C.M.M.; Pereira, A.P.; Moura, J.P. Model of management and decision support systems in the distribution of water for consumption. *Eur. J. Environ. Civ. Eng.* **2011**, *15*, 411–426. [CrossRef]
56. Pacheco, F.A.L.; Van der Weijden, C.H. Weathering of plagioclase across variable flow and solute transport regimes. *J. Hydrol.* **2012**, *420–421*, 46–58. [CrossRef]
57. do Valle Júnior, R.F.; Siqueira, H.E.; Valera, C.A.; Oliveira, C.F.; Sanches Fernandes, L.F.; Moura, J.P.; Pacheco, F.A.L. Diagnosis of degraded pastures using an improved NDVI-based remote sensing approach: An application to the Environmental Protection Area of Uberaba River Basin (Minas Gerais, Brazil). *Remote Sens. Appl. Soc. Environ.* **2019**, *14*, 20–33. [CrossRef]
58. Pacheco, F.A.L.; Szocs, T. "Dedolomitization reactions" driven by anthropogenic activity on loessy sediments, SW Hungary. *Appl. Geochem.* **2006**, *21*, 614–631. [CrossRef]
59. Pacheco, F.A.L. Application of Correspondence Analysis in the Assessment of Groundwater Chemistry. *Math. Geol.* **1998**, *30*, 129–161. [CrossRef]
60. Pacheco, F.A.L.; Landim, P.M.B. Two-Way Regionalized Classification of Multivariate Datasets and its Application to the Assessment of Hydrodynamic Dispersion. *Math. Geol.* **2005**, *37*, 393–417. [CrossRef]
61. Pacheco, F.A.L.; Sousa Oliveira, A.; Van Der Weijden, A.J.; Van Der Weijden, C.H. Weathering, biomass production and groundwater chemistry in an area of dominant anthropogenic influence, the Chaves-Vila Pouca de Aguiar region, north of Portugal. *Water Air Soil Pollut.* **1999**, *115*, 481–512. [CrossRef]
62. Pacheco, F.A.L. Finding the number of natural clusters in groundwater data sets using the concept of equivalence class. *Comput. Geosci.* **1998**, *24*, 7–15. [CrossRef]
63. Hughes, S.J.; Cabecinha, E.; Andrade dos Santos, J.C.; Mendes Andrade, C.M.; Mendes Lopes, D.M.; da Fonseca Trindade, H.M.; dos Santos Cabral, J.A.F.A.; dos Santos, M.G.S.; Lourenço, J.M.M.; Marques Aranha, J.T.; et al. A predictive modelling tool for assessing climate, land use and hydrological change on reservoir physicochemical and biological properties. *Area* **2012**, *44*, 432–442. [CrossRef]
64. Modesto Gonzalez Pereira, M.J.; Sanches Fernandes, L.F.; Barros Macário, E.M.; Gaspar, S.M.; Pinto, J.G. Climate Change Impacts in the Design of Drainage Systems: Case Study of Portugal. *J. Irrig. Drain. Eng.* **2015**, *141*, 05014009. [CrossRef]
65. Miyata, S.; Gomi, T.; Sidle, R.C.; Hiraoka, M.; Onda, Y.; Yamamoto, K.; Nonoda, T. Assessing spatially distributed infiltration capacity to evaluate storm runoff in forested catchments: Implications for hydrological connectivity. *Sci. Total Environ.* **2019**, *669*, 148–159. [CrossRef] [PubMed]

66. Gomi, T.; Sidle, R.C.; Miyata, S.; Kosugi, K.; Onda, Y. Dynamic runoff connectivity of overland flow on steep forested hillslopes: Scale effects and runoff transfer. *Water Resour. Res.* **2008**, *44*. [CrossRef]
67. Gomi, T.; Sidle, R.C.; Ueno, M.; Miyata, S.; Kosugi, K. Characteristics of overland flow generation on steep forested hillslopes of central Japan. *J. Hydrol.* **2008**, *361*, 275–290. [CrossRef]
68. Gomi, T.; Asano, Y.; Uchida, T.; Onda, Y.; Sidle, R.C.; Miyata, S.; Kosugi, K.; Mizugaki, S.; Fukuyama, T.; Fukushima, T. Evaluation of storm runoff pathways in steep nested catchments draining a Japanese cypress forest in central Japan: A geochemical approach. *Hydrol. Process.* **2010**, *24*, 550–566. [CrossRef]
69. Dosskey, M.G.; Helmers, M.J.; Eisenhauer, D.E.; Franti, T.G.; Hoagland, K.D. Assessment of concentrated flow through riparian buffers. *J. Soil Water Conserv.* **2002**, *57*, 336–343.
70. Hénault-Ethier, L.; Larocque, M.; Perron, R.; Wiseman, N.; Labrecque, M. Hydrological heterogeneity in agricultural riparian buffer strips. *J. Hydrol.* **2017**, *546*, 276–288. [CrossRef]
71. Rodrigues, V.; Estrany, J.; Ranzini, M.; de Cicco, V.; Martín-Benito, J.M.T.; Hedo, J.; Lucas-Borja, M.E. Effects of land use and seasonality on stream water quality in a small tropical catchment: The headwater of Córrego Água Limpa, São Paulo (Brazil). *Sci. Total Environ.* **2018**, *622–623*, 1553–1561. [CrossRef]
72. Dai, X.; Zhou, Y.; Ma, W.; Zhou, L. Influence of spatial variation in land-use patterns and topography on water quality of the rivers inflowing to Fuxian Lake, a large deep lake in the plateau of southwestern China. *Ecol. Eng.* **2017**, *99*, 417–428. [CrossRef]
73. Polyakov, V.; Fares, A.; Ryder, M.H. Precision riparian buffers for the control of nonpoint source pollutant loading into surface water: A review. *Environ. Rev.* **2005**, *13*, 129–144. [CrossRef]
74. Damodar Reddy, D.; Subba Rao, A.; Rupa, T.R. Effects of continuous use of cattle manure and fertilizer phosphorus on crop yields and soil organic phosphorus in a Vertisol. *Bioresour. Technol.* **2000**, *75*, 113–118. [CrossRef]
75. Lenka, N.K.; Satapathy, K.K.; Lal, R.; Singh, R.K.; Singh, N.A.K.; Agrawal, P.K.; Choudhury, P.; Rathore, A. Weed strip management for minimizing soil erosion and enhancing productivity in the sloping lands of north-eastern India. *Soil Tillage Res.* **2017**, *170*, 104–113. [CrossRef]
76. Bogunovic, I.; Pereira, P.; Kisic, I.; Sajko, K.; Sraka, M. Tillage management impacts on soil compaction, erosion and crop yield in Stagnosols (Croatia). *Catena* **2018**, *160*, 376–384. [CrossRef]
77. Pacheco, F.A.L.; Varandas, S.G.P.; Sanches Fernandes, L.F.; Valle Junior, R.F. Soil losses in rural watersheds with environmental land use conflicts. *Sci. Total Environ.* **2014**, *485–486*, 110–120. [CrossRef] [PubMed]
78. Shan, N.; Ruan, X.-H.; Xu, J.; Pan, Z.-R. Estimating the optimal width of buffer strip for nonpoint source pollution control in the Three Gorges Reservoir Area, China. *Ecol. Modell.* **2014**, *276*, 51–63. [CrossRef]
79. Bren, L.J. Aspects of the geometry of riparian buffer strips and its significance to forestry operations. *For. Ecol. Manag.* **1995**, *75*, 1–10. [CrossRef]
80. Chang, C.-L.; Hsu, Y.-S.; Lee, B.-J.; Wang, C.-Y.; Weng, L.-J. A cost-benefit analysis for the implementation of riparian buffer strips in the Shihmen reservoir watershed. *Int. J. Sediment Res.* **2011**, *26*, 395–401. [CrossRef]
81. Nigel, R.; Chokmani, K.; Novoa, J.; Rousseau, A.N.; El Alem, A. An extended riparian buffer strip concept for soil conservation and stream protection in an agricultural riverine area of the La Chevrotière River watershed, Québec, Canada, using remote sensing and GIS techniques. *Can. Water Resour. J.* **2014**, *39*, 285–301. [CrossRef]
82. Richit, L.A.; Bonatto, C.; Carlotto, T.; da Silva, R.V.; Grzybowski, J.M.V. Modelling forest regeneration for performance-oriented riparian buffer strips. *Ecol. Eng.* **2017**, *106*, 308–322. [CrossRef]
83. Nigel, R.; Chokmani, K.; Novoa, J.; Rousseau, A.N.; Dufour, P. Recommendations for riparian buffer widths based on field surveys of erosion processes on steep cultivated slopes. *Can. Water Resour. J.* **2013**, *38*, 263–279. [CrossRef]
84. Lin, Y.-F.; Lin, C.-Y.; Chou, W.-C.; Lin, W.-T.; Tsai, J.-S.; Wu, C.-F. Modeling of riparian vegetated buffer strip width and placement. *Ecol. Eng.* **2004**, *23*, 327–339. [CrossRef]
85. Hanowski, J.M.; Walter, P.T.; Niemi, G.J. Effects of prescriptive riparian buffers on landscape characteristics in Northern Minnesota, USA. *J. Am. Water Resour. Assoc.* **2002**, *38*, 633–639. [CrossRef]

Article

A Methodology for the Fast Comparison of Streamwater Diurnal Cycles at Two Monitoring Points

Andrei-Emil Briciu [1,*], Adrian Graur [2], Dinu Iulian Oprea [1] and Constantin Filote [2]

1 Department of Geography, Ștefan cel Mare University of Suceava, Universității 13, 720229 Suceava, Romania; dinuo@atlas.usv.ro

2 Computers, Electronics and Automation Department, Ștefan cel Mare University of Suceava, Universității 13, 720229 Suceava, Romania; Adrian.Graur@usv.ro (A.G.); filote@usm.ro (C.F.)

* Correspondence: andreibriciu@atlas.usv.ro

Received: 23 October 2019; Accepted: 27 November 2019; Published: 29 November 2019

Abstract: There are numerous streamwater parameters that exhibit a diurnal cycle. However, the shape of this cycle has a huge variation from one parameter to another and from one monitoring point to another on the same river. Important variations also occur at the same point during some events, such as high waters. Water level, specific conductivity, dissolved oxygen, oxidation reduction potential, and pH of the Suceava River were monitored for 365 days (2018–2019, hourly sampling frequency) in order to assess the upstream-downstream changes in the diurnal cycle of these parameters, some of these changes being caused by the impact of Suceava city, which is located between the selected monitoring points. The multiresolution analysis of the maximal overlap discrete wavelet transform and the wavelet coherence analysis were combined in a flexible methodology that helped in comparing the upstream and downstream shapes of the diurnal cycle. The methodology allowed for a fast comparison of diurnal profiles during periods of high waters or baseflow. Notable changes were observed in the moments of diurnal maxima and minima.

Keywords: script files; mean diurnal profile; wavelet coherence

---

1. Introduction

The wavelet analysis of streamwater parameters has become more and more popular in the last two decades due to the advantages of this method for the study of non-linear processes [1,2]. Wavelet analysis techniques were applied mostly on stage or discharge time series due to the wide availability of this type of data [3]. It is in the last decade when water quality parameters were involved intensively in wavelet analyses [4,5]. Some physical and chemical properties of streamwaters are proper indicators of water quality and can be used to trace the environmental impact of man. The specific conductivity, dissolved oxygen, oxidation reduction potential (ORP), and pH of a river water can be used to indicate the impact of urban areas on the environment and are sometimes included in wavelet analyses [5–7]. Cities modify the properties of natural waters through various active or passive processes, such as the discharge of stormwater runoff or the creation of an urban heat island, which has multiple effects on urban waters [8]. Urban wastewater treatment plants can alter the diurnal profile of a streamwater chemistry parameter [9,10].

The diurnal cycle in streamwaters is caused by the Earth's rotation around its axis, which leads to the day/night cycle. This cycle modifies evaporation and evapotranspiration in catchments [11,12]. These changes can be measured as variations in the temporal evolution of numerous streamwater parameters, which often have an interdependent behavior with the diurnal oscillation [13]. Natural or anthropogenic events (such as rainfall or pollution events) add transient fluctuations in the diurnal cycle. Finding a relevant shape of the diurnal cycle is needed in order to distinguish the periodic and aperiodic modifications in case study time series. A high resolution diurnal profile is obtained from high frequency measurements. This is why, up-to-date, such profiles are missing in Romania for most water quality parameters. Existing studies on diurnal streamwater profiles have focused on water level or water temperature [10,14].

Streamwater monitoring in Romania is done mainly by the Romanian Waters National Administration, whose data is sometimes used in studies about water chemistry of the Suceava River [15], but this data lacks high frequency sampling and the needed spatial density along a river. The environmental impact of some water contaminants is often assessed for various areas in Romania [16–20] and cities have a traceable impact on streamwater quality [21].

A monitoring system that measures Suceava River water quality upstream and downstream of Suceava city was implemented in 2018 and this study aimed to provide a methodology for a fast analysis of the diurnal cycles of various streamwater parameters at two monitoring points. The wavelet analysis was used to reveal the spatial and temporal variations in the streamwater diurnal cycles. The speedier analysis is needed for the growing sizes of databases. It is often useful to apply an analysis method that acts as an easily adjustable preview of data in order to identify interesting phenomena for further analyses. Also, to our knowledge, this was the first study that computes the average diurnal profiles of specific conductivity, dissolved oxygen, ORP, and pH for the Suceava River.

The methodology proposed in this study was applied to streamwater data from monitoring points located above and below a city in order to discover some details of how the diurnal profile of a river is affected by a city. In this paper, we indicate some links between the observed variations in diurnal profiles and urban elements, such as the urban wastewaters or the urban heat island, that generate the environmental impact of a city.

2. Materials and Methods

2.1. Study Area

The monitoring system. which provided data in this study, was implemented by the University of Suceava and consisted of two monitoring points where water properties are measured every hour (the monitoring points are 11.6 km in a straight line away from each other, with Suceava city in the middle of this distance). Suceava city has a total number of inhabitants fluctuating around 100,000 people.

The study area was located in Suceava Plateau, part of the Moldavian Plateau. Climate is temperate continental with warm and wet summers (the average annual air temperature is 7.86 °C and the annual sum of precipitation is 578 mm [10]). The streambed elevation between the selected points ranged approximately from 280 to 260 m above sea level (a.s.l.) and has a sinuous path, which helps in mixing streamwater so that it records identical values at any point of a transversal section. The upstream monitoring point was at Mihoveni Dam (47.681° N, 26.2° E). This is a quasi-inactive dam aimed at generating run-of-the-river hydroelectricity; it operates only during some high water events for regulating the water level. The downstream monitoring point was placed at Tișăuți (47.618° N, 26.323° E), downstream of an urban wastewater treatment plant and some floodplain landfills (Figure 1). The Suceava River has an average flow rate of 16.87 m^3/s inside Suceava City [10]. The Suceava River has two periods of high waters. The first one occurs during the middle of springtime (April, monthly average flow rate of 29.7 m^3/s) due to snowmelt in the catchment, especially in the mountain area, while the second one happens at the beginning of summer (June, 29.9 m^3/s) as result of heavy rainfalls [10]. During high water events, the Suceava River frequently exceeds 50 m^3/s. The lowest monthly average flow rate is recorded during the winter (January, 6.2 m^3/s) because of the negative air temperatures and little precipitation [10]. The urban tributaries of the Suceava River in the study area had a total discharge of approximately 0.5 m^3/s. The water chemistry of rivers in north-eastern Romania, which includes our study area, was briefly discussed in some studies [22,23], some of them highlighting the intense self-purification processes of Suceava River water inside and downstream of the city [24]. However, these studies were not based on long time series with high frequency of measurements.

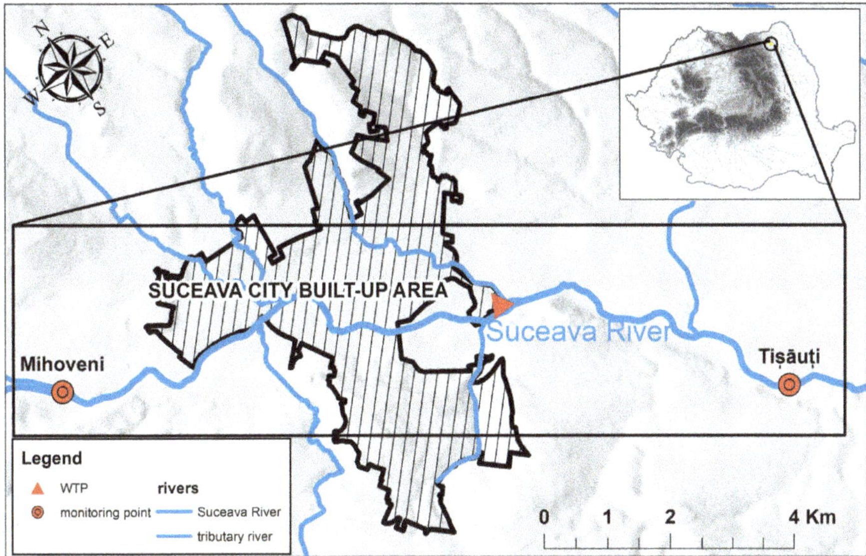

Figure 1. Map of the study area, including the location of the monitoring points and the urban wastewater treatment plant (WTP).

2.2. Instruments

Two AquaTROLL 500 multiparameter probes were used to monitor physical and chemical streamwater properties (paired with two Tube 300R for telemetry). The instruments were equipped with sensors that measured: pressure/level (accuracy: ±0.1% full scale (9 m), resolution: 0.01% full scale), electrical conductivity (automatically converted to specific conductivity (SC); accuracy: ±0.5% of reading +1 μS/cm, resolution: 0.1 μS/cm), dissolved oxygen (DO; accuracy: ±0.1 mg/L, resolution: 0.01 mg/L), oxidation reduction potential (ORP; accuracy: ±5 mV, resolution: 0.1 mV), and pH (accuracy: ± 0.1 pH unit, resolution: 0.01 pH).

2.3. Data

Data analyzed in this study were recorded from 10 October 2018 until 9 October 2019 (with one exception: DO data is missing from November 7 to December 5, 2018). The local standard time (EET) was used for all time series. Corrections were applied to our data as follows: compensations were applied after periodic instrument calibrations, missing singular values were obtained by using linear interpolation, outliers were removed through replacement with values from data smoothed with a moving average filter, and some missing consecutive values were obtained by using regression equations. At the downstream sampling point, the measurements were sometimes affected by residuals from wastewaters and the time series were verified by using independent measurements carried out using a Hach HQ40d portable multiparameter instrument. Data of all parameters can be viewed and downloaded at the project website, http://water.usv.ro/data.php (where a map of the study area can also be inspected).

The yearly average of the specific conductivity was higher downstream (at Tișăuți; 549 μS/cm) than upstream (at Mihoveni; 483.1 μS/cm) and the standard deviation values had the same relationship (97.9 μS/cm versus 72.4 μS/cm). Urban wastewaters were the main cause of the differences; treated and untreated waters are discharged into the Suceava River through the wastewater treatment plant effluent and Cetății Creek [10,23,25]. The difference is much larger after snowfalls, when de-icing measures are applied and/or when high air temperatures lead to important snowmelt and runoff from roads and roofs (see, for example, peaks in SC that were only recorded in the second half of November 2018 downstream after large snowfalls occurred) (Figures 2 and 3). Dissolved oxygen had mean values of 10.6 and 8.82 mg/L (upstream and downstream, with corresponding standard deviations of 2.06 and 2.69 mg/L, respectively). The lower average value downstream of the urban areas was an effect of a warmer and more polluted streamwater. The urban heat island was observed in the study area in a previous study [10].

ORP mean values upstream and downstream were 412.16 and 338.01 mV and standard deviation at these points was 32.11 and 52.97 mV, respectively. The values of pH were 8.45 (upstream) and 8.22 (downstream) and the difference in standard deviations was the greatest of all parameters: 0.09 (upstream) and 0.31 (downstream).

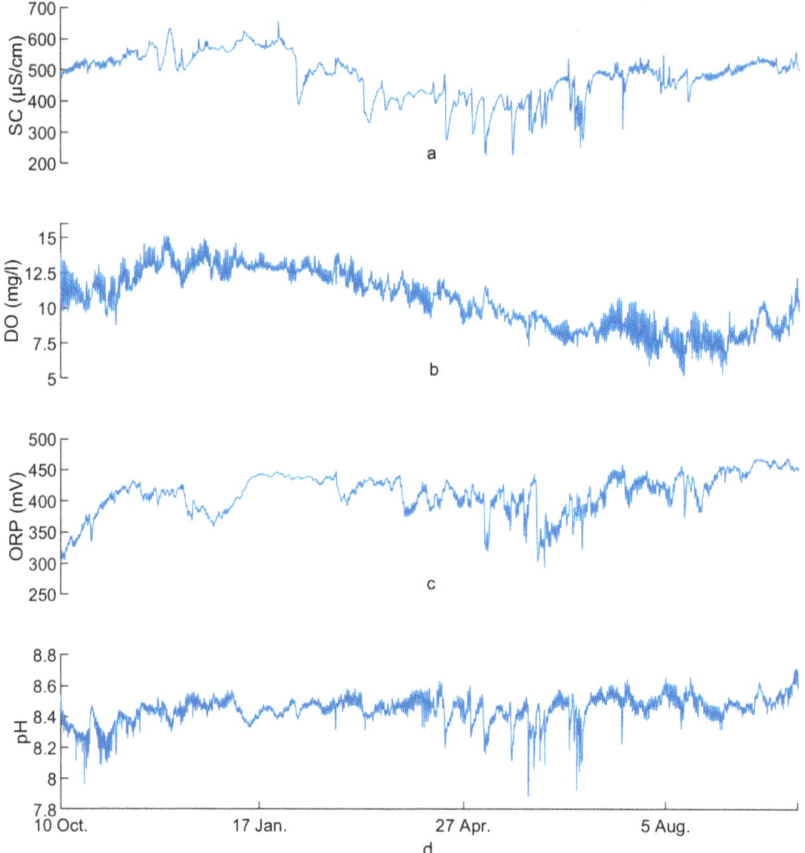

Figure 2. Variations in the studied parameters of the Suceava River at the upstream monitoring point (Mihoveni) from 10 October 2018 to 9 October 2019: (**a**) specific conductivity (SC), (**b**) dissolved oxygen (DO), (**c**) oxidation reduction potential (ORP), and (**d**) pH.

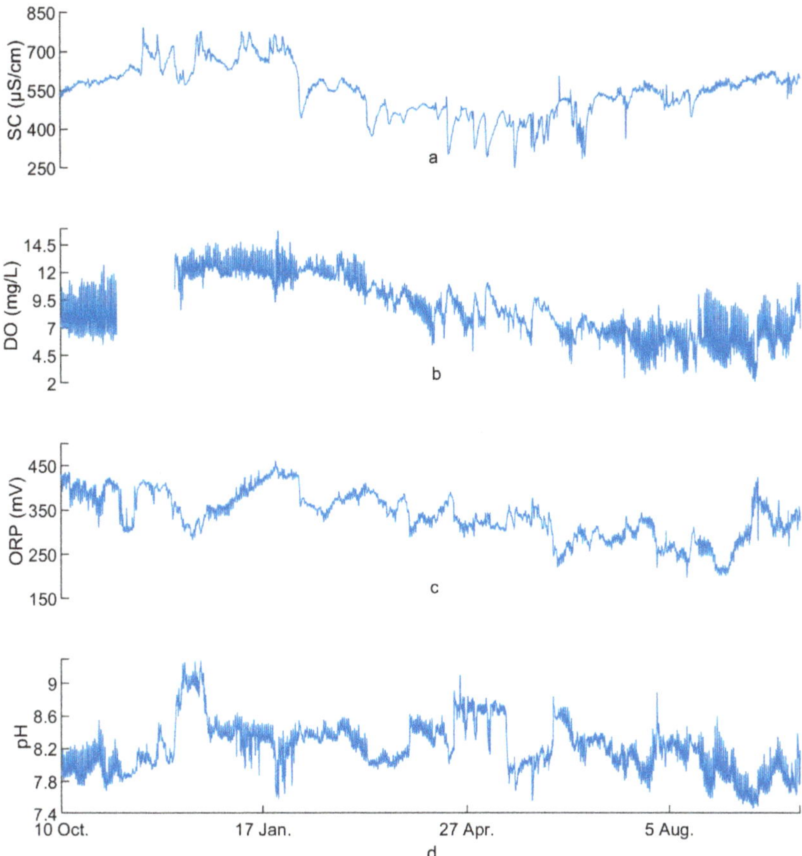

Figure 3. Variations in the studied parameters of the Suceava River at the downstream monitoring point (Mihoveni) from 10 October 2018 to 9 October 2019: (**a**) specific conductivity (SC), (**b**) dissolved oxygen (DO), (**c**) oxidation reduction potential (ORP), and (**d**) pH.

The average diurnal profiles of SC, DO, ORP, and pH indicated differences in the shapes and positions of the diurnal maxima and minima not only between parameters, but also between the values of the same parameter at the two monitoring points (Figure 4). SC upstream had a diurnal maximum during midday (11:00–12:00), while downstream, the maximum was recorded at 03:00. The maximum values of DO recorded were in the late afternoon and early evening at the selected monitoring points; this difference was not caused by the streamwater temperature, which had similar moments of maxima in the study area [10]. Rather, this similarity happens instead of an inverse relationship that should occur theoretically. The average hourly values of downstream ORP created a prolonged interval with maximum values in the first third of the day. The diurnal minimum of upstream pH occurred at 13:00 and might be partly explained by the theoretical inverse relationship between pH and temperature, but the downstream pH maxima was recorded at 15:00.

The average diurnal profile of streamwater level at both monitoring points had irregular fluctuations that do not satisfy the definition of a diurnal cycle. This was due mainly to rainfall events, which have strong responses in the change of the river water level. If the analysis window was restricted only to periods of baseflow, diurnal cycles could be observed for this parameter too (minima in the midday, maxima during late evening). The mean relative streamwater level was

0.786 m upstream of Suceava city and 0.895 m downstream. The standard deviation indicated a higher difference between the two points: 0.121 m upstream and 0.305 m downstream. The increased variability of the water level downstream of Suceava city is to be attributed mainly to water runoff from impervious urban areas because of tributaries that flow into the Suceava River between the two sampling points. These tributaries have small discharges at baseflow and receive some of the pluvial drainage from the metropolitan area (especially Şcheia, Cetăţii, Dragomirna, and Podu Vatafului creeks [9]).

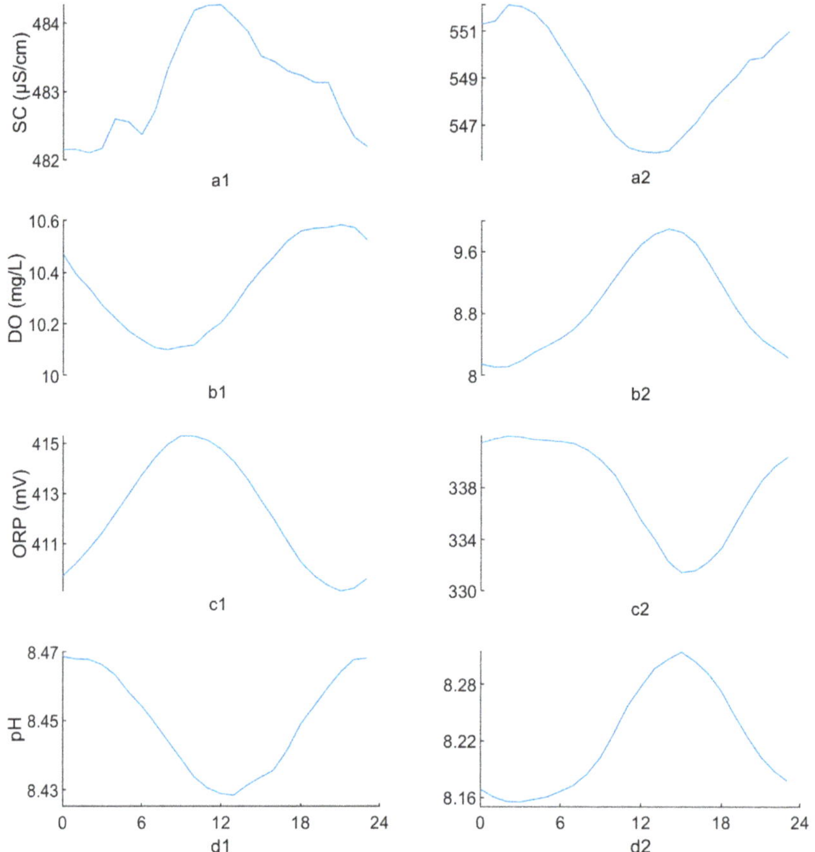

Figure 4. Average diurnal profiles of the selected parameters at the upstream (1) and downstream (2) monitoring points (horizontal axes in hours): (**a**) specific conductivity, (**b**) dissolved oxygen, (**c**) oxidation reduction potential, and (**d**) pH (data from 10 October 2018 until 9 October 2019; with the exception of DO, which was calculated from 6 December 2018 until 9 October 2019 due to missing data).

2.4. Analysis Methods

Variations in flow rates/water level can cause important changes in the diurnal profile of other streamwater parameters. Increases in water level have a clear cause, rainfall, and the slow decline in water level is to be naturally expected until the next rainfall. For this reason, we chose the water level parameter as the control factor when analyzing SC, DO, ORP, and pH. More precisely, we analyzed the water level time series in order to choose the position and length of the analysis windows used for comparing the diurnal profiles of the other parameters. Because the increase in water level is the active change, we decided to split the water level time series into subsets whose ends were given by the minimum values between high waters (Figure 5). A time interval between such minima would be a good case study to observe changes in diurnal cycles induced by peaks in flow rates.

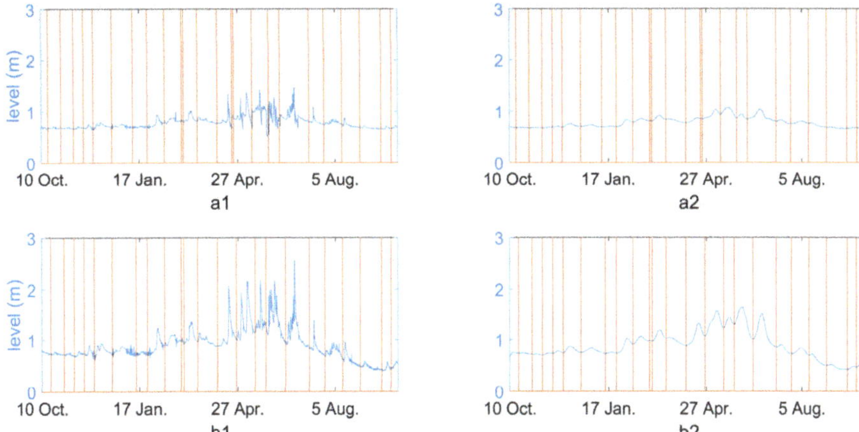

Figure 5. Suceava River water level at the (**a**) upstream and (**b**) downstream monitoring points plotted by using (1) raw data and (2) simplified data computed with the multiresolution analysis. Vertical red lines represent the limits of time intervals between relevant minima (10 October 2018 –9 October 2019).

Data analysis consisted of four major steps: (1) applying the multiresolution and wavelet analysis on raw level data in order to obtain time intervals for case studies; (2) smoothing the time series of water level, SC, DO, ORP, and pH, obtaining the evolution of diurnal cycles with the super-daily trends removed and normalizing the remnant time series for the upstream-downstream comparison; (3) computing the average diurnal profiles for case study time intervals from the normalized time series; and (4) comparing the case study time intervals from the normalized time series through the wavelet coherence analysis.

Prerequisites for the proposed analysis methodology were: MATLAB software (for all steps), the three script files found in the Supplementary Material of this paper (for the first three major steps), the cross wavelet and wavelet coherence toolbox of Aslak Grinsted [26] (https://www.mathworks.com/matlabcentral/fileexchange/47985-cross-wavelet-and-wavelet-coherence), data loaded/created in MATLAB workspace consisting of two variables named "upstream" and "downstream", which were 1D column vectors representing one parameter per monitoring point and having a length equal to a multiple of 24 (because data in this study was sampled 24 times per day).

The first step had four stages:

- Applying the maximal overlap discrete wavelet transform to the water level data from both monitoring points. This was acquired with the function *modwt*, which has the following syntax:

  $$y = modwt(x,wname,n),$$

  where x is the real-valued signal; *wname* is the name of an orthogonal wavelet, which, in our case, was *'haar'*; and n represents the desired number of levels of detail coefficients—7 in this case;
- Executing the multiresolution analysis based on *modwt*, with the function *modwtmra*, which had a similar syntax:

  $$z = modwtmra(y,wname),$$

  where *wname* should be the same as that used with *modwt*;

- Extracting the approximation left from the raw data after the removal of the detail coefficients. This is a vector from the multiresolution analysis matrix and represents the simplified data shown in Figure 5;
- Identifying the local minima on the simplified data (marked with vertical red lines in Figure 5). Note that the positions of minima in the two monitoring points are rarely the same due to differences imposed by geographic location. When selecting the case study time intervals for steps 3 and 4, time intervals delineated by local minima from the two time series were superposed and only the common interval was then selected for being reduced to a size that was multiple of 24 (by starting and ending the case study time series at midnight).

The wavelet decomposition and the maximal overlap discrete wavelet transform were applied previously in a few studies concerning some streamwater parameters, such as water level, DO, and pH [3,5,27,28], which also offer the mathematical description of the mentioned wavelet analyses. Haar wavelet was successfully tested for water parameters [29].

The second step had 3 stages:

- Smoothing the raw time series by using the smooth function, which had the following syntax:

  yy = smooth(x,span,method),

  where the *method* used here was the moving average (*moving*; lowpass filter with coefficients equal to the reciprocal of the span) and the *span* was equal to a day (24). Note that, due to software limitations, span was reduced by 1 to an even number;
- Obtaining the detrended diurnal cycle from raw data after subtracting the smoothed data from it;
- Normalizing the detrended time series between 0 and 1 for the upstream-downstream comparison purposes.

Steps 1 and 2 were computed by executing the *analyze* command line in MATLAB console. This line applies the codes written in the analyze.m script file found in the Supplementary Material. After this command line was applied, two new variables, normalisedU and normalisedD, were created in the workspace and represented the normalized data from upstream and downstream, respectively, obtained at step 2, stage 3 (some other intermediary variables were also created in the workspace, but were handled automatically by codes). The *analyze* command line also generated a graphical output, similar to Figure 5, and was applied to water level data. In order to obtain only normalized data (as in step 2), for the other streamwater parameters, the command line *transform* had to be executed, which used the transform.m script file (in the Supplementary Material) and generated the variables PnormalisedU and PnormalisedD.

All of the new time series from (P)normalisedU and (P)normalisedD were involved in step 3, because they served as the source of data that was cut for case studies by using the time intervals established at the end of step 1. Data selected for case studies had to be included in two new variables, named CSnormalisedU and CSnormalisedD.

Step 3 had only one stage, when the average diurnal profile was computed, and some graphical outputs were generated, as in Figure 6 (using the newly generated variables CSprofileU and CSprofileD). This step was executed with the command line *compare* (with codes written in the compare.m script file, found in the Supplementary Material).

Step 4 had one stage and used the *wtc* function which performs the wavelet coherence analysis between two variables and has the syntax:

wtc(a,b),

where *a* and *b* are values from the upstream and downstream parameters, respectively (e.g., normalisedU, PnormalisedU, or CSnormalisedU). The analysis produced scalograms that indicated how strong two signals co-vary by using a power spectrum and phase arrows. Also, periodicities in signals were tested against the AR1 red noise through a Monte Carlo test and the relevant results were marked with a black line. The wavelet coherence used the Morlet mother wavelet because it is best suited for comparing time series with non-linear processes [26,30]. Wavelet coherence mechanics and results were described by previous studies that applied this type of analysis to hydrological time series [26,31]. The script files used in this methodology are included in the Supplementary Material and described in an accessible manner between the lines of codes contained in the files.

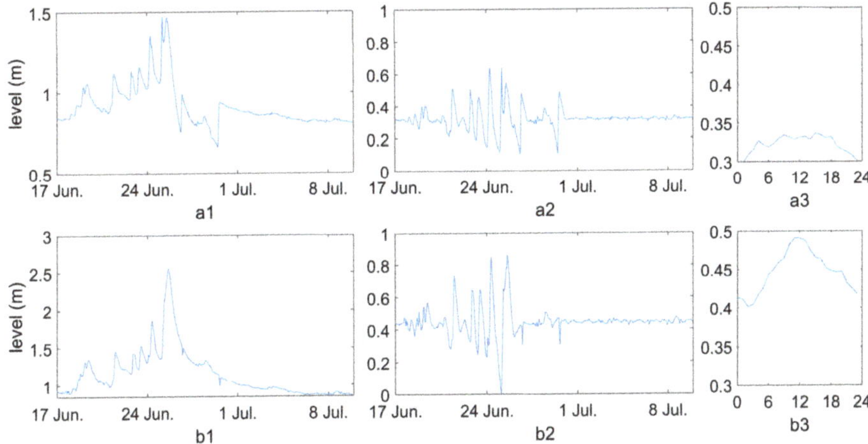

Figure 6. Suceava River water level at the (**a**) upstream and (**b**) downstream monitoring points plotted as (1) raw data, (2) detrended and normalized data, and (3) the average diurnal profile of the normalized data (17 June–9 July 2019).

3. Results and Discussion

The analysis of the major convolutions/waves of water level during a year led us to select two subsets of time series that should serve as relevant case studies of changes in diurnal profiles depending on water level variability. Case study 1 comprised 23 days during a period of intense rainfalls and high waters (17 June–9 July 2019; Figures 6–10), while case study 2 comprised 10 days during the autumn baseflow (20–29 October 2018; Figures 11–15). The length of the subsets with common major evolutions could be increased or decreased depending on the number of levels of detail coefficients of the maximal overlap discrete wavelet transform, which in our case was set to 7.

Depending on the characteristics of water flow or the need for longer/shorter subsets, the number of levels of detail coefficients could be edited in the anlyze.m script file. The smaller the number of levels, the fewer details that were removed from the original time series through decomposition. The subsets were very numerous and small; more levels removed more details and generated and abstracted the approximation of the original data, where only a few subsets could be assessed. Therefore, the choice of decomposition level depends on the user/author [32], as it is the choice of applying the multiresolution analysis only on the water level or on other parameter/all parameters. Available orthogonal wavelets in MATLAB that can replace the Haar wavelet (*'haar'*) option are: Daubechies (*'dbN'*), Symlets (*'sym4'* or *'symN'*), Coiflets (*'coifN'*), and Fejér–Korovkin (*'fkN'*).

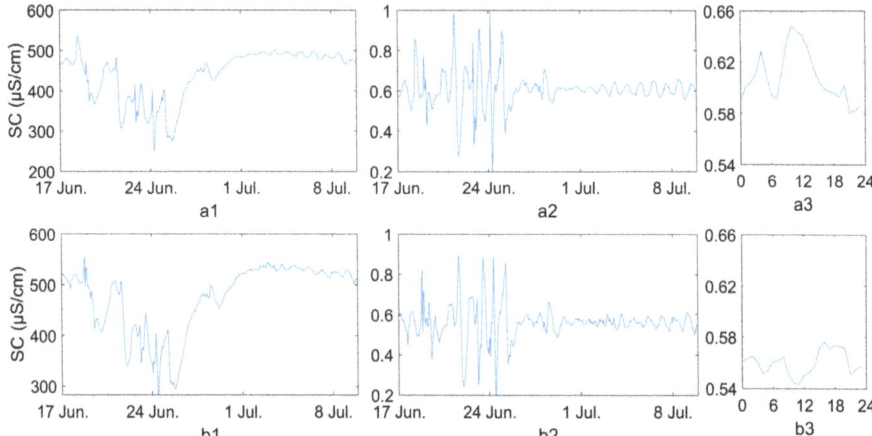

Figure 7. Suceava River SC at the (**a**) upstream and (**b**) downstream monitoring points plotted as (1) raw data, (2) detrended and normalized data, and (3) the average diurnal profile of the normalized data (17 June 17–9 July 2019).

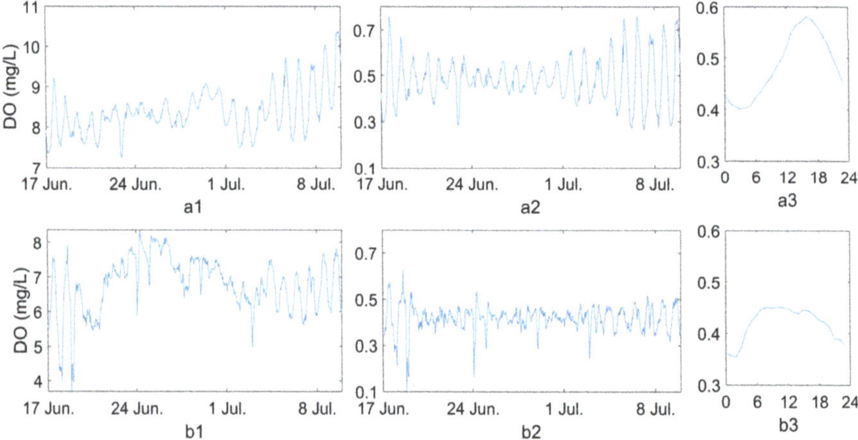

Figure 8. Suceava River DO at the (**a**) upstream and (**b**) downstream monitoring points plotted as (1) raw data, (2) detrended and normalized data, and (3) the average diurnal profile of the normalized data (17 June 17–9 July 2019).

Distinct average diurnal profiles can be observed for any water parameter in case study 1, except for SC (Figure 7), which was very sensitive to water level oscillations during high waters. Such time intervals explain the complex shape of the average diurnal profile of SC calculated for a year (Figure 4a1). Another parameter that had the diurnal cycle easily altered by important changes in water level was pH (Figure 10). Changes in the shapes and positions of the diurnal cycles occurred in the average diurnal profile of DO (Figure 8); at both monitoring points, diurnal maxima moved towards midday or late morning (Figure 4b).

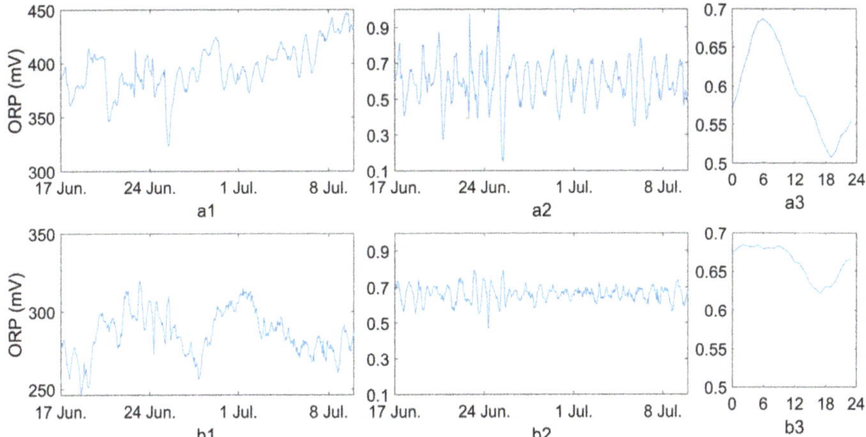

Figure 9. Suceava River ORP at the (**a**) upstream and (**b**) downstream monitoring points plotted as (1) raw data, (2) detrended and normalized data, and (3) the average diurnal profile of the normalized data (17 June 17–9 July 2019).

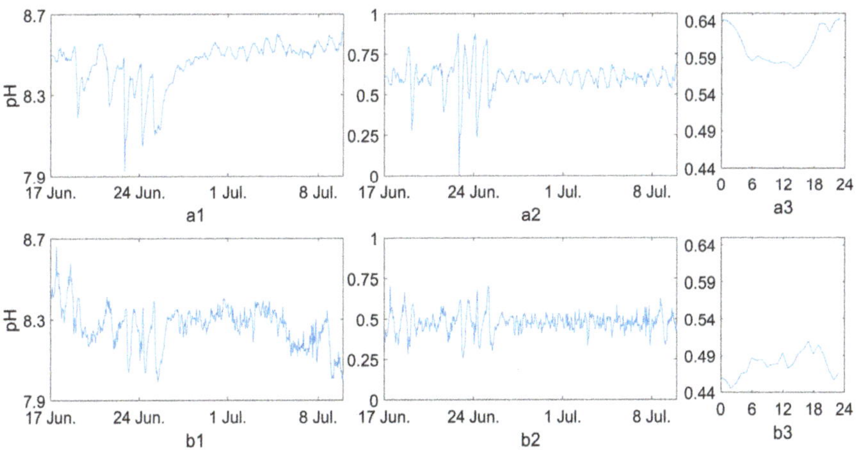

Figure 10. Suceava River pH at the (**a**) upstream and (**b**) downstream monitoring points plotted as (1) raw data, (2) detrended and normalized data, and (3) the average diurnal profile of the normalized data (17 June 17–9 July 2019).

The script files analyze.m and transform.m allowed for editing the smoothing and detrending methods. The smoothing can be done with a filter other than the default moving average. Available filters were: *'lowess'*, a local regression with weighted linear least squares and a 1st degree polynomial model; *'loess'*, the same as the previous, but with a 2nd degree polynomial model; *'rlowess'* or *'rloess'*, robust versions of *'lowess'* or *'loess'* with lower weight for outliers in the regression; *'sgolay'*, a Savitzky–Golay filter. Also, smoothing could be set to be more or less aggressive with the increase or decrease in the span number.

The detrending technique may be different than the proposed difference between the raw and smoothed date. A method with a similar ability to remove general and seasonal trends might show the difference between adjacent elements, if the following syntax is used: *y=diff(x)*.

We inserted a few codes between the lines that execute smoothing and detrending. These are aimed at helping those that intend to obtain detrended values that can no longer be normalized. Instead, a detrended time series with absolute values, similar to those of the raw time series (but lower), will be obtained. It is necessary that this option is applied only to data with inter-diurnal variations significantly higher than the annual oscillation and with weak variability in the diurnal cycle (dedicated variables created in the workspace were differenceX and subtrendX, where X is U (upstream data) or D (downstream data)).

The time interval of case study 2 belonged to the autumn baseflow, where the main aperiodic changes were caused by a few rainfalls. For the first time, SC average diurnal profiles at both monitoring points were similar (Figure 12). This led to the conclusion that, during periods of baseflow, which are longer than the periods affected by intense rainfalls, this similarity is frequent and only these average diurnal profiles might be considered relevant for describing various persistent riverine processes.

Another interesting observation was linked to the diurnal behavior of DO at the downstream monitoring point (Figure 13b). The diurnal cycle had the same relative position of minima as observed in the annual diurnal profile (Figure 4b2), but recorded maxima in two secondary peaks superposed on the diurnal oscillation. Various processes may co-generate this double-peaked shape, including the urban heat island (because the water is measured at the downstream point after passing through the middle of Suceava city). A change also occurred in the hourly position of the maximum value of the DO average diurnal profile: it occurred early in the morning, instead of late evening, as in the annual profile.

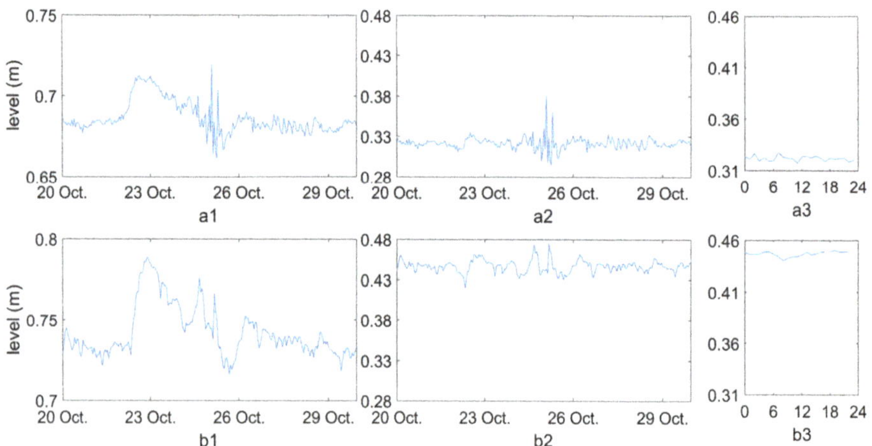

Figure 11. Suceava River water level at the (**a**) upstream and (**b**) downstream monitoring points plotted as (1) raw data, (2) detrended and normalized data, and (3) the average diurnal profile of the normalized data (20–29 October 2018).

The diurnal oscillations of ORP and pH were strongly affected by rainfalls during baseflow at the upstream monitoring point. This was caused by the weak intensity of the diurnal cycle, which became easily modifiable by quasi-random events. In contrast, at Tișăuți, the diurnal cycles of ORP and pH had oscillations stronger than the general trend, which was different at the downstream point than at the upstream point. Given the fact, discussed previously, that the metropolitan area modifies the averages and standard deviations of the measured parameters, we may assume that the stronger diurnal cycle of ORP and pH at the downstream point is regulated by Suceava city. This may also explain why pH values were not lowered by rainfalls, which, by contrast, cause relatively sudden drops at the upstream point. This latter behavior is natural because of the acidic nature of the raindrops, which dissolve gases, such as CO_2, from the atmosphere. The amplitude of the diurnal cycle of pH had amplitudes much greater than 0.2 units, which might induce stress to aquatic fauna. The high variability of pH combined with the low DO during some periods of the year could have caused fish mortality, as reported in June 2018 near the downstream monitoring point (at Lisaura). During the period investigated by us in 2018 and 2019, the minimum DO value was 5.2 mg/L upstream of Suceava city and 2.1 mg/L (critical value) downstream of the city. Pollutants from the wastewater treatment plant of Suceava city contributed to the observed values of the monitored parameters, especially when the plant has difficulties in treating the wastewaters according to specific regulations [10]. The changes in the metropolitan area toward higher imperviousness [10] and the changes in climate toward an increased average air temperature in the study area [33] will generate higher pH oscillations and lower DO concentrations downstream of Suceava city and more fish mortality events are to be expected in the near future.

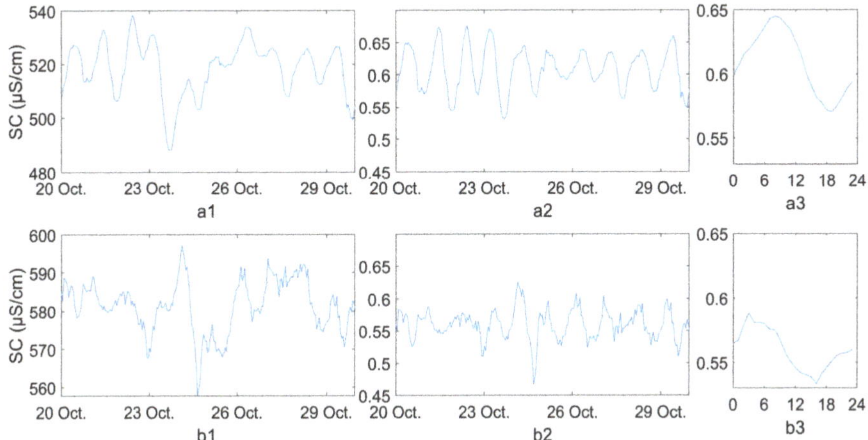

Figure 12. Suceava River SC at the (**a**) upstream and (**b**) downstream monitoring points plotted as (1) raw data, (2) detrended and normalized data, and (3) the average diurnal profile of the normalized data (20–29 October 2018)..

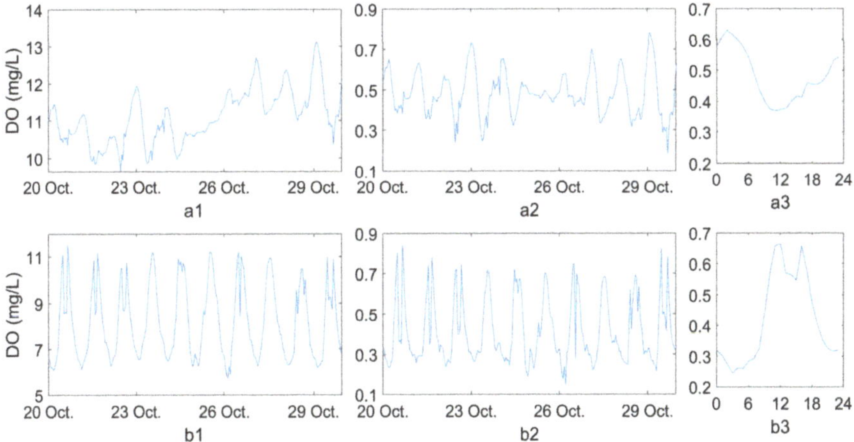

Figure 13. Suceava River DO at the (**a**) upstream and (**b**) downstream monitoring points plotted as (1) raw data, (2) detrended and normalized data, and (3) the average diurnal profile of the normalized data (20–29 October 2018).

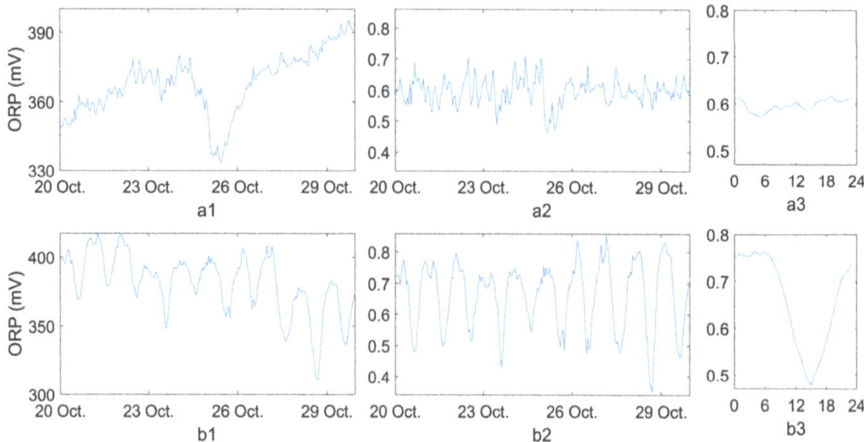

Figure 14. Suceava River ORP at the (**a**) upstream and (**b**) downstream monitoring points plotted as (1) raw data, (2) detrended and normalized data, and (3) the average diurnal profile of the normalized data (20–29 October 2018).

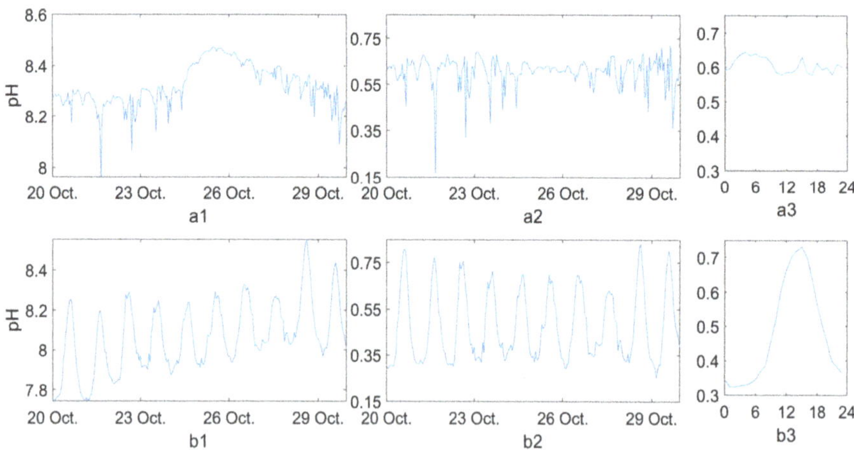

Figure 15. Suceava River pH at the (**a**) upstream and (**b**) downstream monitoring points plotted as (1) raw data, (2) detrended and normalized data, and (3) the average diurnal profile of the normalized data (20–29 October 2018).

The wavelet coherence analysis allowed us to find time intervals when the diurnal cycle was statistically significant (0.95) at both monitoring points at the same time and to observe the temporal changes of the phasing of this cycle (Figures 16–18; arrows pointing right indicate in-phase evolution and anti-phase when pointing left). At a first glance, we observed the high extension of the area with high power (red) on SC scalograms, very similar to that of the water level scalograms. Except for the diurnal band observed for SC in October 2018 (Figure 17a), the high power indicates only very good co-variance over periods greater than 32/64 h; the weak coherence at high frequencies is caused by the disturbing effect of high waters.

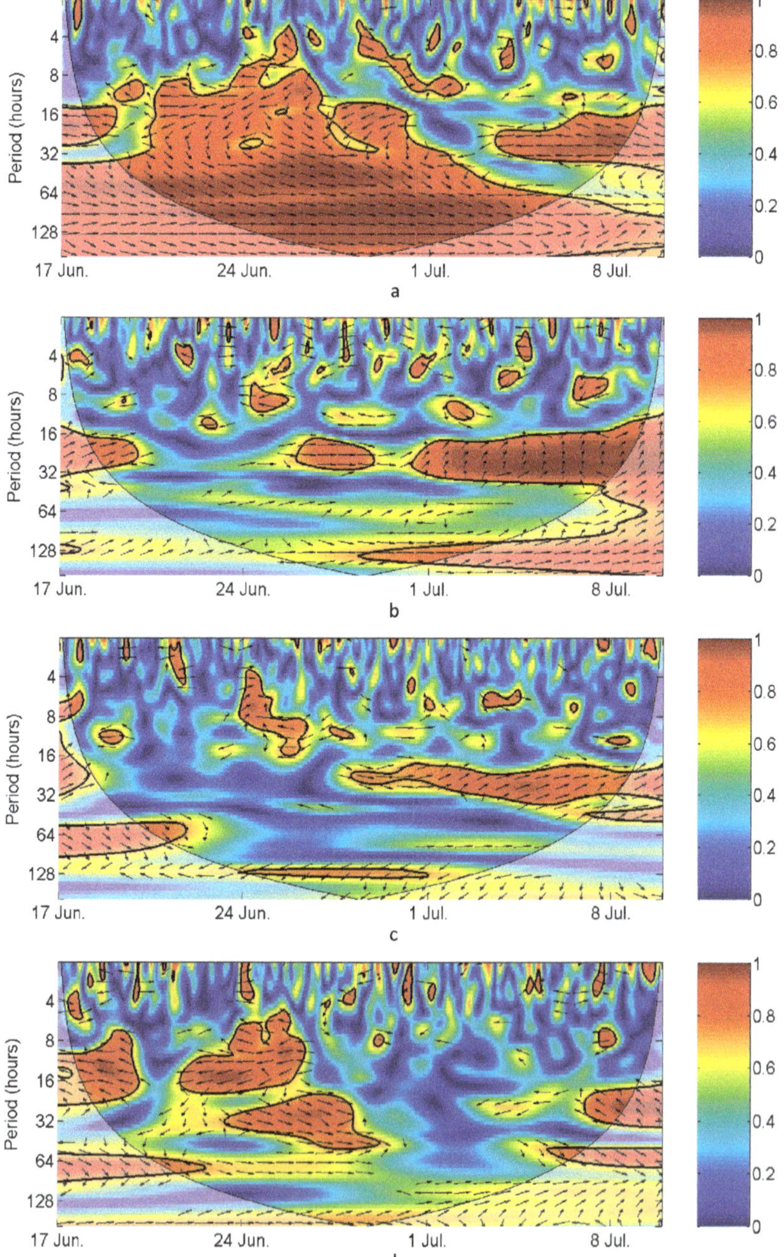

Figure 16. Scalograms of the wavelet coherence analysis of some water parameters measured upstream and downstream during 17 June–9 July 2019: (**a**) SC, (**b**) DO, (**c**) ORP, and (**d**) pH.

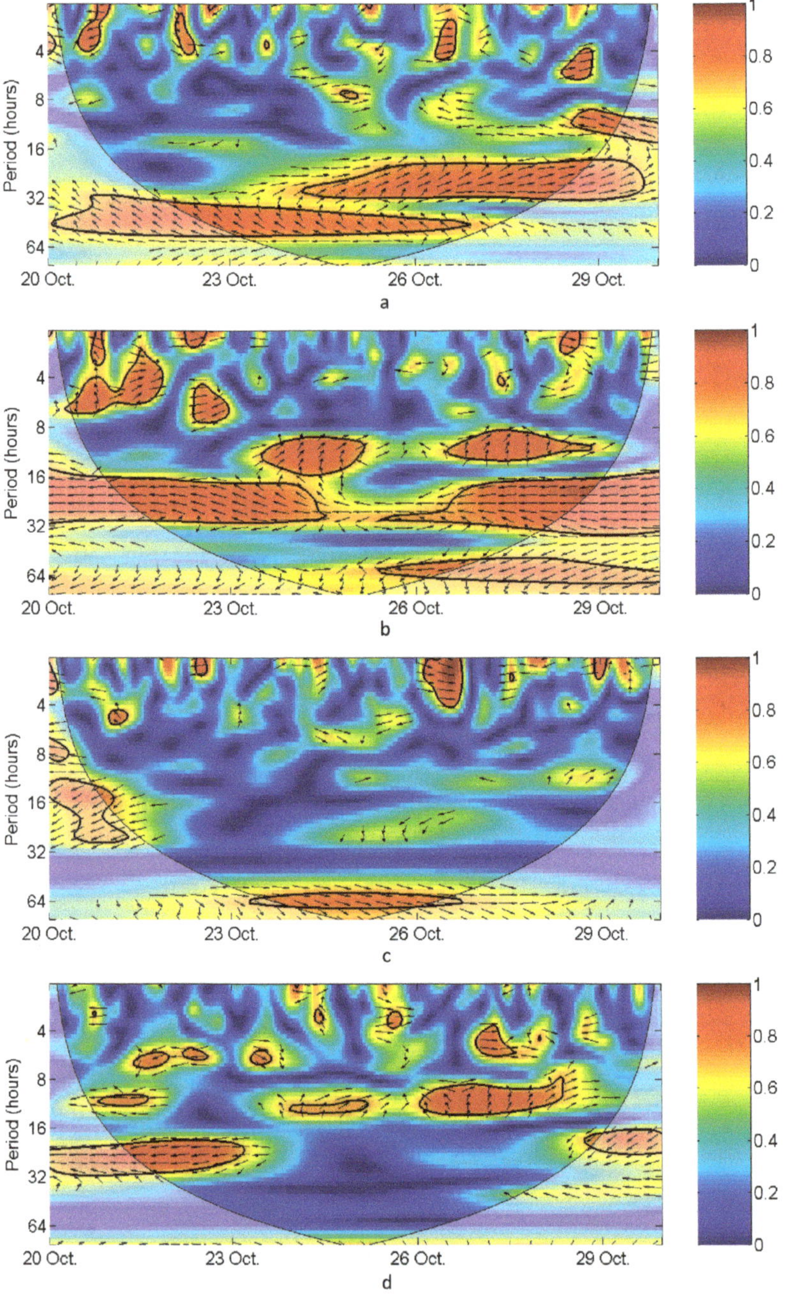

Figure 17. Scalograms of the wavelet coherence analysis of some water parameters measured upstream and downstream during 20–29 October 2018: (**a**) SC, (**b**) DO, (**c**) ORP, and (**d**) pH.

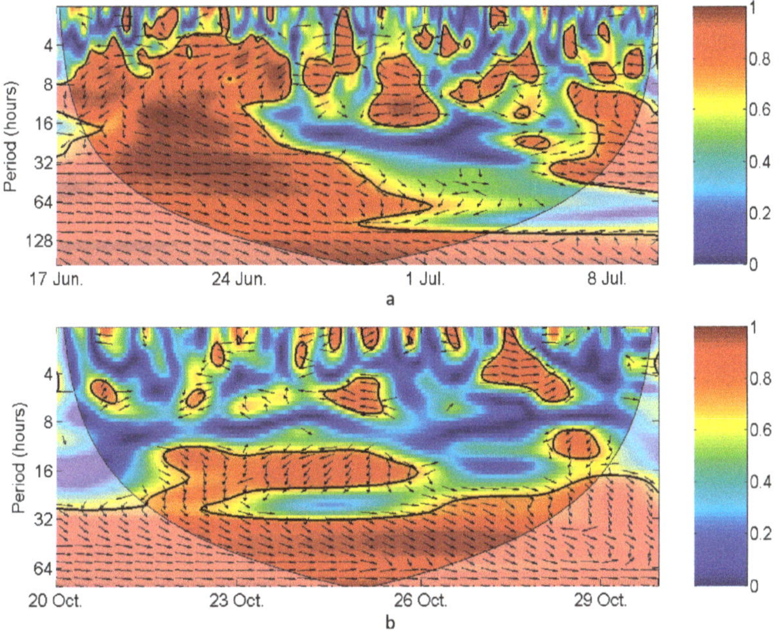

Figure 18. Scalograms of the wavelet coherence analysis of the upstream and downstream water level during: (**a**) 17 June 17–9 July 2019 and (**b**) 20–29 October 2018.

The only persistent coherence between time series recorded upstream and downstream in both case studies was observed for DO (Figures 16b and 17b). During June–July 2019, the diurnal cycles at both monitoring points were in phase or almost in phase, while during October 2018 they were in obvious anti-phase on both sides of the cone of influence (which separates time intervals affected by the edge effects from those that are not affected). Moments when other parameters had coherent evolution of the diurnal cycles in both monitoring points were at the end of June and beginning of July 2019 for ORP and in October 2018 for pH.

Wavelet coherence analyses are not useful only for studying the relationship between two points, but also for investigating the inter-parametric links at the same monitoring point [7]. Climate indices and water quality parameters can also be used for wavelet coherence analyses [34].

4. Conclusions

This study is the first one to use high-frequency (hourly) measurements done over a 365-day interval in the analysis of the temporal and spatial evolution of SC, DO, ORP, and pH of the Suceava River. Raw and normalized time series from case studies, together with average diurnal profiles and scalograms of the wavelet coherence analysis, indicate distinct diurnal evolutions of the selected parameters over time, which cannot be deduced from the annual average diurnal profiles. Important differences in the shape and hourly position of the diurnal cycles of the same parameter occurred between the upstream and downstream monitoring points, despite of the short distance that separated them. Important changes in the monitored water parameters are caused by Suceava city, located between the selected monitoring points.

The case studies time intervals were selected objectively by using the multiresolution analysis of the maximal overlap discrete wavelet transform applied to water level data. The removal of detail coefficients from the raw water level data allowed for detecting the major variations that impose changes in other water parameters. These major variations were generated by increases in water levels

due to rainfalls. The time series of SC, DO, ORP, and pH were detrended prior to comparing their average diurnal profiles at the upstream and downstream monitoring points for a better understanding of the diurnal cycles.

The pair of monitoring points, located upstream and downstream of Suceava city, allowed for measuring the impact of urban wastewaters on the Suceava River. The specific conductivity was higher downstream of the city (yearly averages: 483.1 μS/cm upstream, 549 μS/cm downstream) because of both treated and untreated waters discharged directly or indirectly into the Suceava River. Dissolved oxygen mean annual values were 10.6 mg/L (upstream) and 8.82 mg/L (downstream). The low downstream values are caused by pollutants in water, which consume oxygen in chemical reactions, and because of the urban heat island, which led to warmer waters, less able to dissolve oxygen from the atmosphere.

Three MATLAB script files are provided for a fast comparison process, especially useful when using big data. Codes within the script files are explained and possible variations of the analyses are proposed for a more flexible approach when different data are analyzed. Options include, but are not limited to, variations of the wavelet type, decomposition level, and detrending methods. The proposed methodology is best suited for comparing results from case studies.

Supplementary Materials: The following are available online at http://www.mdpi.com/2073-4441/11/12/2524/s1. Supplementary Material (a zip archive containing three MATLAB .m script files).

Author Contributions: Conceptualization, A.-E.B. and A.G.; Data curation, A.-E.B. and C.F.; Investigation, A.-E.B. and D.I.O.; Methodology, A.-E.B. and C.F.; Resources, A.-E. B., A.G., and D.I.O.; Supervision, A.G.; Visualization, D.I.O. and C.F.; Writing—original draft, A.-E.B.; Writing—review & editing, A.G.

Funding: This research was funded by CNCS - UEFISCDI, project number PN-III-P1-1.1PD-2016-2106.

Acknowledgments: Wavelet coherence software was provided by A. Grinsted. This study was supported by data obtained within the research project SQRTDA (Streamwater Quality Real-Time Data Analysis). This work was supported by a grant of Ministry of Research and Innovation, CNCS - UEFISCDI, project number PN-III-P1-1.1PD-2016-2106, within PNCDI III.

Conflicts of Interest: The authors declare no conflict of interest. The funders had no role in the design of the study; in the collection, analyses, or interpretation of data; in the writing of the manuscript, or in the decision to publish the results.

References

1. Labat, D. Recent advances in wavelet analyses: Part 1. A review of concepts. *J. Hydrol.* **2005**, *314*, 275–288. [CrossRef]
2. Labat, D.; Ronchail, J.; Guyot, J.L. Recent advances in wavelet analyses: Part 2—Amazon, Parana, Orinoco and Congo discharges time scale variability. *J. Hydrol.* **2005**, *314*, 289–311. [CrossRef]
3. Lee, M.; You, Y.; Kim, S.; Kim, K.T.; Kim, H.S. Decomposition of Water Level Time Series of a Tidal River into Tide, Wave and Rainfall-Runoff Components. *Water* **2018**, *10*, 1568. [CrossRef]
4. Evrendilek, F.; Karakaya, N. Monitoring diel dissolved oxygen dynamics through integrating wavelet denoising and temporal neural networks. *Environ. Monit. Assess.* **2014**, *186*, 1583. [CrossRef] [PubMed]
5. Parmar, K.S.; Bhardwaj, R. Wavelet and statistical analysis of river water quality parameters. *Appl. Math. Comput.* **2013**, *219*, 10172–10182. [CrossRef]
6. Baker, M.E.; Schley, M.L.; Sexton, J.O. Impacts of Expanding Impervious Surface on Specific Conductance in Urbanizing Streams. *Water Resour. Res.* **2019**, *55*, 6482–6498. [CrossRef]
7. Rajwa-Kuligiewicz, A.; Bialik, R.J.; Rowiński, P.M. Wavelet Characteristics of Hydrological and Dissolved Oxygen Time Series in a Lowland River. *Acta Geophys.* **2016**, *64*, 649–669. [CrossRef]
8. Somers, K.A.; Bernhardt, E.S.; Grace, J.B.; Hassett, B.A.; Sudduth, E.B.; Wang, S.; Urban, D.L. Streams in the urban heat island: Spatial and temporal variability in temperature. *Freshw. Sci.* **2013**, *32*, 309–326. [CrossRef]
9. Kinouchi, T.; Yagi, H.; Miyamoto, M. Increase in stream temperature related to anthropogenic heat input from urban wastewater. *J. Hydrol.* **2007**, *335*, 78–88. [CrossRef]
10. Briciu, A.-E. *Studiu de Hidrologie Urbană în Arealul Municipiului Suceava (Urban Hydrology Study in Suceava Municipality Area)*; Ştefan cel Mare University Publishing House: Suceava, Romania, 2017.

11. Gribovszki, Z.; Szilágyi, J.; Kalicz, P. Diurnal fluctuations in shallow groundwater levels and streamflow rates and their inter-pretation—A review. *J. Hydrol.* **2010**, *385*, 371–383. [CrossRef]
12. Bond, B.J.; Jones, J.A.; Moore, G.; Phillips, N.; Post, D.; McDonnell, J.J. The zone of vegetation influence on baseflow revealed by diel patterns of streamflow and vegetation water use in a headwater basin. *Hydrol. Process.* **2002**, *16*, 1671–1677. [CrossRef]
13. Nimick, D.A.; Cleasby, T.E.; McCleskey, R.B. Seasonality of diel cycles of dissolved trace-metal concentrations in a Rocky Mountain stream. *Environ. Geol.* **2005**, *47*, 603–614. [CrossRef]
14. Briciu, A.-E.; Oprea-Gancevici, D.I. Diurnal thermal profiles of selected rivers in Romania. *SGEM2015 Conf. Proc.* **2015**, *1*, 221–228.
15. Briciu, A.-E.; Mihăilă, D.; Mihăilă, D. Short, medium and long term stochastic analysis of the Suceava River pollution evolution in the homonymous city. *SGEM2012 Conf. Proc.* **2012**, *3*, 809–816.
16. Teodosiu, C.; Cojocariu, C.; Musteret, C.P.; Dăscălescu, I.G.; Caraene, I. Assessment of human and natural impacts over water quality in the Prut River basin, Romania. *Environ. Eng. Manag. J.* **2009**, *8*, 1439–1450. [CrossRef]
17. Florescu, D.; Ionete, R.E.; Sandru, C.; Iordache, A.; Culea, M. The influence of pollution monitoring parameters in characterizing the surface water quality from Romania southern area. *Rom. J. Phys.* **2011**, *56*, 1001–1010.
18. Batrînescu, G.; Bîrsan, E.; Vasile, G.; Stănescu, B.; Stănescu, E.; Păun, I.; Petrescu, M.; Filote, C. Identification of the aquatic ecosystems integrating variables in the Suceava hydrographic basin and their correlations. *J. Environ. Prot. Ecol.* **2011**, *12*, 1627–1643.
19. Vasilache, V.; Filote, C.; Crețu, M.A.; Sandu, I.; Coisin, V.; Vasilache, T.; Maxim, C. Monitoring of groundwater quality in some vulnerable areas in Botoșani County for nitrates and nitrites based pollutants. *Environ. Eng. Manag. J.* **2012**, *11*, 471–479. [CrossRef]
20. Marusic, G.; Sandu, I.; Vasilache, V.; Filote, C.; Sevcenco, N.; Crețu, M.-A. Modeling of spacio-temporal evolution of fluoride dispersion in "river-type" systems. *Rev. Chim.* **2015**, *66*, 503–506.
21. Iticescu, C.; Georgescu, L.P.; Topa, C.M. Assessing the Danube water quality index in the city of Galați, Romania. *Carpath. J. Earth Environ. Sci.* **2013**, *8*, 155–164.
22. Romanescu, G.; Crețu, M.A.; Sandu, I.G.; Păun, E.; Sandu, I. Chemism of streams within the Siret and Prut drainage basins: Water resources and management. *Rev. Chim.* **2013**, *64*, 1416–1421.
23. Briciu, A.-E.; Toader, E.; Romanescu, G.; Sandu, I. Urban Streamwater Contamination and Self-purification in a Central-Eastern European City. Part I. *Rev. Chim.* **2016**, *67*, 1294–1300.
24. Briciu, A.-E.; Toader, E.; Romanescu, G.; Sandu, I. Urban Streamwater Contamination and Self-purification in a Central-Eastern European City. Part B. *Rev. Chim.* **2016**, *67*, 1583–1586.
25. Florescu, G.; Briciu, A.-E.; Hutchinson, S.M. An assessment of in-channel alluvia (aits) in the Suceava River near Suceava city, NE Romania. *Georeview* **2019**, *29*, 84–97.
26. Grinsted, A.; Moore, J.C.; Jevrejeva, S. Application of the cross wavelet transform and wavelet coherence to geophysical time series. *Nonlinear Process. Geophys.* **2004**, *11*, 561–566. [CrossRef]
27. Seo, Y.; Choi, Y.; Choi, J. River Stage Modeling by Combining Maximal Overlap Discrete Wavelet Transform, Support Vector Machines and Genetic Algorithm. *Water* **2017**, *9*, 525.
28. Chen, Y.; Guan, Y.; Shao, G.; Zhang, D. Investigating Trends in Streamflow and Precipitation in Huangfuchuan Basin with Wavelet Analysis and the Mann-Kendall Test. *Water* **2016**, *8*, 77. [CrossRef]
29. Parmar, K.S.; Bhardwaj, R. Analysis of Water Parameters Using Haar Wavelet (Level 3). *Int. J. Curr. Eng. Technol.* **2012**, *2*, 166–171.
30. Torrence, C.; Compo, G.P. A practical guide to wavelet analysis. *Bull. Am. Meteorol. Soc.* **1998**, *79*, 61–78. [CrossRef]
31. Briciu, A.-E. Wavelet analysis of lunar semidiurnal tidal influence on selected inland rivers across the globe. *Sci. Rep.* **2014**, *4*, 4193. [CrossRef]
32. Yang, M.; Sang, Y.-F.; Liu, C.; Wang, Z. Discussion on the Choice of Decomposition Level for Wavelet Based Hydrological Time Series Modeling. *Water* **2016**, *8*, 197. [CrossRef]

33. Mihăilă, D.; Briciu, A.-E. Actual climate evolution in the NE Romania. Manifestations and consequences. *SGEM2012 Conf. Proc.* **2012**, *4*, 241–252.
34. Hatvani, I.G.; Clement, A.; Korponai, J.; Kern, Z.; Kovács, J. Periodic signals of climatic variables and water quality in a river—Eutrophic pond—Wetland cascade ecosystem tracked by wavelet coherence analysis. *Ecol. Indic.* **2017**, *83*, 21–31. [CrossRef]

Article

An Improved Model for the Evaluation of Groundwater Recharge Based on the Concept of Conservative Use Potential: A Study in the River Pandeiros Watershed, Minas Gerais, Brazil

Marcelo Alvares Tenenwurcel [1], Maíse Soares de Moura [1], Adriana Monteiro da Costa [1], Paula Karen Mota [1], João Hebert Moreira Viana [2], Luís Filipe Sanches Fernandes [3] and Fernando António Leal Pacheco [4,*]

1 Federal University of Minas Gerais, 6620 Antônio Carlos Ave., Pampulha, Belo Horizonte, MG 31270-901, Brazil; m-alvares@hotmail.com (M.A.T.); maisedemoura2013@gmail.com (M.S.d.M.); drimonteiroc@yahoo.com.br (A.M.d.C.); paulakarenmota@gmail.com (P.K.M.)

2 Brazilian Agricultural Research Corporation (Embrapa Maize and Sorghum), Sete Lagoas, MG 35701-97, Brazil; joao.herbert@embrapa.br

3 CITAB—Centre for the Research and Technology of Agro-Environment and Biological Science, University of Trás-os-Montes and Alto Douro, Quinta de Prados Ap. 1013, 5001-801 Vila Real, Portugal; lfilipe@utad.pt

4 CQVR—Chemistry Research Centre, University of Trás-os-Montes and Alto Douro, Quinta de Prados Ap. 1013, 5001-801 Vila Real, Portugal

* Correspondence: fpacheco@utad.pt

Received: 14 February 2020; Accepted: 28 March 2020; Published: 1 April 2020

Abstract: Water resources have been increasingly impacted due to the growth of water demand associated with environmental degradation. In this context, the mapping of groundwater recharge potential has become attractive to water managers as it can be used to direct public policies and conserve this natural asset. The present study modifies (improves) a spatially explicit model to determine groundwater recharge potential at the catchment scale, testing it in the Pandeiros River basin located in the state of Minas Gerais, Brazil. The model is generally based on the water balance approach and the input variables were compiled from institutional sources and processed in a Geographic Information System. The novelty brought by the aforementioned modification relates to the coupling of physical variables (conventional way) and land management practices (introduced here) in the estimation of a percolation factor. The role of land management practices for percolation was assessed by the so-called Conservative Use Potential (PUC) method, which classifies the areas of a river basin in terms of their potential for sustainable use. The results were validated by an independent method, namely the recession curve method based on the interpretation of hydrographs. In general, the groundwater recharge potential is favored in flat to gently undulating areas and forested regions, as well as where the landscape is characterized by well-structured soils, good drainage conditions and large hydraulic conductivity. The map of groundwater recharge potential produced in this study can be used by planners and decision makers in the Pandeiros River basin as a tool to achieve sustainable use of groundwater resources and the protection of recharge areas.

Keywords: groundwater recharge; water resources management; Conservative Use Potential; river basin; geographic information system; water balance

1. Introduction

Clean water resources are becoming scarcer due to the increasing demand for its use and to environmental degradation [1]. It has been estimated that 1/3 of all countries will have to adapt their productive processes by 2025 because water is lacking, and more than 3 billion people might be living in regions under chronic draughts or water stress [2–6]. Owing to its great importance for society, economy and the environment, it is necessary to properly manage water resources. The increasing degradation of surface water resources not only in Brazil but all over the world [7], makes the sustainable management of groundwater become an even more essential activity. In this context, a precise evaluation of groundwater distribution entails the understanding of aquifer refilling processes and the quantification of groundwater inputs. This quantification is called groundwater recharge estimation. In light of this challenge, studies and maps related to the provision of water resources are needed, with the purpose of indicating priority areas for conservation or restoration and to direct public policies to protect this natural resource. In Brazil, groundwater recharge studies that exist are predominantly focused on the river basin scale [8].

The incorrect management of river basins can generate serious threats to water availability, hindering surface and groundwater, because the dynamic of hydrologic systems is vulnerable to human actions [9,10]. An example is the reduced capacity for water infiltration into the soil observed in areas that are occupied with civil constructions or impermeable pavements [10,11]. Other problems, such as the intensification of erosion or floods, can occur in the development of soil permeability declines. In rural catchments, the inadequate uses of the land, among other factors, can amplify the risk of flood occurrence [12] reducing the levels of recharge.

Changes in the dynamics of hydrologic systems are particularly worrying in aquifer recharge zones, which are regions that enable water infiltration and percolation toward an aquifer system, which is defined as the geological system capable of storing and distributing a significant amount of water [13–15]. Moreover, recharge zones can be defined as areas where the soil surface favors the water infiltration and percolation [16,17]. Water can also be retained in the soil and slowly recharge the aquifer [13,18,19].

Some recharge zones are more efficient than others and for that reason are called preferential groundwater recharge zones [20]. The environmental protection of these special areas is important to conserve the quality and quantity of water resources. Thus, detailed information on the groundwater recharge process can aid better land use and cover distribution, and indicate the best areas for agricultural activities with the lowest groundwater contamination risks caused by the release of substances with high polluting potential, such as pesticides.

Despite the importance of sustainable management of aquifer recharge zones [7,21–23], in the state of Minas Gerais, Brazil, this topic has not yet been properly studied. Therefore, a better comprehension about the factors that affect groundwater recharge is necessary, as well as the mapping of recharge areas that consider the sustainable management potential of the basin. These evaluations should be robust and contain physical–environmental factors [10], such as soil characteristics, geology, vegetation cover, climate, and topography. When all these data are evaluated together, they allow for sustainable water use, meaning a use without compromising groundwater recharge [10]. Besides, under these circumstances the volume of water withdrawn from the aquifer system can be defined according to its natural capacity [24].

Numerous groundwater recharge estimation methods exist. However, they all entail some level of uncertainty [25]. In general, the practical and conceptual limitations of recharge estimation models occur because the available hydrological and hydrogeological data are sparse or fragmented, and because the spatial and temporal variations in recharge are significant [26]. This difficulty is significant for semi-arid areas [27], causing recharge estimations in these areas to be even more challenging. There are direct and indirect methods to evaluate the groundwater recharge potential. The direct methods include geological and geophysical explorations, gravimetrical and magnetic models, and perforation tests [10]. The indirect methods include hydrological and hydrogeological models [28,29], using geographical information systems (GIS) combined with field work [30,31]. Other studies have employed different methods to estimate groundwater recharge, which are comprised of tracer methods, water table fluctuation models, lysimeter methods and simple water balance techniques. Some of these studies have used numerical groundwater models or dynamically linked them to hydrological models to estimate recharge variations under different climate and land cover conditions [32–36]. For example, Döll (2008) modeled global groundwater recharge using the WaterGAP Global Hydrological Model (WGHM), which has failed to reliably estimate recharge in semi-arid regions [37]. In that study, the influence of vegetation was not taken into account, even though many studies have showed the importance of this variable for estimating the groundwater recharge [32,38–42]. Moreover, Chowdhury et al. (2010) delineated groundwater recharge zones in West Medinipur district, India, using a GIS approach mixed with remote sensing and multi-criteria decision making techniques [22]. The input variables considered in that study were geomorphology, geology, drainage density, slope and aquifer transmissivity. In general, the choice of a method should consider the precision level needed, the project execution viability, and the available financial resources.

Among methods available in the literature, Costa et al. (2019) proposed one for the evaluation of groundwater recharge potential based on the water balance approach that considers climatic variables, water runoff, and the percolation of water into the soil profile [10]. The authors obtained results for mean annual recharge similar to those calculated by the hydrograph recession curve analysis, which has been used as a validation method. Furthermore, they identified areas with larger recharge potential and suggested management practices to improve groundwater recharge in those areas, making their study a valuable tool for the sustainable use of groundwater and protection of recharge areas [10].

Despite the positive results obtained by Costa et al. (2019), it is worth noting that the land management practices were a consequence of groundwater recharge assessments, not a contributing factor to groundwater recharge included in the model. Indeed, all parameters included in Costa's water balance model were physical, while land management practices had no role, regardless of their potential to dynamically affect groundwater recharge [10]. Thus, the coupling of physical factors and land management practices in a recharge estimation model could be a motivation (and a novelty) for a subsequent study. Before the publication of Costa et al. (2019), Costa et al. (2017) [43] conducted a study in Minas Gerais and developed a method based on multi criteria analysis, which was efficient to map a so-called Conservative Use Potential (PUC). The PUC method weights a considerably large number of variables considering their importance for sustainable land use, including several variables linked to land management practices, such as drainage, soil depth and fertility, erosion potential, and land capability. Hence, one possible route to realize our research motivation would encompass including the PUC, as determined by Costa et al. (2017) [43], within the framework of the Costa et al. (2019) groundwater recharge method [10].

The general purpose of this study is therefore to take that step forward and embed the concept of PUC in the groundwater recharge method of Costa et al. (2019) [10]. In that method, a parameter is defined to measure water percolation based on the soil's effective porosity and hydraulic conductivity. The method presented in this work replaces porosity by three parameters strongly influenced by management practices, which are soil texture, drainage, and profile depth. The replacement has the specific purpose to check whether this set of variables responds more effectively to land use changes than the original variable (porosity). The model was tested in the Pandeiros River Basin (PRB), Brazil, to generate a spatially explicit map of groundwater recharge potential, at regional scale (1:100.000). This map has the potential to subsidize indications of preferred areas for restoration, recovery, and protection, ensuring a more sustainable water resources management in this basin.

2. Materials and Methods

2.1. Study Area

The Pandeiros River basin (PRB) is located in the northern state of Minas Gerais, Brazil, and has an area of 396,028 ha, which encompasses part of Januária, Bonito de Minas, and Cônego Marinho municipalities (Figure 1). The climate of the region is predominantly dry with mean annual temperature of 24.6 °C, which occasionally reaches a maximum of 33 °C in October, and a minimum of 14 °C in

July [44]. Rainfall is concentrated in April to September, since the climate is semiarid, with mean annual rainfall of approximately 1,050 mm year [45–47]. The longest and largest tributaries of the Pandeiros River are the Catolé, Suçuarana, Borrachudo and Macaúbas streams.

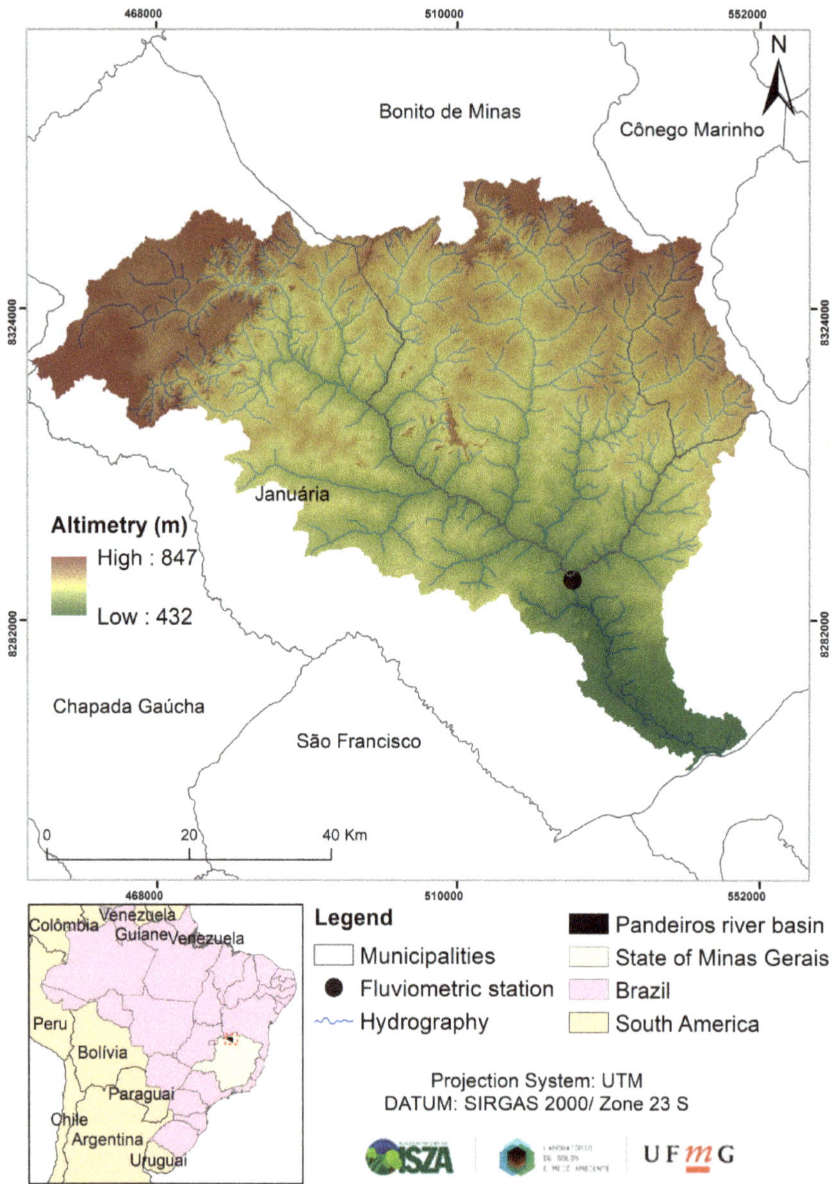

Figure 1. Location of the Pandeiros River basin in the state of Minas Gerais, Brazil.

According to the Brazilian Institute for Geography and Statistics (2018), the municipalities comprising the PRB have a total population of 86,311 inhabitants, most of them living in Januária. The largest part of this population live in rural areas, especially in the municipalities of Bonito de Minas and Cônego Marinho (Table 1) [48].

Table 1. Area and population living in the municipalities of Pandeiros River basin. Source: Brazilian Institute for Geography and Statistics. The population numbers refer to 2018.

Municipality	Total Area (ha)	Urban Population (2018)	Rural Population (2018)	Total Population (2018)
Januária	665,700	41,322	24,141	67,628
Bonito de Minas	390,200	2209	7464	11,088
Cônego Marinho	164,100	1915	5186	7595
Total	1,220,000	45,446	36,791	86,311

The PRB is in an area with transitional vegetation, presenting phytophysionomies of Cerrado and Caatinga biomes [49]. This ecotone is characterized by swamp regions that contain the springs of the São Francisco River. These springs are responsible for the reproduction of most fishes that live between the Três Marias (MG) and Sobradinho (BA) dams [50].

The relief is predominantly flat. The plain was formed by the filling of the São Francisco Depression with sediments that were sourced from the erosion of rocks from the São Francisco Plateau [51]. This process is also responsible for a small proportion of steep slope areas in the basin, as shown in Figure 2 and quantified in Table 2 [52].

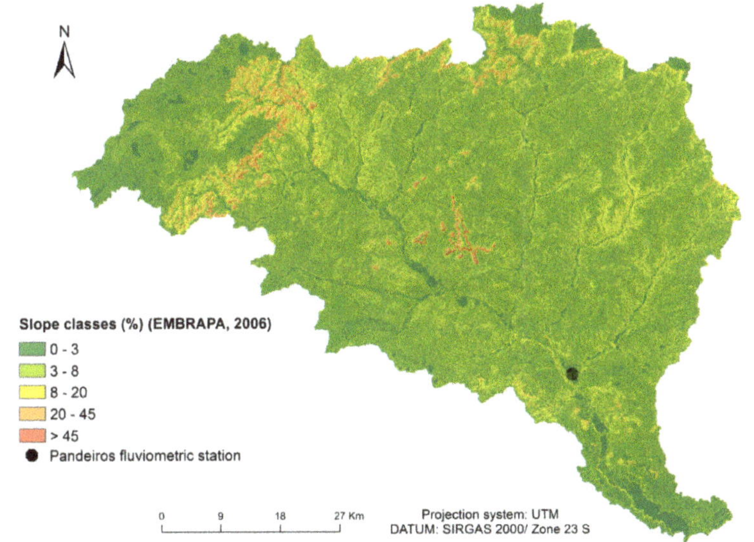

Figure 2. Slope map of the Pandeiros River basin.

Table 2. Area of slope classes in the Pandeiros River basin.

Slope (%)	Relief Type	Area (ha)	%
0 to 3	Plain	122.048,53	30.83
3 to 8	Slightly wavy	206.106,85	52.04
8 to 20	Wavy	59.179,5	14.94
20 to 45	Strongly wavy	7.806,56	1.97
Above 45	Mountainous to scarped	886,85	0.22
Total	396,028.29	100	100

As regards the local geology, there is the presence of alluvial deposits over the main drains in the area and in wetland regions (Veredas), which are the results of natural and anthropogenic erosion processes with consequent transport and deposition of sediments. Figure 3 and Table 3 show the spatial distribution and proportion of lithotypes in the studied basin, respectively.

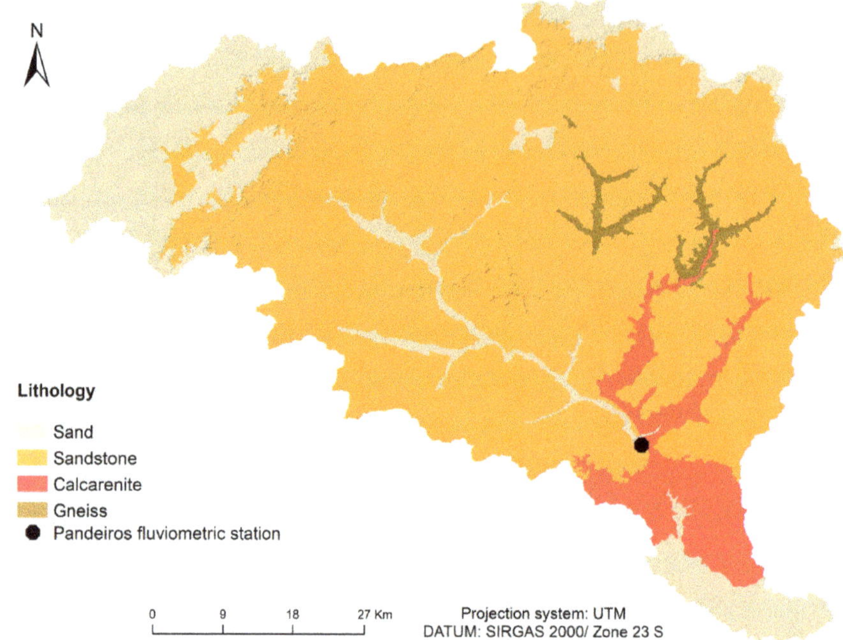

Figure 3. Lithological map of the Pandeiros River basin.

Table 3. Spatial distribution and proportion of lithotypes in the Pandeiros River basin.

Lithotype	Area (ha)	%
Sand	70,886.74	17.90
Sandstone	290,804.78	73.43
Calcarenite	27,570.98	6.96
Gneiss	6,765.81	1.71
Total	396,028.29	100

The Soil Map of Minas Gerais [53] presents an area predominantly composed of red–yellow latosols, which cover more than 87% of the hydrographic basin, followed by fluvic neosols (5.48%). The occurrence of quartzarenic neosols, litolic neosols, cambisols and melanic gleysols are related to much smaller areas (Figure 4 and Table 4).

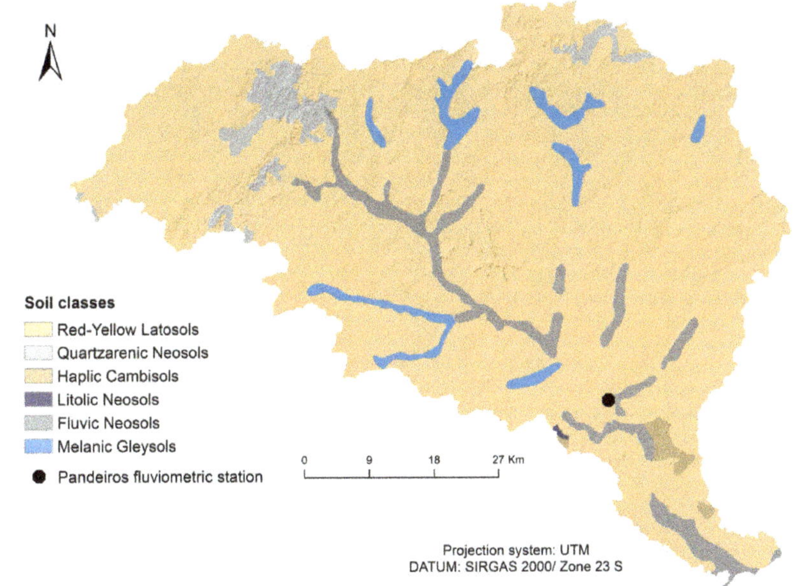

Figure 4. Soil map of the Pandeiros River basin.

Table 4. Spatial distribution and proportion of soil classes in the Pandeiros River basin.

Soil Classes	Area (ha)	%
Haplic cambisols	2,810.25	0.71
Melanic gleysols	11,856.63	2.99
Red–yellow latosols	346,549.01	87.51
Fluvic neosol	21,706.62	5.48
Litolic neosol	176.09	0.04
Quartzarenic neosol	12,929.70	3.26
Total	396,028.29	100

The spatial distribution of the land use and cover classes in the basin (Figure 5) shows the predominance of a typical Cerrado vegetation (savanna) of low to medium size [54], which is usually associated with the occurrence of latosols and sandstone regions. These are found mainly in the central part, towards the northern and northeast of the Pandeiros River basin. This phytophysiognomy covers 183,719.88 ha in the basin, representing 46.3% of its area (Table 5). The second most significant land use and cover class in the study area is the dense Cerrado (21.5%), followed by the sparse Cerrado (10.42%).

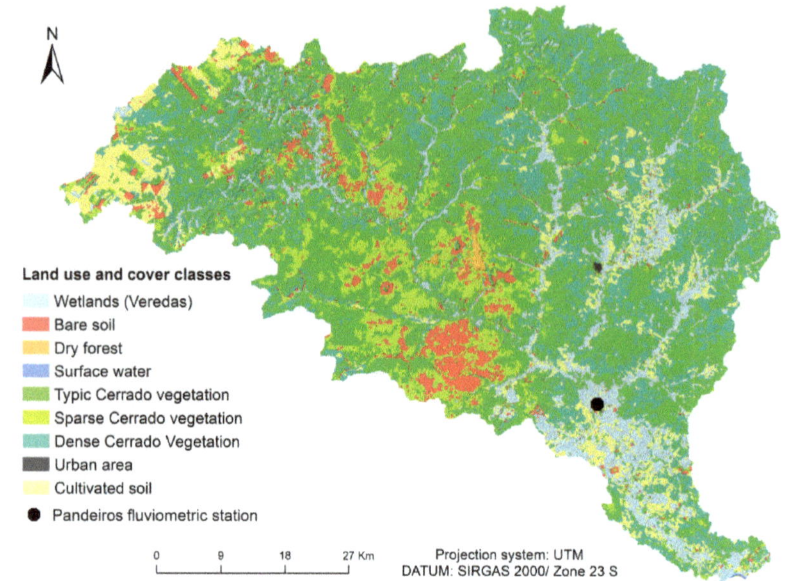

Figure 5. Land use and cover map of the study area, Pandeiros River basin.

Table 5. Spatial distribution and proportion of the land use and cover classes in the Pandeiros River basin.

Class	Area	%
Cultivated soil	33,138.21	8.37
Urban area	121.19	0.03
Dense Cerrado vegetation	85,203.09	21.51
Sparse Cerrado vegetation	41,246.36	10.42
Typical Cerrado vegetation	183,719.88	46.39
Surface water	70.08	0.02
Dry forest	1,102.50	0.28
Bare soil	18,351.75	4.63
Wetland regions (Veredas)	33,075.23	8.35
Total	396,028.29	100

2.2. Material

The materials used in this study (Table 6) consisted of (i) a digital elevation model (ALOS PALSAR), with a spatial resolution of 12.5 meters; (ii) a land use and cover map (scale of 1: 25.000); (iii) the soil map of Minas Gerais state (scale of 1: 600.000); (iv) values of groundwater recharge calculated by the PUC method and the hydraulic conductivity of each soil class in the basin; (v) rainfall and evapotranspiration data from meteorological stations located in municipalities near the basin; and vi) flow data records from the station code 45,250,000 located near the mouth of the Pandeiros River.

Table 6. Material used in the groundwater recharge evaluations.

Data Type	Use in the Work	Web Site Page
Digital elevation model	Calculation of slope length and steepness factor	https://www.asf.alaska.edu
Land use and cover map	Calculation of RF, the runoff factor	https://www.earthexplorer.usgs.gov/
Soil map	Calculation of PF, the percolation factor	https://www.dps.ufv.br
Rainfall and evapotranspiration data	Calculation of recharge potential	https://www.inmet.gov.br
Streamflow data	Validation of recharge potential data (analysis of hydrograph recession curve)	https://www.snirh.gov.br/hidroweb

2.3. Methodology

The spatially explicit groundwater recharge potential in the Pandeiros River basin was evaluated using the methodology proposed by Costa et al. (2019) [10], with the adjustments described below. The workflow was divided into 5 main steps: (i) acquisition of a land use and cover map, digital elevation model, soil type map, and climate data (rainfall and evapotranspiration); (ii) calculation of surface runoff using the slope length and steepness factor, and runoff coefficients for the land use and cover types; (iii) calculation of water percolation based on the PUC method [43] (replacing the effective porosity of Costa's approach and representing the proposed methodological improvement) and adapted hydraulic conductivity values with fuzzy logic [55,56]; (iv) calculation of groundwater recharge in different points of the basin and a mean value for the whole basin, using a geographical information system; (v) validation of results by comparing the previously calculated mean groundwater recharge with the value estimated by the hydrograph recession curve analysis (Figure 6).

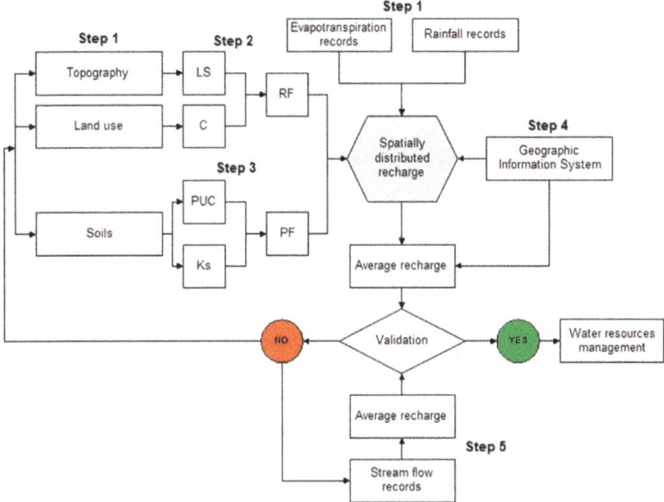

Figure 6. Flowchart for the groundwater recharge potential calculation and model validation. Adapted from Costa et al. (2019) [10]. Abbreviations: LS—slope length and steepness factor [57]; C—runoff coefficient (Table 7); RF—runoff factor (Equation (1)); PUC—Conservative Use Potential [43]; Ks—soils' hydraulic conductivity (Table 8); PF—water percolation factor (Equation (2)).

In the first step, the mean annual rainfall and evapotranspiration of municipalities near the study area were estimated using data from the Brazilian Institute for Meteorology (INMET) relative to the 2009–2018 period, which were obtained from meteorological stations in Arinos (MG), Januária (MG), Montes Claros (MG), Salinas (MG), Cariranha (BA), Espinosa (MG), Formoso (MG), Posses (GO), and Brasília (DF). Longer temporal series would be more adequate for estimating groundwater recharge. However, these records were not available. The flaws in the temporal series were resolved by using the regional weighting method, and the information was interpolated through inverse distance weighting (IDW) raised to the power of two [58]. This method was used because it raises the importance of closer stations in the interpolation. The data were spatialized and trimmed up to the study area limits.

In the second step, the land use and cover map and the topographical information from the digital elevation model (slope length and steepness factor—LS factor) were used to calculate the surface runoff factor, based on the method proposed by Böhner and Selige (2002) [59]. The runoff factor was calculated according to Equation (1), reproduced from Costa et al. 2019 [10].

$$RF = 1 - (C + LS_{FUZZY}) \tag{1}$$

where *RF* is the runoff factor (dimensionless); C is the runoff coefficient (dimensionless values adopted from Table 7); LS_{FUZZY} is the slope length and steepness factor estimated using the method of Desmet and Govers (1996) [57], but recast to the 0–1 range using a fuzzy logic algorithm (the steeper the slope the closer to 1).

Table 7. Runoff coefficients for each land use and cover class in the Pandeiros River basin. Adapted from the ASCE—American Society of Civil Engineers, 1969, and Costa et al., 2019 [10,60].

Land and Cover Class	Runoff Coefficient
Anthropized areas	0.50
Urban areas	0.85
Forest formation	0.10
Exposed soil and ground vegetation	0.60
Wetland regions ("Veredas")	0.09

In the third step, the water percolation was calculated considering the soil classes in the basin [53]. A systematic literature review was performed to set up hydraulic conductivity values for each soil class, at the second category level of Brazilian soil classification system (SiBCS) [61]. The hydraulic conductivity values were established according to Freire et al. (2003), Costa et al. (2019), Pedron (2011) and Amaral (2017) [10,55,56,62].

Table 8. Soil classes and respective Ks_{Fuzzy} values. The scores of groundwater recharge parameter were set up by the PUC method in the Pandeiros River basin [10,55,56,62].

Soil Class	PUC Score	Ks_{fuzzy}
Haplic cambisols	0.4	0.3
Melanic gleysols	0.1	0.01
Red–yellow latosols	0.7	1
Fluvic neosols	0.3	0.02
Litolic neosols	0.3	0.06
Quartzarenic neosols	0.3	0.61

Different from the method proposed by Costa et al. (2019) [10], the water percolation factor in this model was calculated using the groundwater recharge category of the PUC method, developed by Costa et al. (2017) [43]. The PUC is a method that allows for mapping areas of a basin based on their limitations and potentialities for conservationist land use, through the combined assessment and weighting of several environmental variables (soils, geology, and geomorphology) [43].

The PUC assigns values from 1 to 5 to the different classes of lithology, slope and soil in the watersheds from the state of Minas Gerais, Brazil. The analyses are focused on groundwater recharge, agricultural use potential and soil resistance to erosion in the catchments [43]. For the identification of lithologies, slopes and soil classes existing in Minas Gerais, the available official databases are used.

The attribution of grades for the different types of lithologies took into consideration their potential to provide nutrients (greater weight for rocks with higher absolute content of essential macro elements to plants) and their susceptibility to weathering processes (considering the main mineral constituents and scored according to their resistance to weathering, based on Goldish's stability [63]). Regarding the slope parameter, the same weights were attributed for groundwater recharge potential for agricultural use and resistance to erosion. For groundwater recharge, it was considered that the slope has a direct relationship with the flow velocity and the opportunity time for water infiltration. The higher the slope, the higher the water velocity and the shorter the time for water infiltration. In this context, the mountainous relief received a weight 1 and the flat relief a weight 5.

For the attribution of grades to different soils, the variables texture, drainage, effective depth and fertility were considered. For groundwater recharge, the attribute fertility was disregarded and the classes of soils characterized as favorable to water infiltration and percolation received greater weight.

The recharging potential for each soil type was obtained by the simple average of the values of texture, drainage and effective depth, normalized so that the final scale was in the range of 1 to 5.

In this work, only the soil parameter of groundwater recharge was used, which assigns the basin´s soil classes a score from 1 to 5. This parameter takes into account the effective depth, texture and drainage of each soil class at the first category level of SiBCS [61]. Thus, to the soil that presents considerable effective depth, satisfactory drainage and texture favorable to infiltration, a higher score regarding groundwater recharge is attributed.

However, in order to adapt the PUC method to the recharge model, a rescaling process was implemented, whereby the ratings from 1 to 5 were recast to the range 0 to 1. The values used in the model are represented in Table 8. The percolation factor was evaluated using Equation (2), reproduced from Costa et al. 2019 [10]:

$$PF = PUC \times Ks_{FUZZY} \quad (2)$$

where PF is the water percolation factor (dimensionless); Ks_{FUZZY} (dimensionless) is the soil's hydraulic conductivity fitted to the 0 to 1 range by the fuzzy logic algorithm (the higher the soil's hydraulic conductivity in the class the closer to 1); PUC are the scores of groundwater recharge set up by the PUC method but fitted to the 0 to 1 range.

In the fourth step, the groundwater recharge potential was calculated for each point in the basin, using Equation (3), according to Costa et al. 2019 [10]:

$$RPot = [(P - ET_R) \times RF \times PF] \times 10 \quad (3)$$

where $RPot$ is the groundwater recharge potential (m^3 ha^{-1} year^{-1}); P is the mean annual rainfall depth (mm year^{-1}); ET_R is the mean evapotranspiration (mm year^{-1}); RF is the surface runoff factor; and PF is the water percolation factor.

The results were validated in the fifth step through comparison of the calculated mean groundwater recharge with a homologous median value estimated by the hydrograph recession method based on the Maillet equation [64,65]. The hydrograph for the Pandeiros River was drawn from daily average stream flow data measured at the hydrometric station nº 4425000 located in the Pandeiros River mouth (Pandeiros River Dam), compiled from the Hidroweb portal [66]. Streamflow data from 2013 to 2018 were the most recent and more continuous in the historical series, showing no flaws or absent data. Thus, these data were used to calculate the Maillet equation (Equation (4)). To separate and analyze the recession curve and the recession days, the methodology proposed by Barnes (1939), Dewandel et al. (2003) and Kovacs et al. (2005) was used [64,65,67–69]:

$$Q_T = Q_0 \times e^{-\alpha t} \quad (4)$$

where Q_T is the flow at time t (m^3 s^{-1}); Q_0 is the flow at the beginning of a recession (m^3 s^{-1}); α is the coefficient of recession; t is the time (days); and e is the basis of Neperian logarithm (2.71828).

Thus, the coefficient of recession can be determined numerically, based on the logarithmic form of Equation (4), represented and rearranged in Equation (5):

$$\alpha = \frac{LogQ_0 - LogQ_t}{0.4343t} \quad (5)$$

Subsequently, the groundwater recharge volume was calculated using Equation (6):

$$V = \frac{Q_0 \times t'}{\alpha} \quad (6)$$

where V is the recharge volume (m^3); Q_0 is the flow at the beginning of recession (m^3 s^{-1}); t' is the converter of the t unit (days into seconds; 86,400); α is the coefficient of recession (dimensionless).

The constant of recession (α; Equation (5)) is dependent on the aquifer characteristics and therefore should not vary significantly from year to year. On a Q versus t plot (hydrograph), where the Q values are represented in logarithmic scale and the values are in linear scale, the baseflow within a hydrologic year (from the recharge period to the end of recession) should define as a straight line, the slope to which is related in terms of α. If t_{cycle} is the time of a log cycle for discharge, meaning the time for discharge to change from 1 to 10 m^3/s, from 10 to 100 m^3/s, and so forth, then $LogQ_0 - LogQ_t = 1$ and $\alpha = 1/(0.4343t_{cycle})$. This simplified representation of Equation (5) is frequently used in the calculation of α and will be adopted in the present study. Conversely, the values of Q_0 can vary in response to the annual variations of precipitation. In this case, a value of Q_0 should be calculated for each hydrologic year, while mean ± standard deviation values are derived therefrom.

3. Results

The interpolation of precipitation data from 2009 to 2018, obtained from climatologic stations close to the Pandeiros River basin, showed mean rainfall depths ranging from 904.7 to 1056.3 mm $year^{-1}$ (Figure 7).

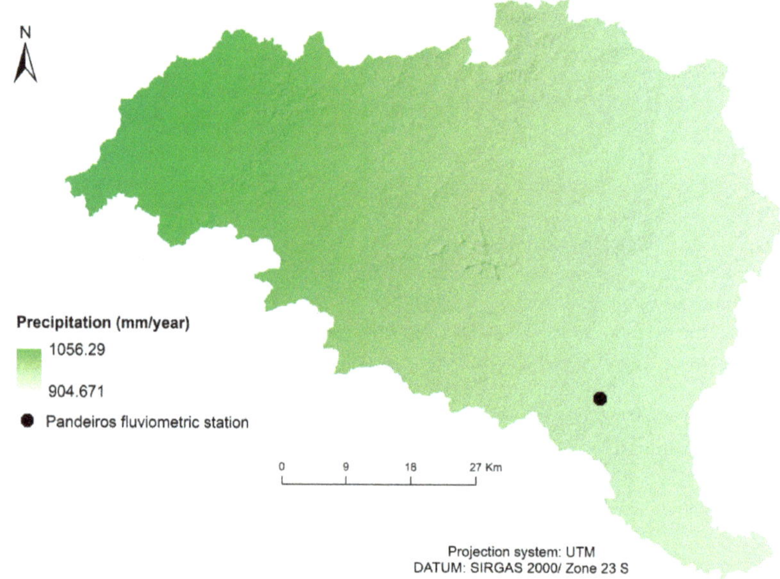

Figure 7. Spatial distribution of rainfall in the Pandeiros River basin.

The evapotranspiration ranged from 729.6 to 856.5 mm $year^{-1}$ (Figure 8). The spatial distribution of this variable was similar to rainfall; the lowest values were found in the southeast, east, and northeast regions, whereas the highest values were found in the northwest region of Pandeiros River basins.

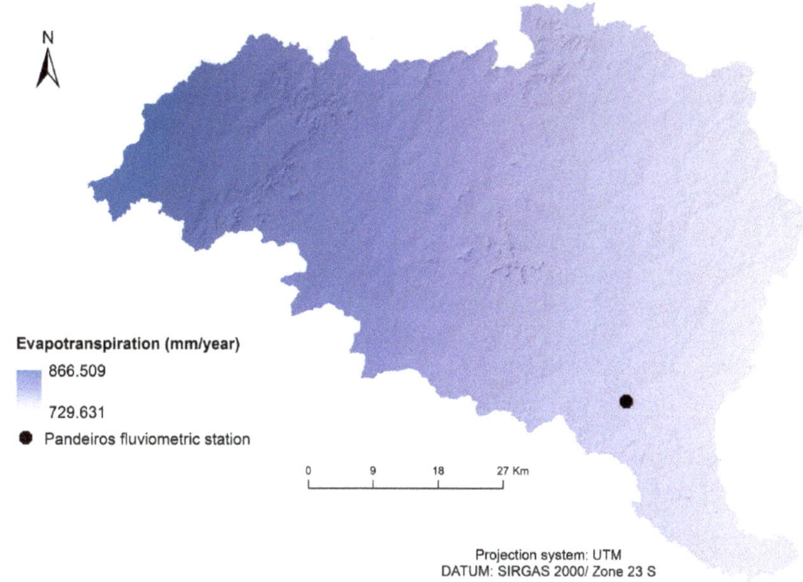

Figure 8. Spatial distribution of evapotranspiration in the Pandeiros River basin.

The evaluation of the runoff factor exposes a strong effect of runoff coefficients on the RF, as observed when the land use and cover map (Figure 5) is compared with the runoff map (Figure 9). The map of Figure 9 showed a lower groundwater recharge in urban areas, which have a higher runoff. In contrast, higher potential for groundwater recharge takes place in dense vegetation areas, indicating that the presence of vegetation decreases surface runoff.

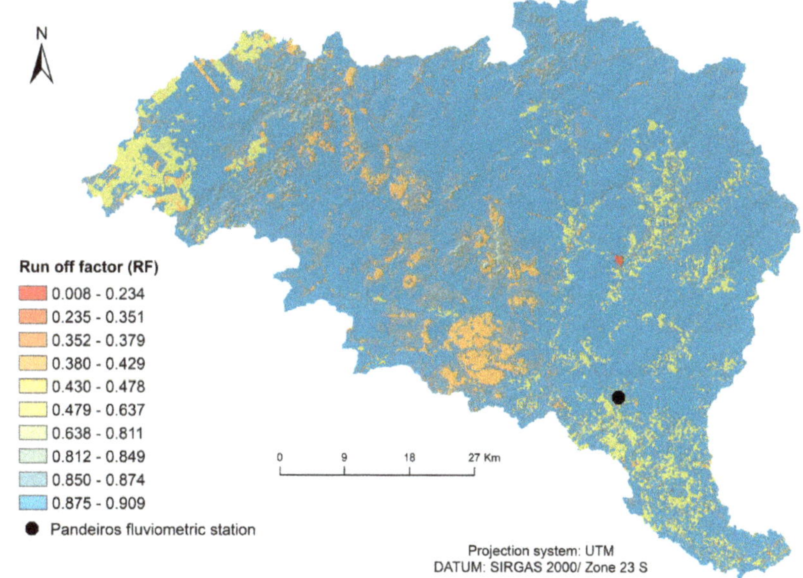

Figure 9. Spatial distribution of runoff factor in the Pandeiros River basin.

The water percolation factor ranged from 0.001 to 0.7 (dimensionless), expressing the combined variation of hydraulic conductivity values and PUC scores fitted to the 0–1 range (Figure 10 and Table 8). Areas with melanic gleysols and fluvic neosols presented the lowest water percolation factors, decreasing the groundwater recharge in these areas because of their low hydraulic conductivity.

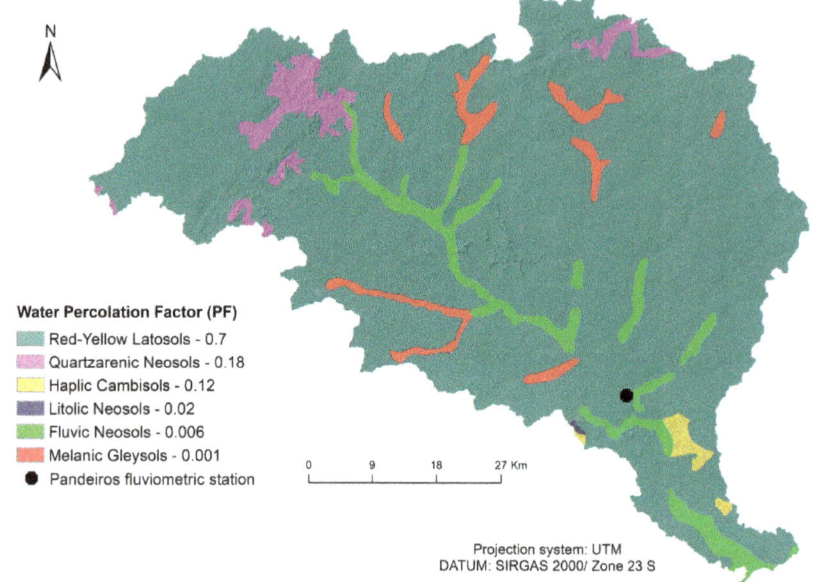

Figure 10. Spatial distribution of water percolation in the Pandeiros River basin.

The map of Figure 11 shows the groundwater recharge potential of the Pandeiros River basin ranging from 0 to 122.7 mm year^{-1}, with a mean value of 93.99 mm year^{-1}. The areas with higher groundwater recharge potential are located in regions with dense tree vegetation cover, areas with flat or slightly wavy relief, and areas with developed and structured soils, where the porosity and hydraulic conductivity allow water percolation to the water table. These areas are distributed throughout the basin and are found in the three municipalities that encompass the basin.

The areas with lowest groundwater recharge potential are represented in Figure 11 with a red color and are located in the urban area and in areas with the presence of melanic gleysols and fluvic neosols. The reduced potential is due to the soil sealing process occurring in urban areas, and to low hydraulic conductivities that decrease water percolation and consequently recharge in the aforementioned soil types.

The validation of PUC-based recharge estimates is depicted in Table 9. The mean groundwater recharge is 93.99 mm year^{-1} which is close to the median value obtained with the hydrograph recession method (87.2 mm year^{-1}). The difference between the two values is just 6.6 mm year^{-1} or 7.3%. The hydrograph used to calculate groundwater recharge based on base flows is shown in Figure 12. Hydrographs describe the succession of peaks representing the watershed response to a precipitation event, which are separated by baseflow segments that describe the aquifer response to drainage. Hydrograph recession curves are usually separated in the quick flow stage, which depicts the runoff and water infiltration towards the saturated zone, and the baseflow stage when only the saturated zone discharges [68]. The baseflow discharge is the most representative feature of an aquifer's global response because it is less influenced by the temporal and spatial variations on infiltration [69]. Generally, hydrographs are analyzed together with rainfall data. The peak of a hydrograph rising curve shows the highest values of stream flows, which take place in the months when rainfall values are the highest. In these periods the superficial runoff also reaches its highest values and it decreases during the recession time, marked by the red segments in Figure 12.

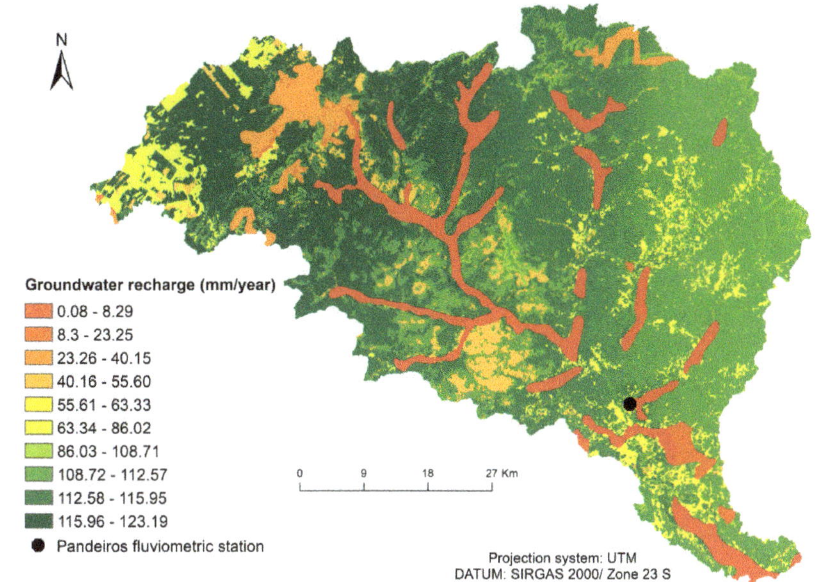

Figure 11. Spatial distribution of groundwater recharge potential in the Pandeiros River basin.

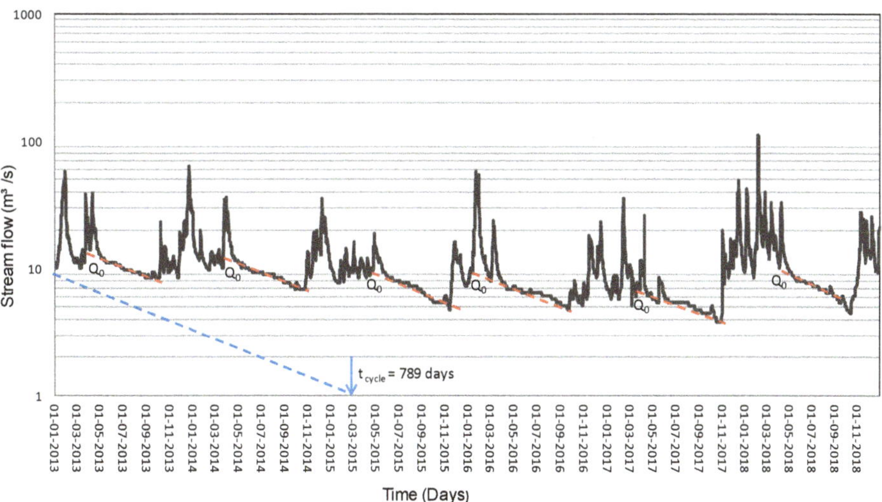

Figure 12. Hydrograph of Pandeiros River basin, state of Minas Gerais, Brazil, used to calculate the recession constant value.

Table 9. Groundwater recharge in the Pandeiros River basin, estimated by the spatially distributed PUC-based and hydrograph recession analysis methods.

Method	Mean Groundwater Recharge (mm year^{-1})	Average Difference (mm year^{-1})	Average Difference (%)
Spatialization	93.99 ± 36.92	4.8	5.2%
Hydrograph recession curve analysis	98.77 ± 20.32		

In the assessment of recharge using the method of hydrograph recession analysis, we looked for straight line segments corresponding to base lows. These segments allowed us to draw the corresponding fitting lines (dashed red lines). As would be expected, the lines are all virtually parallel because the line slope is solely dependent on the aquifer characteristics and dimension, which are invariant at the timescale of a few years (present case). For all years, the point in the graph where the fitting line intercepted the hydrograph at the upper flows (left edge point) was defined as the Q_0,t_0 point, i.e., the point where the recession period began. On average $Q_0 = 10.8 \pm 1.8$ m^3 s^{-1}. The coefficient of recession was estimated by the simplified version of Equation (5), $\alpha = 1/(0.4343t_{\text{cycle}})$. In the Pandeiros River basin, the estimated t_{cycle} is 789 days, and therefore α = 0,002918. Using Q_0 and α in Equation (6) results in the recharge value of V =98.8 ± 20.3 mm year^{-1}.

Figure 13 shows the spatialization of groundwater recharge within the municipalities that constitute the studied basin. Although the municipality of Januária encompasses a large area of the Pandeiros River basin, approximately 212,956.8 ha, this municipality presents a higher proportion of melanic gleysols, fluvic neosols, and litolic neosols, which are soils with the lowest water percolation factor (PF) assessed by the model. Moreover, it presents a considerable proportion of exposed soil, sparse Cerrado vegetation, and cultivated soils in the northwest region of the basin, which increases the runoff potential. For these reasons, this municipality presented a mean annual groundwater recharge of 90.30 mm year^{-1}.

The municipality of Cônego Marinho presented the highest mean annual groundwater recharge (105 mm year^{-1}). This municipality occupies a smaller area of the basin (approximately 26,083.86 ha) than the other municipalities, but it is in a region where soil and topography favor groundwater recharge.

Bonito de Minas presented mean annual groundwater recharge of 97.18 mm year^{-1}. The area of the basin within this municipality is approximately 156,987.95 ha. Despite the considerable presence of fluvic neosols and melanic gleysols in this region of the basin, the proportion is lower than that of Januária. The urban area of Bonito de Minas within the basin is small, representing 0.03% of the total area of the basin, which does not significantly affect the mean annual groundwater recharge.

The land use and cover map (Figure 5) and the groundwater recharge map (Figure 11) showed that the areas with forest cover presented the best recharge values. This type of land use and cover combined with flat or slightly wavy relief and areas overlaid with red–yellow latosols proved preferable for groundwater recharge.

Contrastingly, regions presenting such land use and cover as urban areas, exposed soil, ground vegetation, poorly structured soils and low hydraulic conductivity (fluvic neosols, litolic neosols, haplic cambisols, melanic gleysols), combined with steeper areas, presented lower recharges.

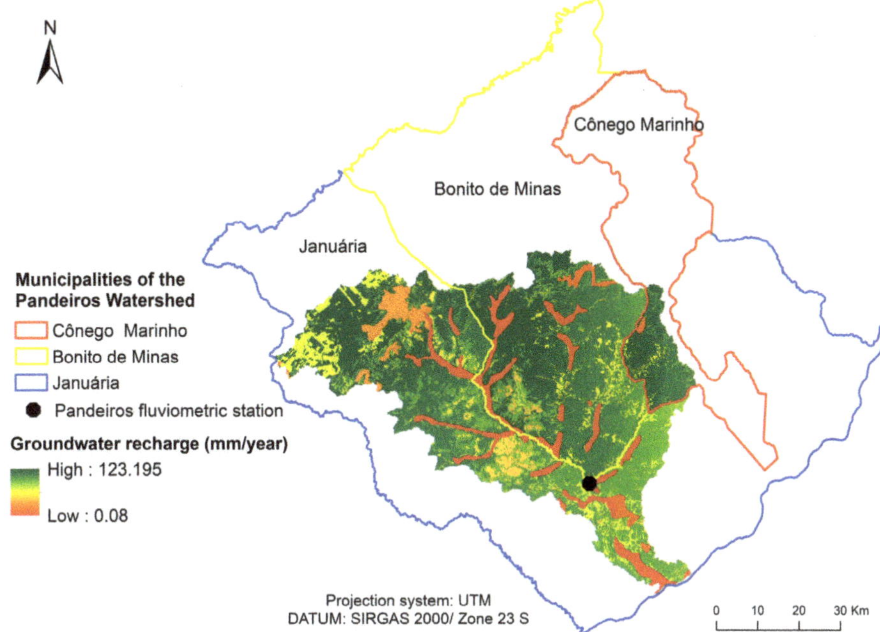

Figure 13. Spatial distribution of groundwater recharge potential in the municipalities of Pandeiros River basin.

4. Discussion

The rainfall depths (Figure 7) are consistent with Neves (2011), who evaluated a historical series (1931–1990) in the same region and reported a mean rainfall depth of 1050 mm year^{-1} [45]. However, the Brazilian Institute for Geography and Statistics (2014) considers values below 800 mm year^{-1} to define the semiarid region of Brazil [70], denoting that the rainfall depths of the study area were relatively high for a semiarid region.

Yang et al. (2016) showed that the rainfall depth can be a determinant for evapotranspiration and temperature values [71]. Therefore, practically all water that leaves the system (output) in regions with dry climate and low rainfall, such as the Brazilian semiarid regions, is due to evapotranspiration [72]. According to Loos, Gayler, and Priesack (2007), evapotranspiration is one of the most critical variables, with a high impact on water loss in similar regions [73].

The spatial distribution of runoff factor showed that cover plants are essential to maintaining the water cycle and to protect the soil against the impacts of raindrops. Moreover, the presence of vegetation increases soil porosity and permeability by the action of roots, thereby reducing runoff, and keeping soil moisture in the vicinity of organic colloids [74].

Soils under forest are characterized by expressive plant residue layers (litter fall) and by an A horizon rich in organic matter, which enables a higher soil aggregation, preserving its porosity [75]. Soils under forest usually present significant porosity, mainly macropores due to dead roots and animal holes, which are important to facilitate water infiltration and recharge. Therefore, water infiltration capacity is usually more expressive in areas with forest vegetation [75,76] than in pastureland or cropland, as found in the present work, resulting in a lower surface runoff.

Costa et al. (2019) [10] used a groundwater recharge model and reported a higher surface runoff in urban areas than in vegetated areas. Urban areas have drainage systems, street pavements, and infrastructures that hinder infiltration of rainwater into the soil, decreasing the recharge potential [10].

Comparing the first model employed by Costa et al. (2019) with the present application, it is interesting to see how including the groundwater recharge parameter from the PUC method improved the spatialization of groundwater recharge, mainly because the first model [10] used solely total porosity values for each soil class in the Jequitiba River basin [10]. This parameter is not the best to assess drainage and water percolation, because according to Silva et al. (2013), the micro and macro porosity better reflect the movement of water through the soil profile [77]. Moreover, the PUC method uses a management approach that was not taken into consideration in the PF in the first model. Another important difference from the first model to the present one is the separation until the second category level of the neosols class, which has shown more reliable spatialization and values for the different category levels of this soil class. The differences between the physical characteristics (structure, texture and effective depth) of fluvic neosols, litolic neosols and quartzarenic neosols) make it questionable to give the same values of hydraulic conductivity and porosity to this soil class.

Regarding the validation stage of the work, the higher the slope of the recession curve, the faster the depletion of the water table reserves and the higher the demand of this system to regulate the surface system; whereas the lower the slope, the lower the time of depletion of water resources. The surface and groundwater form a system with mutual contribution, thus, any change in one will affect the other in the short or long term.

The comparison of the methods (spatialization; recession curve analysis) showed a small difference (5.2%) because of their particularities, since each method has inherent and inevitably uncertainty levels [78]. The spatialization method probably has a higher uncertainty in groundwater recharge values because the number of parameters involved with the calculations is much larger [10]. However, the recession curve analysis provides only an average groundwater recharge estimate (mean) for the entire basin, whereas the spatialization method provides one estimate for each point in the basin.

The small difference between the results obtained with the spatially distributed PUC-based and hydrograph recession methods attributes to geology a limited role in the estimation of recharge in the Pandeiros River basin, and that this process is predominantly controlled by the infiltration capacity and profile depth of the soil. The effect of vadose zone thickness on the recharge was recently reported in China [79]. Information on the characteristics of a river basin and its potentialities and limitations are essential to an adequate management of water resources [10,80,81].

The potentialities found for the municipality of Cônego Marinho include the prevalence of red–yellow latosols, which are soils that favor groundwater recharge because of their structural characteristics, such as the occurrence of macropores that increase hydraulic conductivity [43]. Moreover, the region has few cultivated areas and has a predominance of typical Cerrado vegetation, which also favor groundwater recharge, according to the model. Thus, the mean annual groundwater recharge was higher in this municipality than in the other two in the basin.

The considerable overlay of fluvic neosols and melanic gleysols in Bonito de Minas is compensated for by a high proportion of typical Cerrado and dense Cerrado vegetation and by the presence of few cultivated areas, which decreased the runoff potential: Consequently, the mean annual water recharge in this municipality was higher than that in Januária. The presence of native vegetation favors the groundwater recharge, because the forest reduces the surface runoff, favors water percolation, and maintains the soil physical and mechanical stability, assisting in the storage of water and the supply of groundwater [10]. Several studies evaluated the effect of land use and cover and the benefits of forested areas to groundwater recharge [42,82–88].

Considering these results, the method used in the present work provides aid in the better management of river basins by considering their potentialities and limitations, not generalizing a mean value for the whole area evaluated. Therefore, preferred areas can be identified for recharge, thus directing public policies and conservationist actions for each area, according to their needs.

5. Conclusions

The groundwater recharge potential was higher in areas covered with forests, located in plains or slightly wavy relief areas, or overlaid with red–yellow latosol. These are the areas to protect in the watershed management plan if recharge is to be favored or restored. The soil classes and their structural attributes, as well as the land use and cover types were considered as key factors for groundwater recharge. The results showed that areas with higher groundwater recharge potential were concentrated in the municipality of Cônego Marinho, followed by Bonito de Minas, and Januária. Areas with a presence of melanic gleysols and fluvic neosols presented the worst responses in the model.

As made evident from the results, this study should be used as a tool for the management of water resources in the Pandeiros River basin, because the preferred recharge areas could be successfully identified. It is urgent therefore that public policies and conservationist actions are enforced in these areas to improve natural groundwater recharge, and hence increase the accessible water volume to the local population. The adjustment of irrigation methods, adoption of soil preservation practices to improve water infiltration, seasonal storage of surface water in areas of low recharge potential and the preservation of forest vegetation, are examples of feasible actions. Moreover, this work provided subsidies for further studies that seek methods for the spatialization of groundwater recharge potential in river basins with a key role assigned to management practices.

Author Contributions: Conceptualization, M.A.T., M.S.d.M. A.M.d.C., P.K.M., J.H.M.V.; methodology, M.A.T., M.S.d.M., A.M.d.C. and J.H.M.V.; validation, M.A.T and F.A.L.P.; resources, A.M.d.C.; writing—original draft preparation, M.A.T. and M.S.d.M.; writing—review and editing, M.A.T., M.S.d.M., A.M.d.C., P.K.M., F.A.L.P and L.F.S.F.; supervision, A.M.d.C.; funding acquisition, A.M.d.C. All authors have read and agreed to the published version of the manuscript.

Funding: This study was funded by the research project "Sustentabilidade da Bacia do Rio Pandeiros" (Sustainability of Pandeiros River Basin), sponsored by grant APQ-03773-14 from FAPEMIG – Fundação de Amparo à Pesquisa do Estado de Minas Gerais that included a research scholarship for the author Maíse Soares de Moura. The manuscript translation from Portuguese to English was financed by the CAPES – Coordenação de Aperfeiçoamento de Pessoal de Nível Superior. For the author integrated in the CITAB research centre, the research was financed by National Funds of FCT–Portuguese Foundation for Science and Technology, under the project UIDB/04033/2020. For the author integrated in the CQVR, the research was supported by National Funds of FCT–Portuguese Foundation for Science and Technology, under the projects UIDB/00616/2020 and UIDP/00616/2020.

Acknowledgments: This study was conducted within the work plan of research group "GEISS – Grupo de Estudos Integrado em Solos e Sustentabilidade", in its research line on "Gestão Sustentável de Recursos Hídricos e Segurança Hídrica". We wish to thank the support of the team of Soil and Environment Laboratory of the Federal University of Minas Gerais for the help in the analyses made in the study.

Conflicts of Interest: The authors declare no conflict of interest. The funders had no role in the design of the study; in the collection, analyses, or interpretation of data; in the writing of the manuscript or in the decision to publish the results.

Nomenclature

The list of mathematical symbols and their measurement units as used in the present article are listed below in alphabetical order:

A	recession coefficient;
C	runoff coefficient (dimensionless);
E	basis of Neperian logarithm (2.71828)
ET_R	mean evapotranspiration (mm year^{-1});
Ks_{FUZZY} (dimensionless)	soil hydraulic conductivity fitted to the 0 to 1 range by the fuzzy logic algorithm;
LS_{FUZZY}	slope length and steepness factor;
P	mean annual rainfall depth (mm year^{-1});
PF	water percolation factor (dimensionless);
PUC	Conservative Use Potential;
Q_T	stream flow discharge at time t (m^3 s^{-1});
Q_0	stream flow discharge at the beginning of a recession (m^3 s^{-1});
RF	runoff factor (dimensionless);
$RPot$	groundwater recharge potential (m^3 ha^{-1} year^{-1});
T	time (day);
t'	converter of t measurement unit (days into seconds; 86,400)

References

1. Grey, D.; Sadoff, C.W. Sink or Swim? Water security for growth and development. *Water Policy* **2007**, *9*, 545–571. [CrossRef]
2. Hanjra, M.A.; Qureshi, M.E. Global water crisis and future food security in an era of climate change. *Food Policy* **2010**, *35*, 365–377. [CrossRef]
3. Carrão, H.; Naumann, G.; Barbosa, P. Mapping global patterns of drought risk: An empirical framework based on sub-national estimates of hazard, exposure and vulnerability. *Glob. Environ. Chang.* **2016**, *39*, 108–124. [CrossRef]
4. Ahmadalipour, A.; Moradkhani, H.; Castelletti, A.; Magliocca, N. Future drought risk in Africa: Integrating vulnerability, climate change, and population growth. *Sci. Total Environ.* **2019**, *662*, 672–686. [CrossRef] [PubMed]
5. Blakeslee, D.; Fishman, R.; Srinivasan, V. Way down in the hole: Adaptation to long-term water loss in rural India. *Am. Econ. Rev.* **2020**, *110*, 200–224. [CrossRef]
6. Calicioglu, O.; Flammini, A.; Bracco, S.; Bellù, L.; Sims, R. The future challenges of food and agriculture: An integrated analysis of trends and solutions. *Sustainability* **2019**, *11*, 222. [CrossRef]
7. Scanlon, B.R.; Keese, K.E.; Flint, A.L.; Flint, L.E.; Gaye, C.B.; Edmunds, W.M.; Simmers, I. Global synthesis of groundwater recharge in semiarid and arid regions. *Hydrol. Process.* **2006**, *20*, 3335–3370. [CrossRef]
8. Bressers, H.; Kuks, S. Integrated Governance and Water Basin Management. In *Integrated Governance and Water Basin Management: Conditions for Regime Change and Sustainability*; Bressers, H., Kuks, S., Eds.; Springer: Heidelberg, Germany, 2004; pp. 247–265. ISBN 978-1-4020-2482-5.
9. Lerner, D.N. Groundwater recharge in urban areas. In *Hydrological Processes and Water Management in Urban Areas*; Massing, H., Packman, J., Zuidema, F.C., Eds.; International Association of Hydrological Sciences: Wallingford, Oxfordshire, England, 1990; pp. 59–65. ISBN 0-947571-82-5.
10. da Costa, A.M.; de Salis, H.H.C.; Viana, J.H.M.; Leal Pacheco, F.A. Groundwater Recharge Potential for Sustainable Water Use in Urban Areas of the Jequitiba River Basin, Brazil. *Sustainability* **2019**, *11*, 2955. [CrossRef]
11. Yang, J.L.; Zhang, G.L. Water infiltration in urban soils and its effects on the quantity and quality of runoff. *J. Soils Sediments* **2011**, *11*, 751–761. [CrossRef]
12. Caldas, A.M.; Pissarra, T.C.T.; Costa, R.C.A.; Neto, F.C.R.; Zanata, M.; Parahyba, R.d.B.V.; Fernandes, L.F.S.; Pacheco, F.A.L. Flood vulnerability, environmental land use conflicts, and conservation of soil and water: A study in the Batatais SP municipality, Brazil. *Water* **2018**, *10*, 1357. [CrossRef]
13. Nimmo, J.R.; Schmidt, K.M.; Perkins, K.S.; Stock, J.D. Rapid measurement of field-saturated hydraulic conductivity for areal characterization. *Vadose Zone J.* **2009**, *8*, 142–149. [CrossRef]
14. Dawoud, M.A.; Darwish, M.M.; El-Kady, M.M. GIS-based groundwater management model for Western Nile Delta. *Water Resour. Manag.* **2005**, *19*, 585–604. [CrossRef]
15. Saraf, A.K.; Choudhury, P.R.; Roy, B.; Sarma, B.; Vijay, S.; Choudhury, S. GIS based surface hydrological modelling in identification of groundwater recharge zones. *Int. J. Remote Sens.* **2004**, *25*, 5759–5770. [CrossRef]

16. CHOWDHURY, R. *Slope Analysis*, 1st ed.; Elsevier: Chatswood, Australia, 1978; ISBN 9780444601391.
17. YEH, J.T.C.; GELHAR, L.W.; WIERENGA, P.J. Observations of spatial variability of soil water pressure in a field soil. *Soil Sci.* **1986**, *142*, 7–12. [CrossRef]
18. Allison, G.B.; Gee, G.W.; Tyler, S.W. Vadose-zone techniques for estimating groundwater recharge in arid and semiarid regions. *Soil Sci. Soc. Am. J.* **1994**, *58*, 6–14. [CrossRef]
19. Delin, G.N.; Healy, R.W.; Lorenz, D.L.; Nimmo, J.R. Comparison of local- to regional-scale estimates of ground-water recharge in Minnesota, USA. *J. Hydrol.* **2007**, *334*, 231–249. [CrossRef]
20. Yeh, H.F.; Cheng, Y.S.; Lin, H.I.; Lee, C.H. Mapping groundwater recharge potential zone using a GIS approach in Hualian River, Taiwan. *Sustain. Environ. Res.* **2016**, *26*, 33–43. [CrossRef]
21. Singh, A.; Panda, S.N.; Kumar, K.S.; Sharma, C.S. Artificial groundwater recharge zones mapping using remote sensing and gis: A case study in Indian Punjab. *Environ. Manag.* **2013**, *52*, 61–71. [CrossRef]
22. Chowdhury, A.; Jha, M.K.; Chowdary, V.M. Delineation of groundwater recharge zones and identification of artificial recharge sites in West Medinipur district, West Bengal, using RS, GIS and MCDM techniques. *Environ. Earth Sci.* **2009**, *59*, 1209–1222. [CrossRef]
23. Jiménez-Martínez, J.; Skaggs, T.H.; van Genuchten, M.T.; Candela, L. A root zone modelling approach to estimating groundwater recharge from irrigated areas. *J. Hydrol.* **2009**, *367*, 138–149. [CrossRef]
24. Valle Junior, R.F.; Varandas, S.G.P.; Sanches Fernandes, L.F.; Pacheco, F.A.L. Environmental land use conflicts: A threat to soil conservation. *Land Use Policy* **2014**, *41*, 172–185. [CrossRef]
25. Spannenberg, J.; Atangana, A.; Vermeulen, P.D. New approach to groundwater recharge on a regional scale: Uncertainty analysis and application of fractional differentiation. *Arab. J. Geosci.* **2019**, *12*, 1–12. [CrossRef]
26. Cherkauer, D.S. Quantifying Ground Water Recharge at Multiple Scales Using PRMS and GIS. *Ground Water* **2004**, *42*, 97–110. [CrossRef] [PubMed]
27. Conrad, J.; Nel, J.; Wentzel, J. The challenges and implications of assessing groundwater recharge: A study—Northern Sandveld, Western Cape, South Africa. *Water SA* **2004**, *30*, 75–81. [CrossRef]
28. Pacheco, F.A.L. Regional groundwater flow in hard rocks. *Sci. Total Environ.* **2015**, *506–507*, 182–195. [CrossRef]
29. De Luca, D.A.; Lasagna, M.; Gisolo, A.; Morelli di Popolo e Ticineto, A.; Falco, M.; Cuzzi, C. Potential recharge areas of deep aquifers: An application to the Vercelli–Biella Plain (NW Italy). *Rend. Lincei* **2019**, *30*, 137–153. [CrossRef]
30. Blasch, K.W.; Bryson, J.R. Distinguishing sources of ground water recharge by using δ2H and δ18O. *Ground Water* **2007**, *45*, 294–308. [CrossRef]
31. Sukhija, B.S.; Reddy, D.V.; Nagabhushanam, P.; Hussain, S.; Giri, V.Y.; Patil, D.J. Environmental and injected tracers methodology to estimate direct precipitation recharge to a confined aquifer. *J. Hydrol.* **1996**, *177*, 77–97. [CrossRef]
32. Mohan, C.; Western, A.W.; Wei, Y.; Saft, M. Predicting groundwater recharge for varying land cover and climate conditions-a global meta-study. *Hydrol. Earth Syst. Sci.* **2018**, *22*, 2689–2703. [CrossRef]
33. Aguilera, H.; Murillo, J.M. The effect of possible climate change on natural groundwater recharge based on a simple model: A study of four karstic aquifers in SE Spain. *Environ. Geol.* **2009**, *57*, 963–974. [CrossRef]
34. Ali, R.; McFarlane, D.; Varma, S.; Dawes, W.; Emelyanova, I.; Hodgson, G. Potential climate change impacts on the water balance of regional unconfined aquifer systems in south-western Australia. *Hydrol. Earth Syst. Sci.* **2012**, *16*, 4581–4601. [CrossRef]
35. Herrera-Pantoja, M.; Hiscock, K.M. The effects of climate change on potential groundwater recharge in Great Britain. *Hydrol. Process.* **2008**, *22*, 73–86. [CrossRef]
36. Sanford, W. Recharge and groundwater models: An overview. *Hydrogeol. J.* **2002**, *10*, 110–120. [CrossRef]
37. Döll, P.; Fiedler, K. Global-scale modeling of groundwater recharge. *Hydrol. Earth Syst. Sci.* **2008**, *12*, 863–885. [CrossRef]
38. Pang, Z.; Huang, T.; Chen, Y. Diminished groundwater recharge and circulation relative to degrading riparian vegetation in the middle Tarim River, Xinjiang Uygur, Western China. *Hydrol. Process.* **2010**, *24*, 147–159. [CrossRef]
39. Oliveira, P.T.S.; Leite, M.B.; Mattos, T.; Nearing, M.A.; Scott, R.L.; de Oliveira Xavier, R.; da Silva Matos, D.M.; Wendland, E. Groundwater recharge decrease with increased vegetation density in the Brazilian cerrado. *Ecohydrology* **2017**, *10*, 1–8. [CrossRef]

40. Jackson, R.B.; Carpenter, S.R.; Dahm, C.N.; McKnight, D.M.; Naiman, R.J.; Postel, S.L.; Running, S.W. Water in a changing world. *Ecol. Appl.* **2001**, *11*, 1027–1045. [CrossRef]
41. Kim, J.H.; Jackson, R.B. A Global Analysis of Groundwater Recharge for Vegetation, Climate, and Soils. *Vadose Zone J.* **2012**, *11*. [CrossRef]
42. Scanlon, B.R.; Reedy, R.C.; Stonestrom, D.A.; Prudic, D.E.; Dennehy, K.F. Impact of land use and land cover change on groundwater recharge and quality in the southwestern US. *Glob. Chang. Biol.* **2005**, *11*, 1577–1593. [CrossRef]
43. da Costa, A.M.; Viana, J.H.M.; Evangelista, L.P.; De Carvalho, D.C.; Pedras, K.C.; Horta, I.D.M.; Salis, H.H.d.C.; Pereira, M.P.R.; Sampaio, J.L.D. Ponderação de variáveis ambientais para a determinação do Potencial de Uso Conservacionista para o Estado de Minas Gerais. *Geografias* **2017**, *14*, 118–133.
44. INMET—Instituto Nacional de Meteorologia (Brazilian Institute for Meteorology). The Database on Climatic Data. Available online: http://www.inmet.gov.br/portal/ (accessed on 1 January 2020).
45. Neves, W.V. *Avaliação da Vazão em Bacias Hidrográficas com Veredas, em Diferentes Estádios de Conservação, na APA do Rio Pandeiros—MG*; Universidade Federal de Minas Gerais: Belo Horizonte, Brazil, 2011.
46. Alvares, C.A.; Stape, J.L.; Sentelhas, P.C.; Gonçalves, J.L.D.M.; Sparovek, G. Köppen's climate classification map for Brazil. *Meteorol. Z.* **2013**, *22*, 711–728. [CrossRef]
47. Silva, C.G. *Da Caracterização Física e Ambiental da Bacia Hidrográfica do rio Pandeiros- MG em Eventos de El Niño-Oscilação Sul*; Universidade Federal Rural do Rio de Janeiro: Seropédica, Brazil, 2018.
48. IBGE—Instituto Brasileiro de Geografia e Estatística (Brazilian Institute for Geography and Statistics). The database on demography. Available online: https://www.ibge.gov.br/ (accessed on 12 December 2019).
49. Coutinho, L.M. O conceito de bioma. *Acta Bot. Brasilica* **2006**, *20*, 13–23. [CrossRef]
50. Nunes, Y.R.F.; Azevedo, I.F.P.; Neves, W.V.; Veloso, M.d.D.M.; Souza, R.d.A.; Fernandes, G.W. Pandeiros: O Pantanal Mineiro. *MG Biota* **2009**, *2*, 4–17.
51. Rezende, É.A. *O Papel da Dinâmica Espaço-Temporal da rede Hidrográfica na Evolução Geomorfológica da Alta/Média Bacia do Rio Grande, Sudeste Brasileiro*; Universidade Federal de Ouro Preto: Ouro Preto, Brazil, 2018.
52. EMBRAPA. *Sumula da X Reunião Técnica de Levantamento de Solos*; Palmieri, F., Rodrigues, T.E., dos Santos, H.G., Motchi, E.P., de Freitas, F.G., Matos, M.D.M., Eds.; Empresa Brasileira de Pesquisa Agropecuaria: Brasília, Brazil, 1979.
53. UFV—Universidade Federal de Viçosa (Federal University of Viçosa). The database on Minas Gerais Soil Map. Available online: https://www.ufv.br/ (accessed on 20 December 2019).
54. Eiten, G. The cerrado vegetation of Brazil. *Bot. Rev.* **1972**, *38*, 201–341. [CrossRef]
55. Pedron, F.d.A.; Fink, J.R.; Rodrigues, M.F.; Azevedo, A.C. de Condutividade e retenção de água em Neossolos e saprolitos derivados de arenito. *Rev. Bras. Cienc. Solo* **2011**, *35*, 1253–1262.
56. Amaral, J.R. *Do Caracterização Físico-Hídrica dos Solos da Bacia do Córrego Marinheiro, Sete Lagoas—MG*; Universidade Federal de Minas Gerais: Belo Horizonte, Brazil, 2017.
57. Desmet, P.J.J.; Govers, G. A GIS procedure for automatically calculating the USLE LS factor on topographically complex landscape units. *J. Soil Water Conserv.* **1996**, *51*, 427–433.
58. de Oliveira, L.F.C.; Fioreze, A.P.; Medeiros, A.M.M.; Silva, M.A.S. Comparação de metodologias de preenchimento de falhas de séries históricas de precipitação pluvial anual. *Rev. Bras. Eng. Agrícola Ambient.* **2010**, *14*, 1186–1192. [CrossRef]
59. Böhner, J.; Selige, T. Spatial prediction of soil attributes using terrain analysis and climate regionalisation. *Gött. Geogr. Abh.* **2002**, *115*, 13–28.
60. Joint Committee of the ASCE and the Water Pollution Control Federation. *Design and Construction of Sanitary and Storm Sewers*; American Society of Civil Engineers: New York, NY, USA, 1969.
61. dos Santos, H.G.; Jacomine, P.K.T.; dos Anjos, L.H.C.; de Oliveira, V.A.; Lumbreras, J.F.; Coelho, M.R.; de Almeida, J.A.; de Araujo Filho, J.C.; de Oliveira, J.B.; Cunha, T.J.F. *Sistema Brasileiro de Classificação de Solos, 5 ed.*; Embrapa Solos: Rio de Janeiro, Brazil, 2018; ISBN 85-85864-19-2.
62. Freire, M.B.G.d.S.; Ruiz, H.A.; Ribeiro, M.R.; Ferreira, P.A.; Alvarez, V.V.H.; Freire, F.J. Condutividade hidráulica de solos de Pernambuco em resposta à condutividade elétrica e RAS da água de irrigação. *Rev. Bras. Eng. Agríc. Ambient.* **2003**, *7*, 45–52. [CrossRef]
63. Goldich, S.S. A Study in Rock-Weathering. *J. Geol.* **1938**, *46*, 17–58. [CrossRef]
64. Pacheco, F.A.L.; Van der Weijden, C.H. Integrating topography, hydrology and rock structure in weathering rate models of spring watersheds. *J. Hydrol.* **2012**, *428–429*, 32–50. [CrossRef]

65. Santos, R.M.B.; Sanches Fernandes, L.F.; Moura, J.P.; Pereira, M.G.; Pacheco, F.A.L. The impact of climate change, human interference, scale and modeling uncertainties on the estimation of aquifer properties and river flow components. *J. Hydrol.* **2014**, *519*, 1297–1314. [CrossRef]
66. ANA—Agência Nacional de Águas (Brazilian Agency for Water). The data base on hydrologic information and stream flow records. Available online: https://capacitacao.ana.gov.br/ (accessed on 2 December 2019).
67. Barnes, B.S. The structure of discharge-recession curves. *Eos Trans. Am. Geophys. Union* **1939**, *20*, 721–725. [CrossRef]
68. Dewandel, B.; Lachassagne, P.; Bakalowicz, M.; Weng, P.; Al-Malki, A. Evaluation of aquifer thickness by analysing recession hydrographs. Application to the Oman ophiolite hard-rock aquifer. *J. Hydrol.* **2003**, *274*, 248–269. [CrossRef]
69. Kovács, A.; Perrochet, P.; Király, L.; Jeannin, P.Y. A quantitative method for the characterisation of karst aquifers based on spring hydrograph analysis. *J. Hydrol.* **2005**, *303*, 152–164. [CrossRef]
70. IBGE—Instituto Brasileiro de Geografia e Estatística (Brazilian Institute for Geography and Statistics). The database on the "Semiárido Brasileiro" (Brazilian semiarid region). Available online: https://www.ibge.gov.br/geociencias/cartas-e-mapas/mapas-regionais/15974-semiarido-brasileiro.html?=&t=sobre (accessed on 5 January 2020).
71. Yang, Z.; Zhang, Q.; Hao, X. Evapotranspiration Trend and Its Relationship with Precipitation over the Loess Plateau during the Last Three Decades. *Adv. Meteorol.* **2016**, 1–10. [CrossRef]
72. Saxton, K.E.; Mc Guinness, J.L. Evapotranspiration. In *Hydrologic Modeling of Small Watersheds*; Haan, C.T., Johnson, H.P., Brakenstek, D.L., Eds.; American Society of Agricultural Engineers: St. Joseph, Michigan, EUA, 1982; pp. 229–258.
73. Loos, C.; Gayler, S.; Priesack, E. Assessment of water balance simulations for large-scale weighing lysimeters. *J. Hydrol.* **2007**, *335*, 259–270. [CrossRef]
74. Hartanto, H.; Prabhu, R.; Widayat, A.S.E.; Asdak, C. Factors affecting runoff and soil erosion: Plot-level soil loss monitoring for assessing sustainability of forest management. *For. Ecol. Manag.* **2003**, *180*, 361–374. [CrossRef]
75. Cheng, J.D.; Lin, L.L.; Lu, H.S. Influences of forests on water flows from headwater watersheds in Taiwan. *For. Ecol. Manag.* **2002**, *165*, 11–28. [CrossRef]
76. Best, A.; Zhang, L.; Mc Mahon, T.; Western, A.; Vertessy, R. *A Critical Review of Paired Catchment Studies with Reference to Seasonal Flows and Climatic Variability*; Murray-Darling Basin Commission: Canberra, Australia, 2003; ISBN 1-876-830-57-3.
77. da Silva, A.J.P.; Coelho, E.F. Estimation of water percolation by different methods using TDR. *Rev. Bras. Ciência Solo* **2014**, *38*, 73–81. [CrossRef]
78. Scanlon, B.R.; Healy, R.W.; Cook, P.G. Choosing appropriate techniques for quantifying groundwater recharge. *Hydrogeol. J.* **2002**, *10*, 18–39. [CrossRef]
79. Cao, G.; Scanlon, B.R.; Han, D.; Zheng, C. Impacts of thickening unsaturated zone on groundwater recharge in the North China Plain. *J. Hydrol.* **2016**, *537*, 260–270. [CrossRef]
80. Dessu, S.B.; Melesse, A.M.; Bhat, M.G.; McClain, M.E. Assessment of water resources availability and demand in the Mara River Basin. *Catena* **2014**, *115*, 104–114. [CrossRef]
81. McMahon, P.B.; Plummer, L.N.; Böhlke, J.K.; Shapiro, S.D.; Hinkle, S.R. A comparison of recharge rates in aquifers of the United States based on groundwater-age data. *Hydrogeol. J.* **2011**, *19*, 779–800. [CrossRef]
82. Brauman, K.A.; Freyberg, D.L.; Daily, G.C. Land cover effects on groundwater recharge in the tropics: Ecohydrologic mechanisms. *Ecohydrology* **2012**, *5*, 435–444. [CrossRef]
83. Ilstedt, U.; Bargués Tobella, A.; Bazié, H.R.; Bayala, J.; Verbeeten, E.; Nyberg, G.; Sanou, J.; Benegas, L.; Murdiyarso, D.; Laudon, H.; et al. Intermediate tree cover can maximize groundwater recharge in the seasonally dry tropics. *Sci. Rep.* **2016**, *6*, 21930. [CrossRef]
84. Zomlot, Z.; Verbeiren, B.; Huysmans, M.; Batelaan, O. Spatial distribution of groundwater recharge and base flow: Assessment of controlling factors. *J. Hydrol. Reg. Stud.* **2015**, *4*, 349–368. [CrossRef]
85. Maréchal, J.C.; Varma, M.R.R.; Riotte, J.; Vouillamoz, J.M.; Kumar, M.S.M.; Ruiz, L.; Sekhar, M.; Braun, J.J. Indirect and direct recharges in a tropical forested watershed: Mule Hole, India. *J. Hydrol.* **2009**, *364*, 272–284. [CrossRef]
86. Le Maitre, D.C.; Scott, D.F.; Colvin, C. A review of information on interactions between vegetation and groundwater. *Water SA* **1999**, *25*, 137–152.

87. Zhang, H.; Hiscock, K.M. Modelling the impact of forest cover on groundwater resources: A case study of the Sherwood Sandstone aquifer in the East Midlands, UK. *J. Hydrol.* **2010**, *392*, 136–149. [CrossRef]

88. Allen, A.; Chapman, D. Impacts of afforestation on groundwater resources and quality. *Hydrogeol. J.* **2001**, *9*, 390–400. [CrossRef]

Article

The Assessment of Hydrological Availability and the Payment for Ecosystem Services: A Pilot Study in a Brazilian Headwater Catchment

Mariana Bárbara Lopes Simedo [1,6], Teresa Cristina Tarlé Pissarra [1,6], Antonio Lucio Mello Martins [2], Maria Conceição Lopes [2,6], Renata Cristina Araújo Costa [6], Marcelo Zanata [3,6], Fernando António Leal Pacheco [4,6,*] and Luís Filipe Sanches Fernandes [5,6]

1 Faculdade de Ciências Agrárias e Veterinárias, Universidade Estadual Paulista (UNESP), Via de Acesso Prof. Paulo Donato Castellane, s/n, Jaboticabal 14884-900, Brazil; mariana_blopes@hotmail.com (M.B.L.S.); teresa.pissarra@unesp.br (T.C.T.P.)

2 Polo Regional Centro Norte, Departamento de Descentralização do Desenvolvimento—APTA, Secretaria de Agricultura e Abastecimento—SAA, Rodovia Washington Luis, Km 371, s/n, Pindorama 15830-000, Brazil; lmartins@apta.sp.gov.br (A.L.M.M.); conceicao@apta.sp.gov.br (M.C.L.)

3 Instituto Florestal do Estado de São Paulo, Divisão de Florestas e Estações Experimentais, Rodovia Cândido Portinari, km 347, Horto Florestal, Batatais 14300-000, Brazil; Marcel_zanata@hotmail.com

4 CQVR—Centro de Química de Vila Real, Universidade de Trás-os-Montes e Alto Douro, Ap. 1013, 5001-801 Vila Real, Portugal

5 CITAB—Centro de Investigação e Tecnologias Agroambientais e Biológicas, Universidade de Trás-os-Montes e Alto Douro, Ap. 1013, 5001-801 Vila Real, Portugal; lfilipe@utad.pt

6 POLUS—Grupo de Política de Uso do Solo, Universidade Estadual Paulista (UNESP), Via de Acesso Prof. Paulo Donato Castellane, s/n, Jaboticabal 14884-900, Brazil; renata.criscosta@hotmail.com

* Correspondence: fpacheco@utad.pt

Received: 24 August 2020; Accepted: 27 September 2020; Published: 29 September 2020

Abstract: The assessment of water availability in river basins is at the top of the water security agenda. Historically, the assessment of stream flow discharge in Brazilian watersheds was relevant for dam dimensioning, flood control projects and irrigation systems. Nowadays, it plays an important role in the creation of sustainable management plans at the catchment scale aimed to help in establishing legal policies on water resources management and water security laws, namely, those related to the payment for environmental services related to clean water production. Headwater catchments are preferential targets of these policies and laws for their water quality. The general objective of this study was to evaluate water availability in first-order sub-basins of a Brazilian headwater catchment. The specific objectives were: (1) to assess the stream flow discharge of first-order headwater sub-basins and rank them accordingly; (2) to analyze the feasibility of payment for environmental services related to water production in these sub-basins. The discharge flow measurements were conducted during five years (2012 to 2016), in headwaters in a watershed on the São Domingos River at the Turvo/Grande Watershed, represented as the 4th-largest hydrographic unit for water resources management—UGRHI-15 in São Paulo State, Brazil. A doppler velocity technology was used to remotely measure open-channel flow and to collect the data. The discharge values were obtained on periodic measurements, at the beginning of each month. The results were subject to descriptive statistics that analyzed the temporal and spatial data related to sub-basins morphometric characteristics. The discharge flows showed space–time variations in magnitude between studied headwater sub-basins on water availability, assessed based on average net discharges. The set of ecological processes supported by forests are fundamental in controlling and recharging aquifers and preserving the volume of water in headwater in each sub-basin. The upstream inflows influence downstream sub-basins. To avoid scarcity, the headwater rivers located in the upstream sub-basins must not consider basin area as a single and homogeneous unit, because that may be the source of water conflicts. Understanding this relationship in response to conservationist practices installed uphill influenced by anthropic actions is crucial for water security assessment. The headwaters should be considered a great potential for ecosystem services, with respect to the "provider-receiver" principle, in the context of payments for environmental services (PES).

Keywords: flow; water discharge ecosystem services; payments for environmental services; land use; riparian forest

---

1. Introduction

The study of water availability in watersheds is fundamental for the demonstration of water potential and hydrological behavior of a region. It also helps with increasing the capacity of a population to safeguard access to adequate quantity and acceptable quality of water to sustain well-being and the environment. Surface and groundwater reserves are considered strategic and essential components of ecosystems and are fundamental for economic, social, and sustainable development [1,2]. The availability, quality, management, and governance of water resources are currently at the center of technical and scientific discussions [3–6].

Although most of the planet Earth's surface is occupied by water, 97.5% of available water is salty, and only 2.5% is fresh water. From the percentage of fresh water, 68.9% is concentrated in glaciers, polar ice caps, or mountainous regions, 29.9% in groundwater, 0.9% in other reservoirs, and only 0.3% make up the portion of surface fresh water present in rivers and lakes [7]. Brazil has 13% of the world volume of fresh water available, with a higher concentration in the Amazon region, where there is less population and less demand, according to the National Water Agency (ANA) [8]. The State of São Paulo, which is the most populous, has 1.6% of Brazilian fresh water [9]. The multiple uses of water resources are diversified—public supply, food production, hydroelectricity generation, navigation and industrial development, and their intensity is related to social, agricultural, and industrial development [1].

In the agriculture sector, approximately 70% of total fresh water is used in production activities. This consumption tends to increase with population growth and demand for food [10,11]. Thus, the need to obtain information and metrics that contribute to the establishment of planning measures and future management policies for water security is evident. Water security involves ensuring water quantity and quality acceptable for livelihood. It is related to the development of solutions to manage and mitigate impacts of scarcity and possible risks regarding environmental conditions and climate uncertainties, in order to guarantee human well-being, socioeconomic development and preservation of ecosystems [12–17].

The management of water resources in an integrated planning policy is increasingly needed to ensure water security in the future [3,15,16,18]. Studies showed that more than half of the global population has experienced severe water shortages for at least one month in a year, and climate change is affecting the reliability of supplies and infrastructures available in many regions [15,16,19]. According to scenario-based studies from the Fifth Assessment of the Intergovernmental Panel on Climate Change (IPCC), and data modeled on WaterGAP3, it is estimated that by 2050 urban water demand will increase by 80% [20].

Brazil experiences annual imbalances between water supply and demand. A remarkable episode was in 2013–2014, in the southeastern region, in which drought affected approximately 80 million people in Minas Gerais, São Paulo, and Rio de Janeiro States [18,21–23]. There were considerable problems for public water supply, food production, energy and navigation, causing the loss of 5000 jobs and millions of tons of non-transported materials [18]. Water crises generate significant socioeconomic and environmental impacts. Thus, the development of strategies for adapting to change that represent improvements and efficiency in future availability of water in hydrographic basins is welcome.

In water security research, the working territorial unit is a watershed. That is the unit where the land use and land cover are best managed [24–28]. Research studies on water availability [29–35] and flow distribution [36–42] within this unit contribute to manage water scarcity or stress, understand availability and demand, establish grants, identify environmental impacts, and even implement public policies [43–48].

The information on flow (discharge) metrics of Brazilian drainage networks is relevant for preparing policies, projects for new dams and flood control, as well as creating sustainable management plans within the scope of land use and land cover management in the hydrographic basin. The basin is a place where several interactions between the elements of an ecosystem occur, and thus it can be considered that this space has an ecosystem function considering water and nutrient cycle, in regulation of gases, in transfer of energy, and in the water infiltration/discharge relationship. The ecosystem services generated by these functions trigger a series of benefits, directly or indirectly, that humans can be appropriate.

A single ecosystem service can be a product of two or more functions, and a single function can generate more than one ecosystem service. Discharge flow controlling management practices can enhance environmental services or minimize the impact of human activities in a territory, or even promote economic incentives aimed at conserving ecosystems and increasing these services, which is of great value [49–51]. Ecosystem services promoted by maintenance and conservation of water resources have extremely high economic and social value. Valuation of those services and their diversity have gained scientific and economic attention, and environmental service benefits and control functioning of ecosystems must provide human well-being [52–55].

The Payment for Environmental Services (PES) instrument is considered an efficient program, as it rewards those who produce or maintain environmental services and encourages those who would not promote these services in the absence of monetary stimulus [53,55,56]. The monetary compensation and reduction of tax charges can be applied, or even, use of public resources from municipal and state funds, charging water users, environmental compensation, and carbon credit [54,57,58].

In Brazil, PES initiatives are centered on projects related to water and watershed, carbon storage programs, biodiversity, and landscape protection. Some pioneering and developing examples in Brazil are: Water Producer Program from National Water Agency–ANA (https://www.ana.gov.br/programas-e-projetos/programa-produtor-de-agua); Atlantic Forest Connection Project, financed by different development agencies (https://conexaomataatlantica.mctic.gov.br/cma/o-projeto/o-que-e); Conservative Water Project of Extrema, State of Minas Gerais (https://www.extrema.mg.gov.br); Ecocredit Program in the Montes Claros State of Minas Gerais (https://portal.montesclaros.mg.gov.br).

The assessment of hydrological availability in the Brazilian watershed on headwater sub-basins should therefore be recognized and promoted as a primary strategy to support the continued provision of ecosystem services in watersheds. The potential for ecosystem services provision, considering the principle of "provider-receiver", to establish PES must be investigated in the sequel. The costs to protect and manage those areas are substantial, and PES programs are a promising strategy. The landowner will receive a financial support as compensation to guarantee the provision of water ecosystem service to a willing buyer or beneficiary (e.g., Water Supplies Companies).

The general objective of this study was to evaluate water availability on sub-basins of first order and their contribution to water security assessment on Brazilian watersheds. The specific objectives were: (1) to assess the hydrological discharge of headwater sub-basins; (2) comprehend which first-order sub-basins are higher on water flow discharge; and (3) analyze the feasibility of payment for environmental services for the water-producing sub-basins.

2. Study Area

The experimental study area was the Olaria Stream Watershed, a tributary of the São Domingos river. This river contributes with flow to the Turvo/Grande watershed, the 4th-largest water resources management unit—UGRHI-15 in the State of São Paulo, Brazil (Figure 1), according to the Hydrographic Basins Committee of the Turvo and Grande Rivers—CBHTG (http://www.comitetg.sp.gov.br). The Olaria Stream Watershed covers 9.45 km^2 and is located between latitudes 21°05′47″ S and

21°19′35″ S and longitudes 49°03′02″ W and 48°42′52″ W Gr., expressed in UTM—Universal Transverse Mercator Projection System, Zone 22K, with altitudes varying from 507 to 616 m.

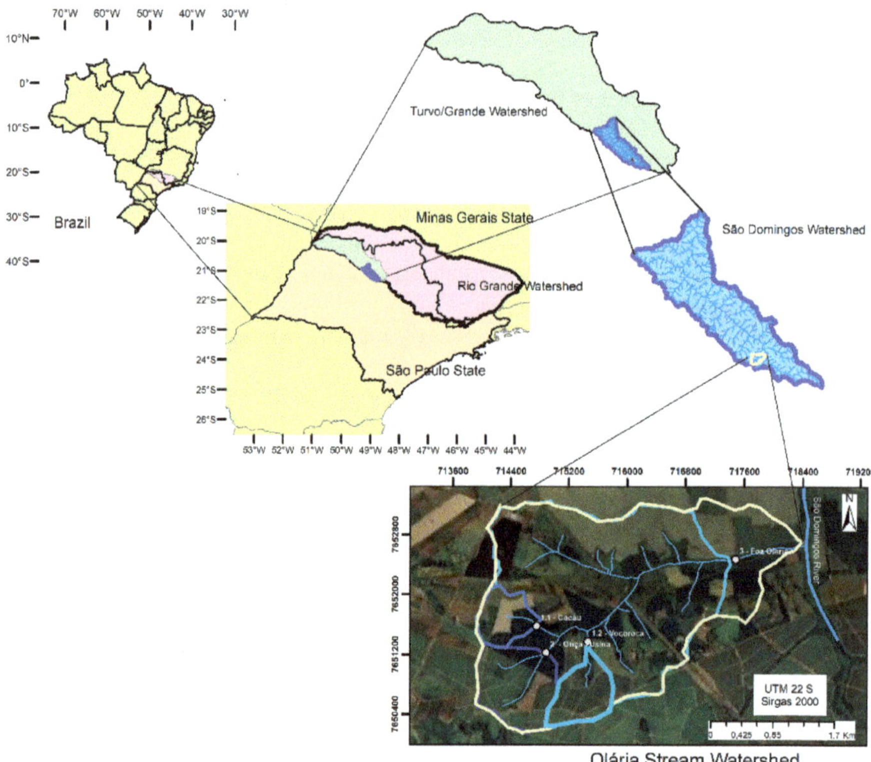

Figure 1. Olaria Stream Watershed, São Domingos River watershed, Turvo/Grande watershed, UGRHI-15, São Paulo State, Brazil.

The river flow supplies water to the municipalities of Catanduva, Catiguá, Tabapuã, and Uchoa, with a total of 169,232 inhabitants, according to data from the Brazilian Institute of Geography and Statistics—IBGE (https://cidades.ibge.gov.br). The area is socioeconomic-important for regional development, namely, the agricultural and industrial productive system. The sub-basin areas belong to Polo Regional Centro Norte, São Paulo Agribusiness Technology Agency—APTA (https://www.apta.sp.gov.br), a Department of the State of Agriculture and Supply Secretariat, São Paulo, Brazil. This unit carries out agricultural research and experimentation with annual and perennial crops, in order to meet technological demand on agribusiness production chains.

The climate is classified as Cwa, defined as a subtropical dry winter (with temperatures below 18 °C) and hot summer (with temperatures above 22 °C), with an average annual air temperature of 20–22 °C [59]. The local water balance (Figure 2), according to 30 year data (1961 to 1990) from the Brazilian Agricultural Research Corporation—EMBRAPA [60], is characterized by: average temperature of 22.8 °C; total annual precipitation of 1388 mm; average annual precipitation of 116 mm; potential evapotranspiration of 1134 mm; soil water storage of 788 mm; real evapotranspiration of 1054 mm; water deficiency of 80 mm; and water surplus of 334 mm.

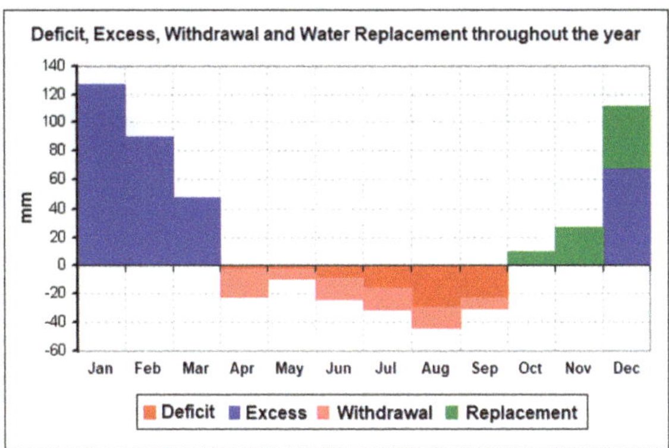

Figure 2. Water balance in the APTA-Pindorama, State of São Paulo, Brazil. Adapted from Brazil's climate database, available at the Brazilian Agricultural Research Corporation [60].

The watershed soil was classified as Red-Yellow Argisols, sandy/medium texture, with wavy and smooth wavy relief [61]. The predominant natural vegetation land cover was classified as semi-deciduous seasonal tropical forest, Atlantic Forest biome [62]. Land use was mainly distributed by urban areas, industries, agro-industries and agriculture, with a predominance of sugarcane and crop production systems such as citrus, rubber, grains, livestock, pasture, and eucalyptus [63]. The sub-basins uphill have an intense agricultural crop production and areas of native forests, considered as a Permanent Preservation Area (PPA) by Environmental Brazilian Laws.

3. Materials and Methods

Flow measurements in natural watercourses were carried out in order to determine the value of the surface runoff of a basin, its temporal variability, and the characteristics of the runoff. The activities of monitoring the amount of water in the Córrego da Olaria watershed in Pindorama-SP were initiated to understand the water in the hydrological processes and water production capacity of the sub-basins, obtaining stream flow data and the elaboration of quantitative analysis of flows, which will be indispensable for future actions and policies on land and water uses.

3.1. Discharge Flow Monitoring in Sub-Basins

The volume of water resources (e.g., discharge, flow) monitored in the Olaria stream watershed began in January 2012, as an activity of a research project entitled "Recovery of headwaters of the Polo Regional Centro Norte (Pindorama Experimental Station)". Subsequently, in 2013, it continued with the project "Monitoring water resources to assess changes associated to land use and soil management at Olaria stream watershed". The data acquisition was carried out in three sub-basins located at the headwaters of Olaria catchment and near its mouth where the stream debouches into the São Domingos river (Figures 3 and 4).

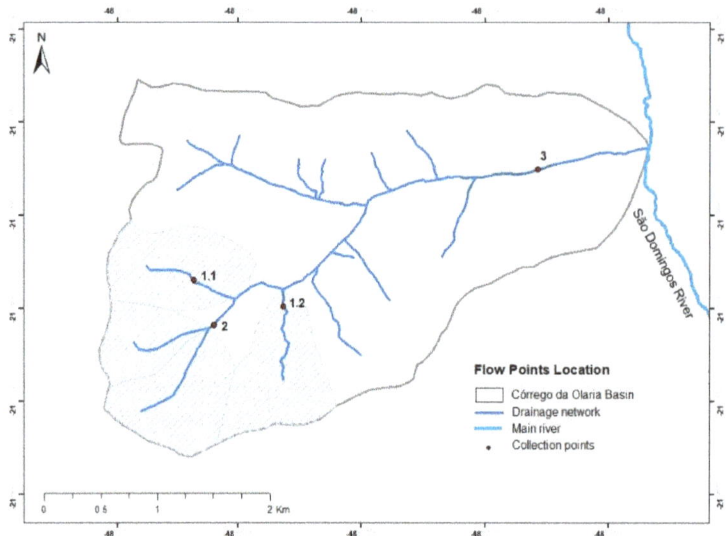

Figure 3. Drainage network of Olaria stream watershed, with identification of monitored sub-basins (1.1; 1.2; 2; 3).

Figure 4. Location of discharge flow monitoring points: (**a**) sub-basin 1.1; (**b**) sub-basin 1.2; (**c**) sub-basin 2; (**d**) sub-basin 3.

The uphill of sub-basin 1.2 (Figure 3) comprises the recovery and reforestation of a degraded area. This specific area had an aggravated erosion process due to inadequate soil management (Argisol—susceptible to erosion) (Figure 5). For decades, the coffee production system was installed uphill and later, pasture and cattle raising, with no soil conservationist practices [64]. The lack of vegetation for soil protection and its incorrect land use have severely degraded the sub-basin hills, resulting in a severe erosion process of a 700 m long gullet and some stretches, up to 15 m deep (Figure 5a) [64,65].

In order to contain the erosion process, recover the source and biodiversity, four uneven reservoirs were built (Figure 5b), interconnected by drainage channels (spillways on concrete stairs) (Figure 5c), which allowed water to pass through with a controlled flow. In addition, surrounding agricultural experimentation areas were managed with conservationist and maintenance treatments within natural vegetation land cover [64–66]. In 2011, reforestation of the surrounding stream was made within an Agroforestry System (AS), under a different management on the uphill of the sub-basin (Figure 5d). The purpose was to restore permanent preservation areas (PPA) and complement the other environmental recovery practices [66,67].

Figure 5. History of recovery and reforestation of degraded area in sub-basin 1.2: (**a**) intense erosion and gullies; (**b**) formation of four uneven reservoirs; (**c**) drainage channels; and (**d**) implementation of reforestation and the Agro-forestry System.

The selected points 1.1 and 1.2 belong to sub-basins of the 1st order, according to the characterization of hierarchical position of fluvial systems and topological ordering of hydrographic basins developed by [68]. Sub-basin 2 occurs in a micro-basin of 2nd order and sub-basin 3 on a streams junction of 3rd order (Figures 3–5). The discharge flow monitoring took place from January 2012 to December 2016. The data were collected monthly on the same point and hour sequence, from 7:00 a.m. to 11:00 a.m.

The methods used for flow measurements in watercourses are necessary to evaluate the passage of water, to obtain the measurement of the average speed of the flow in the section area and the net discharge [69–71]. The measurement of water flow in a watershed can be performed using different methods and instruments. Conventional methods include immersed current meters, which obtain the average flow velocity of the watercourse section. The velocity is then multiplied by the corresponding area, and the sum of these products will result in the average flow of the watercourse.

The measuring of flow with immersed current meters as used in natural watercourses or in artificial channels is frequent in research in Brazil. Structures are built on the riverbed, such as openings of geometric shape, as was done for this research, through which the water will flow, facilitating the measurement of the flow in the determined section [69–71]. The spillways can be: rectangular, trapezoidal, triangular, with a thin or thick threshold [69–71].

Another way of measurement could be by the artificial tracking method with the use of chemical or radioactive tracers [70,72–74]. The use of these tracers is appropriate for small watercourses, as they are considered low cost, easy to handle, presenting satisfactory results. In the past, radioactive tracers such as tritium have been used in large rivers, while tracers with fluorescents (dyes) have been commonly explored in the USA [70], especially in studies focused on groundwater [74]. However, the use of tracers introduces some problems, if the water is turbid. Suspended sediments can easily absorb some tracers and, in addition, there are normative restrictions regarding radiological protection (e.g., with the use of tritium) [70,72,73].

In this study, we opted for the electronic flow meter method (model ISCO 2150), represented as a "smart" probe". This sensor uses the concept of digital electronics. The analog level is digitized inside the sensor itself to avoid electromagnetic interference (https://www.clean.com.br/Produto/Detalhe/56).

The instrument allows the monitoring of the flow at various points along the cross section of a river, in a short period, since the communication of the device with microcomputers is direct, with the transfer of the calculated flow data automatically done, substantially reducing the time necessary to fill in spreadsheets in the field and the digitization of these data. The major disadvantage of these instruments is the acquisition cost.

The discharge flow was measured with an ISCO 2150 portable flow meter (Figure 6), which allows the measurement of total flow (m^3), flow speed (m/s), and level (m) of water. This equipment has doppler continuous wavelength technology to measure the average flow velocity (https://www.clean.com.br/Produto/Detalhe/56). The equipment has a cable with a sensor (flow meter) and a communicator cable for data transmission to a notebook. The flow was calculated using Isco Flowlink 5.1 software. The input parameters were configured considering the shape of each stream channel: round, U-shaped, rectangular, trapezoidal or elliptical; and by width section measurement (cm) of the watercourse (Figure 6a). The sensor was placed in each stream (Figure 6b,c) for six minutes in each sub-basin. The software automatically stored data every 30 s, and after finishing flow reading, the data were exported and saved (Figure 6d).

Figure 6. Flow data collection with the portable ISCO 2150 m: (**a**) width section measurement (cm) of a watercourse; (**b**) input parameters; (**c**) sensor in a watercourse; (**d**) reading flow and data storage.

In order to define the sampling points for monitoring and collecting the flow data, the area was characterized as natural sections. The 1.1 (a), 2 (c), and 3 (d) points (Figure 4) were carefully selected to determine water drainage in stream tributaries, that is, specific points where there was no separation or streambed dispersions of water flow for different parts or directions.

The flow results were subject to descriptive statistics comprising temporal and spatial analyses. The averages were showed in boxplots, with graphical representations of discharge flow distribution with a numerical summary of water flow (m^3/s). For statistical treatment, the data were processed in software R, considered a tool for statistical computing and graph production (https://www.r-project.org). The pluviometric data were collected from the Integrated Center for Agrometeorological Information—CIIAGRO (http://www.ciiagro.sp.gov.br), which provides daily data by location in online system.

3.2. Thematic Maps

The land use and land cover (LULC) thematic maps of the sub-basins were elaborated using geographic information systems (GIS) and remote sensing techniques by using a visual interpretation of LULC on high-spatial-resolution orbital images.

The ESRI's ArcMap and Google Earth Pro GIS software packages were used to view, edit, create, and analyze geospatial data in hydrological and environmental studies in watersheds [26,27,72,73]. The maps' base information was compiled from spatial databases, using maps published by the

Brazilian Institute of Geography and Statistics (https://ww2.ibge.gov.br) at a scale of 1:100,000, and a digital elevation model (DEM) that was obtained from ASTER GDEN V2 satellite image.

The geoprocessing resources served as a basis for the mapping of the compositions of land use and land cover and identification of the points of collection of the water flow and of the watershed under study. The interpretative analysis of each land use was carried out based on the exposure of colors for different spectral waves of in image, with the representation and definition of the classes in regard to sub-basins' soil cover. The classification procedure was carried out based on research works developed by [27,54] and followed the characterization of environments that reflect the watershed ecosystem upstream of the water catchment point.

The location of the sub-basins, the segmentation of drainage networks and topographic dividers were carried out, and the polygons were vectorized on each LULC. The points, lines, and polygons were plotted on Google Earth Pro software, and, subsequently, the units mapped in KML format were transferred to ArcGis®, using the KML to Layer command conversion tool. The calculation of the areas for each LULC was performed in ArcMap software using the Area tool, in order to tabulate the differences of areas in percentage, in each sub-basin.

4. Results and Discussion

The headwater volumetric flow rate of water was characterized in terms of volume of fluid per unit of time (L/s) and space (sub-basins), which passed through the stream cross-section of each drainage sampling point from sub-basins: 1.1, 1.2, 2, and 3 of the Olaria stream watershed (Figure 3). Values on descriptive statistics of discharge can be seen in Table 1.

The discharge showed a range of values from 0.09 L/s (sub-basin 1.2) to 107.99 L/s (sub-basin 3), with mean values of 2.95 L/s (sub-basin 1.1); 0.92 L/s (sub-basin 1.2); 4.84 L/s (sub-basin 2); and 13.91 L/s (sub-basin 3). The standard deviation (SD) indicates the high values spread out over a wider range (1.32 to 20.23 L/s), in a high coefficient of variation (CV), positive asymmetry ($\gamma 1 > 0$) and kurtosis coefficient.

The analysis of morphometric characteristics of each sub-basin provided a quantitative description of geometric aspects and slope of each sub-basin (Table 2).

Table 1. Descriptive statistics of discharge (L/s) of sub-basins, from 2012 to 2016.

Sub-Basin	Max	Min	Mean	Med	SD	CV (%)	AC	KC
1.1	8.50	0.18	2.95	2.50	1.57	53.2	0.803	0.0001
1.2	8.42	0.09	0.92	0.41	1.32	144.1	2.542	0.012
2	17.80	1.44	4.84	3.97	2.85	58.8	1.605	0.005
3	107.99	1.01	20.47	13.91	20.23	98.8	2.327	0.010

Max—maximum value; Min—minimum value; Mean—mean value; Med—median value; SD—standard deviation; CV—coefficient of variation; AC—asymmetry coefficient; KC—kurtosis coefficient.

Table 2. Morphometric characteristics of sub-basins at the Olaria stream watershed.

Sub-Basins	Area (ha)	Perimeter (km)	L (km)	W (km)	Sl (km)	Ha (m)	La (m)	S (%)
1.1	80.9	3.54	1.11	1.1	0.82	607	554	6.4
1.2	77.8	3.97	1.22	0.84	0.80	605	547	5.7
2	111.2	4.42	1.96	1.4	2.72	615	554	7.4
3	794	11.9	4.46	2.55	7.35	615	520	5.4

Symbols: L—maximum length; W—maximum width; Sl—stream length; Ha—high altitude; La—low altitude; S—slope.

The data obtained from dimensional morphometry (Table 2) showed that 1st-order sub-basins (1.1 and 1.2) have similarity values in terms of area (80.9 ha and 77.8 ha), perimeter (3.54 km and 3.97 km), and stream length (0.82 and 0.80 km). The highest altitude (H) was identified in sub-basin 2 (615 m), and the lowest was on point 3 (520 m). The greatest slope (S) was on sub-basin 2 (7.4%), followed by sub-basin 1.1 (6.4%), sub-basin 1.2 (5.7%), and sub-basin 3 (5.4%). The 2nd-order sub-basin had an area of 111.2 ha and a perimeter of 4.42 km.

The headwaters (sub-basins 1.1; 1.2, and 2) form an essential link in the hydrological cycle of the Olaria stream watershed. There are several approaches to topological ordering of streams based on distance from water source (headwaters) upstream to downstream to sub-basin 3. From the highest points to the lowest points, there is a hierarchical position of 1.1 and 1.2 (first order), 2 (second order) in the river system (Table 2; Figure 3), after each stream confluence. Considering the fluvial system as a continuation of the main channel (sub-basin 3), as observed in Figure 3, the headwaters are the smaller streams, of 1st-order of magnitude [68], which are the positive integer used in geomorphology and hydrology to indicate level of branching in a river system.

The headwater upstream points of first order (sub-basins 1.1, 1.2, and 2) are the extreme tributaries on the Olaria Watershed. The Strahler order is designed to morphology of a watershed and forms the basis of important hydrographic indicators of its structure. Its base is initial discharge flow line, which characterizes spring water.

The sub-basins of 1st order of magnitude (1.1 and 1.2), with similar areas (80.9; 77.8 ha), presented different hydrological behavior regarding amplitude of volumes over time (along month and years) and space (Figure 7).

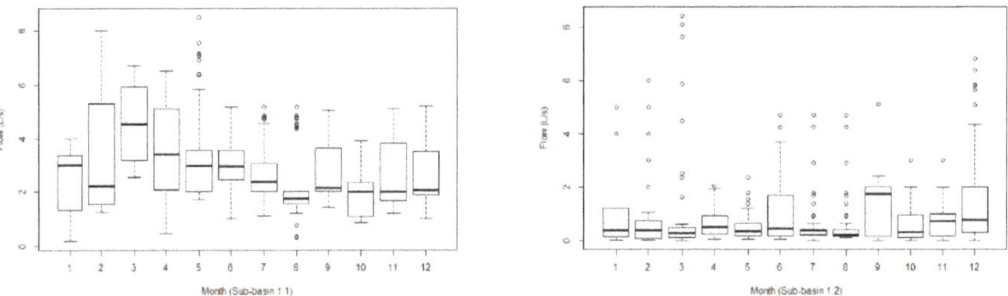

Figure 7. Sub-basins 1.1 and 1.2 discharge flow in the Olaria stream watershed, State of São Paulo, Brazil.

The minimum, first quartile (Q1), median, third quartile (Q3), and maximum values from sub-basins 1.1 and 1.2 showed that the water source of sub-basin 1.1 has higher volumetric flow rate of water transported through the given stream cross-sectional area than sub-basin 1.2, which showed the mostly outlier values along the monthly years, mainly in March (Month 3) (Figure 7).

The distribution, its central value, and its variability showed a different behavior of concentration of water flow on sub-basin 2, which has two stream branches and is along the drainage network (sub-basin 3) over the monthly monitored period (Figure 8).

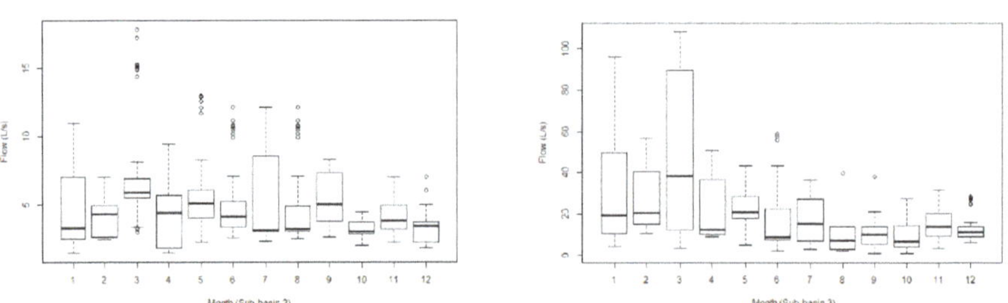

Figure 8. Sub-basins 2 and 3 discharge flow in the Olaria stream watershed, State of São Paulo, Brazil.

The distribution of annual flows in sub-basins 1.1, 1.2, 2, and 3 (Figures 7 and 8) in the Olaria stream watershed indicates the values measured on an interval scale (1.1; 1.2; interval 0–8 L/s); (2; interval

0–15 L/s); (3; interval 0–100 L/s), from month 1, January, to month 12, December. Those headwaters are considered perennial, that is, they produce water throughout the year, but with flow rates varying throughout it. If two discharges from the same order stream are merged (Figure 8, sub-basin 2), the resulting discharge flow increases from 8 L/s to 15 L/s. The 2nd order of magnitude in which two headwaters from two first-order watercourses merge (sub-basin 2) showed the average value of 4.84 L/s (Table 1). Based on volume of water from confluence (the point where two 1st-order rivers merge), a minimum value of 1.44 L/s and a maximum of 17.80 L/s were observed over 5 years.

The freshwater provided from the Olaria stream on point 3 (Figure 8; sub-basin 3) showed an average flow rate of 20.47 L/s that will be used for consumption to urban areas downstream of the São Domingos River, crop irrigation systems, and multiple uses. Through the Olaria stream watershed, the discharge flow is the environmental transport on each open channel (i.e., free surface conduits of water). The water storage and control of sub-basins within headwaters (points 1.1, 1.2, and 2) determine a constant condition of water production over time.

The water discharge recognizes water source on a watershed, which allows defining the main flow and its contribution to environmental services considering the drainage networks of 1st order of magnitude. The valuation of discharge in headwaters as an ecosystem service will benefit and control the sub-basins in a watershed functioning. In this sense, discharge can be a variable that allows development of inductive and not only repressive public policies and can be considered in the importance of ecosystem services (ES) promoted by maintenance and conservation of water resources, mainly on changing the principle of "polluter pays" to "provider-receiver", in payments for environmental services (PES) schemes [53,54].

As mentioned, this program presents an importance of balancing the dynamics of habitats and ecosystems and favors maintaining, recovering, and improving environmental conditions [54,57]. So, there is a need to recognize the anthropogenic landscapes on distribution of land use and land cover (LULC) of uphill on headwaters areas. The percentage values of LULC varied in each sub-basin, with predominance for cropping and forest cover on sub-basins 1.1, 1.2, and 2 (Figure 9).

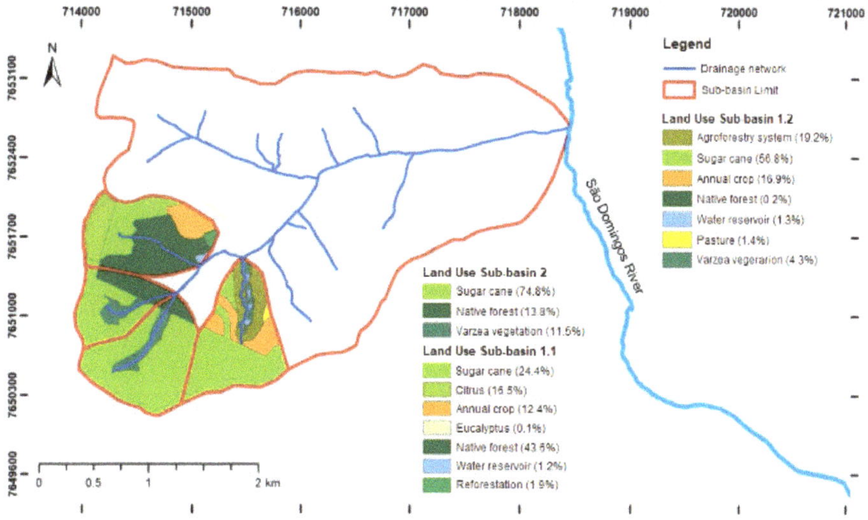

Figure 9. Land use land cover (LULC) at Olaria stream sub-basins, State of São Paulo, Brazil.

Agricultural land use was noticeable in all sub-basins (Figure 9). Sugarcane represented 74.8% (sub-basin 2), 56.8% (sub-basin 1.2), and 24.4% (sub-basin 1.1), annual crops were 16.9% (sub-basin 1.2) and 12.4% (sub-basin 1.1), citrus culture indicated 16.5% (sub-basin 1.1), and pasture 1.4% (sub-basin 1.2). Therefore, the sub-basins have significant anthropogenic influence. In addition to considerable crop areas, sub-basins are substantially covered with native forest, 43.6% (sub-basin 1.1), 13.8% (sub-basin 2). It is important to highlight the Agroforestry System (AS) 19.2% (sub-basin 1.2), which was implemented in 2011, aiming to restore the Permanent Preservation Area (PPA) on site and enable a better functioning of ecosystems, as well as promoting environmental services.

The characterization of water availability (upstream) allows taking into account the spatiotemporal heterogeneity of the Olaria Stream watershed, regarding physical characteristics (morphometric), LULC density, and topography. The headwaters' upstream sub-basins do not consider the crop area as a single and homogeneous unit, but regions that respond to the conservationist practices installed in an agricultural system.

The monthly average flows in sub-basins 1.1 and 1.2 (1–60 months) and the runoff coefficient (C) showed different hydrological behavior over time. The runoff coefficient (C) expresses the different relationships between the amount of runoff and the amount of precipitation received (Figures 10 and 11). The total rainfall values varied each year, (2012: 1152.1 mm; 2013: 1429.4 mm; 2014: 961.2 mm; 2015: 1367.8 mm; 2016: 1376.6 mm). The lowest value occurred in 2014 and the highest value in 2013.

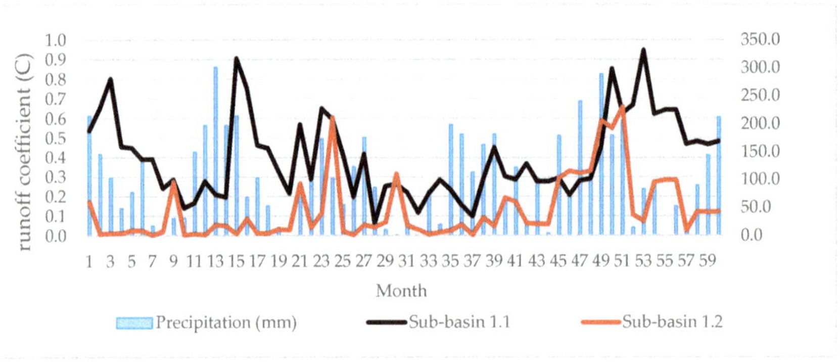

Figure 10. Precipitation (mm) and sub-basin 1.1 and 1.2 runoff coefficient (C) from January 2012 (Month 1) to December 2016 (month 60) in the Olaria stream watershed.

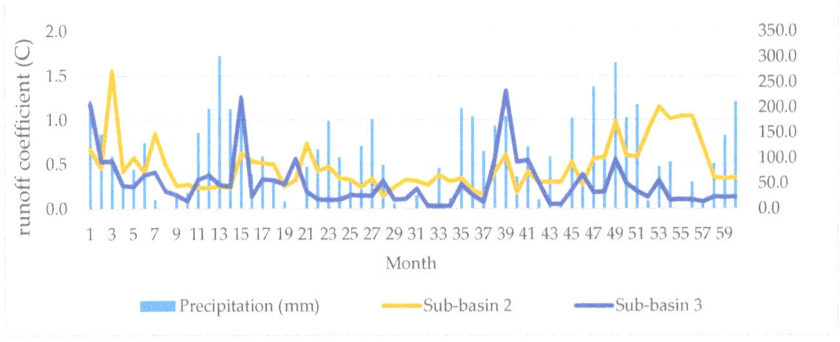

Figure 11. Precipitation (mm) and sub-basin 2 and 3 runoff coefficient (C) from January 2012 (month 1) to December 2016 (month 60) in the Olaria stream watershed.

It is important to highlight that even in months where there was no rain or there was little rainfall (May, June, and July), the headwaters continued to produce water volume in the watershed. The sub-basins' overland surface runoff after a higher precipitation shows the water running downhill over the landscape of the Olaria watershed (Figures 10 and 11). Stream discharges are directly influenced by seasonality, years (Table 3), and climatic variability; thus, critical periods in terms of water availability must be assessed in order to guarantee a safety margin for planning and management activities [74]. Discharge can be conceptualized as a result of hydrological processes of interaction between regional precipitations, infiltration, defluvium (Figures 10 and 11 and Table 3), and physiographic conditions of the watershed (Table 2), in which man's actions directly influence proper management of biodiversity conservation and water production [75]. When precipitation (mm) falls onto each sub-basin, it starts moving according to the laws of gravity, downhill to retain water in the streams.

Table 3. Sub-basins 1.1., 1.2, 2, and 3 flows (m^3/s) in the Olaria stream watershed.

	Year	Flow (L/s)		
		Average	S.D.	
Sub-basin 1.1	2012	2.9	1.4	c
	2013	3.4	1.7	b
	2014	1.7	0.8	d
	2015	2.1	0.7	d
	2016	4.5	1.2	a
Sub-basin 1.2	2012	0.3	0.6	d
	2013	0.8	1.4	bc
	2014	0.4	0.6	cd
	2015	1.2	0.9	b
	2016	1.9	1.9	a
Sub-basin 2	2012	5.4	3.6	b
	2013	4.4	1.6	c
	2014	2.9	0.7	d
	2015	3.9	1.6	c
	2016	7.7	3.1	a
Sub-basin 3	2012	28.2	21.3	a
	2013	23.4	22.7	a
	2014	10.4	7.8	b
	2015	26.6	25.6	a
	2016	13.7	10.9	b

The water storage in the basin system has increased, and the response of the basin has become more direct as the sub-basin area increases. On sub-basin 1.2, there is an area of old gully, recovered by soil conservation practice, with the implantation of a permanent preservation area and agroforestry system, in early 2011. It is important to note that, in this place, an increase in water production can be clearly seen from September 2013 (month 47, Figure 11), a period in which soil conservation practices were consolidating along the headwaters. This result demonstrates the importance of growth of the riparian forest area, and practices of terracing, no-till, and level planting that occurred in the uphill watershed. The effect of vegetation and forests contribute to regulation of the hydrobiological cycle and greater quantity and quality of water in watersheds [52,76–79].

The water availability values of headwaters (1.1, 1.2, and 2) showed the lowest average flow in sub-basin 1.2 (0.09 L/s). Non-forested springs present several problems, such as flood events, silting of water courses by uncovered soil, in addition to low water infiltration in the hydrological system [30,80]. The lack of forest cover drastically prevents water from infiltrating the soil, thereby supplying springs, impairing proper functioning of the hydrological regime [30,81]. The headwater (1.2) is located in a large percentage of agricultural areas and an important reforestation implantation in 2011 (Figure 11). In this place, there is a clear increase in water production in 2013, confirming results of the importance of growth of the riparian forest and a better water ecosystem service promoted within a set of environmental benefits [52,55].

The results indicate the importance of a forest to increase water production and base flow (Qbf), since forest floor provides greater water infiltration in the soil (IF), intercepts (Ic) the precipitated water (P) by means of crowns, trunks and roots, which infiltrate porous soil, percolating to the deepest layers of soil, supplying the water table levels, aquifers and, consequently, springs and riverbeds, favoring the regularization of the hydrological regime. Statements like these were made by several authors, among them [69,75,82–85]. In an area with forest cover, there is less direct runoff (Qds), avoiding sediment transport, erosion, river sedimentation, flooding, and decreased loss of nutrients from soil [69,86–89].

All headwaters are perennial in an intermittent streams condition. Even in the dry season, the volume of water remained regulated, with no water outages or water crises (Table 1, Figures 7–10). In rainy months (November to April), there were no flood peaks or episodes of maximum flows, which shows the great importance of native forest in regulating discharge flow and water availability for springs, in agreement with several studies conducted [52,69,75,82–85,90,91].

The groundwater occurs within sub-basins, and the water moves slowly on sub-basin 1.2 (Table 3). The precipitation is absorbed in the watershed, where it flows and becomes part of the surface water in different ways, as shown in Figures 10 and 11. The streams flow from 1.1 and 1.2 of the watershed, join the stream on point 3, and to the São Domingos river. Thus, groundwater flows toward a stream, and the sub-basins are used as the basic hydrologic unit for both surface water and groundwater planning purposes [79–83,88,89].

Most groundwater and surface water are interconnected, but in some environments, such as sub-basin 1.2, there was not enough groundwater discharge to maintain stream flow like in 1.1. The water availability on sub-basins of first order contribute to water security assessment on the São Domingos river watershed, which contributes water flow to the Turvo/Grande river, an important Brazilian watershed.

The water security assessment framework should consider a flow discharge on a basin-scale analysis, using an indicator, mainly on first-order sub-basins, as 1.1 and 1.2. The discharge from headwaters is a dimension of water flow and can be considered as an indicator of water security. The public policies should be elaborated to consider this point of view in making the water assessment framework on the basis of headwater flow discharge. Therefore, the water availability in watersheds is a measure of how the discharge will have a bearing on water security to a better land use land cover (LULC) governance.

Throughout the period, the sub-basins functioned as an impermeable container, returning the water received by precipitation and retaining part of that water in the water storage in the dam system. The water retention capacity is influenced by several factors, among which, the forest cover, physical characteristics and geomorphological factors, topography, hydraulic works present in the basin and conservationist planting practices [26,45,47,55,67,69].

The forest cover has the function of interception, storage, and reduction of runoff [67]. Concerning sub-basins 1.1 and 1.2, morphometric characteristics at the Olaria stream watershed (Table 2) and the stream flow along the months (Figure 11), the water retention in the basin's hydrological system varied according to geomorphologic factor, as the lower slope of the terrain, soil formation, and implementation of native tree species land cover along the dams. The agroforestry system on sub-basin 1.1 was consolidated on natural vegetation cover on the basin area system along the dam, and it was observed that the flow, over time, was more stable. However, in sub-basin 1.2, in the period of growth of vegetation cover (2011 to 2013), there was a lower volume of runoff. After the consolidation of the permanent preservation area (after 2013), the volume of flow water increases gradually, with a considerable increase after 2015. According to [89], the water absorption by the root system in the growth period of the vegetation cover increased the time of permanence of water in the hydrographic basin, causing events of less volume of water flowing to the basin from the evapotranspiration processes.

With identical amounts of precipitation, the sub-basins produced varying amounts of flow, due to different physical characteristics of the hydrographic basin and vegetation cover by area. Another factor is the topography of the terrain, which may have influenced the water storage capacity on these. In areas with a higher slope (sub-basin 2; 7.6%), there was less storage capacity than flatter areas, sub-basin 1.1 (6.4%), and sub-basin 1.2 (5.7%).

The discharge flows showed space–time variation in magnitude between studied headwaters sub-basins 1.1 and 1.2, on water availability, assessed based on average net discharges (Figure 11). Another factor that can be considered is the hydraulic works present in the basin, intended to contain the runoff (dams), sub-basins 1.1 and 1.2, which result in a reduction in the maximum flow of a basin, in view of the water storage in dams. In the Olaria watershed, the lowest flow was always observed in sub-basin 1.2. It is important to note that, in this place, there are dams that were built to stabilize a gull and to recover the local spring. These dams have drainage channels to control the speed of the water in each weir.

The new Brazilian Forest Code, Law No. 12,651 of 2012 [92], currently in force, Art. 41., establishes that the Federal Executive Government will consider the "Support and incentive program for conservation of environment" for ecologically sustainable development. However, it does not declare a deadline for such action and effectiveness. In Art. 41., mentioned above, it is reported the actions that the legislation intends to apply, as well as the monetary compensations and incentives, Items I and II, namely:

I—payment or incentive for environmental services such as remuneration, monetary or not, for the activities of conservation and improvement of ecosystems and generate environmental services, such as: sequestration, conservation, maintenance and increase of stock and reduction of flow carbon, conservation of scenic beauty and biodiversity, conservation of water and water and soil services, climate regulation, cultural enhancement and traditional ecosystem knowledge, maintenance of Permanent Preservation Areas, Legal Reserves, and restricted use.

II—Compensation for the environmental conservation measures necessary for fulfillment of the objectives of this Law, namely: Obtaining agricultural credit (with lower interest rates, limits and longer terms than in the market); Hiring of agricultural insurance; Granting of tax credits (by deducting the Permanent Preservation Areas, Legal Reserve and restricted use of the calculation basis for the Tax on Rural Territorial Property—ITR); Financing lines for the recovery of degraded areas, projects for the preservation of native vegetation, protection of species of native flora threatened with extinction, forest management and sustainable agroforestry carried out on the property or rural areas; Tax exemption for the main inputs and equipment.

Therefore, Brazil is awaiting regulation from the Federal Executive Government to have a Program for Payments for Environmental Services in force in the Law. However, compensation for the generation of ecosystem services already occurs and is distributed throughout the national territory, through actions promoted in programs and projects, with federal, state, and municipal government support, or support from private-sector companies.

Some notable examples that can be mentioned are: Water Producer Program of the National Water Agency—ANA (https://www.ana.gov.br/programas-e-projetos/programa-produtor-de-agua), being executed to initiatives by city halls, basin committees, or sanitation companies. The rural producers interested in conserving springs on their property and other priority areas for water production should seek out these institutions for applications on registration in the program and financial contemplation.

Another recognized program is the Conexão Mata Atlântica project (Project for the recovery and protection of climate and biodiversity services in the southeastern corridor of the Brazilian Atlantic Forest), which benefits, through the PES financial mechanism, rural owners who adopt conservation actions native forest, recover degraded areas, and implement sustainable production practices. This program is supported by the federal government, through the Ministry of Science, Technology, Innovations and Communications (MCTIC), and by the governments of the states of Rio de Janeiro, São Paulo, and Minas Gerais. The resources are made available by the Global Fund for the Environment (Global Environmental Facility—GEF), in order to implement actions to encourage the recovery of conservation of ecosystem services. For participation, interested parties must submit the required documentation through a public selection notice (https://conexaomataatlantica.mctic.gov.br/cma/o-projeto/o-que-e).

There are also the Payment for Environmental Services Projects, provided for the Forest Remnants Program of the State of São Paulo—Brazil, which comply with State Law 13,798/2009 (State Policy on Climate Change) and State Decree 55,947/2010 [93,94]. These projects cover various types of related environmental services: the conservation of forest remnants; recovery of riparian forests and implantation of native vegetation for the protection of springs; planting seedlings of native species and/or implementing practices that favor natural regeneration for the formation of biodiversity corridors; reforestation with native species or with native species intercropped with exotic species for sustainable exploitation of wood and non-wood products; implementation of agroforestry and silvopastoral systems that include the planting of at least 50 individuals of native tree species per hectare; implantation of commercial forests in areas contiguous to the remnants of native vegetation; and management of forest remnants to control competing species

Other initiatives comprise conservation-planting practices. Conservation practices in agricultural crops include the use of terracing, or monitoring of level curves, to direct runoff (reducing slope) and avoid erosion and damage to crops, contribute to better conditions infiltration of water in the system, in low- or medium-intensity rains [89]. Sub-basins 1.2 have a percentage of agricultural areas that receive conservationist planting treatments, which contributes to greater water infiltration into the hydrological system. The water moves within the hydrographic basin in an infinite number of superficial and/or underground trajectories, in which there is an accumulation of groundwater and less lateral flow.

The watershed ecosystem is a fundamental unit to provide an environmental service considering natural resources. The PES programs should be applied in headwaters on sub-basins in a watershed. The scheme will bring several benefits for society, balancing the dynamics of habitats and ecosystems, and favoring maintenance, recovery, or improvement of environmental conditions [54,57,95].

The sale of a best cropping production considering best management practices and an agricultural production conducted within sustainability (e.g., agroforestry production systems) brings benefit to a collectivity. The "provider-receiver" principle is based on a landowner that will offer an environmental service which generates a benefit of a better quality and quantity of water to society. Therefore, the producer that produces in a headwater sub-basin considering agricultural practices will have the right to be remunerated for maintaining soil and land on its best quality, decreasing the erosion process and preserving the natural resources as soil and water in a better quality.

The control instrument of the "provider-receiver" principle should be a land use policy. The landowner must be motivated to include the best management practices uphill headwaters as an environmental service in their decision-making regarding a better land use land cover (LULC). The conservation of land should be a financially attractive option. The productive practices with a soil and water better management and conservation should be considered as income generation on furnishing environmental services. The PES is planning a legal framework to define the economic activity and it is an instrument of public policies at the least cost to society. The ES is determined by a market established between private and public institutions to establish a market for ES compensation.

5. Conclusions

The hydrological discharge of headwater sub-basins showed space–time variation in magnitude. The water flow discharge of forested protected headwater on 1st order of magnitude produced a better quantity of water. The Olaria stream watershed has water availability, assessed based on average net discharges (flow) and indicates a potential for environmental services payment programs, with respect to the "provider-receiver" principle, and water security schemes.

The hydrological discharge of the headwater sub-basins presented a flow range from 0.09 L/s (sub-basin 1.2) to 107.99 L/s (sub-basin 3) or from 8 to 9330 m^3/day. Sub-basin 1.1 was the one that remained stable throughout the monitoring period, with an average flow of 2.95 L/s or 255 m^3/day. Sub-basin 1.2 showed an increase in volume over time, ranging from 0.3 to 1.9 L/s or 25.9 to 164.1 m^3/day.

Gully recovery management, reforestation with native trees, and agroforestry system in sub-basin 1.2, concomitant with the annual increase in flow during the recovery works, is an example of an activity indicated for receiving payment for the environmental service provided, discharging more water and quality for the owner downstream of the watershed.

This study contributes with an example of water security assessment in Brazilian watersheds, for the feasibility of paying for environmental services for the water-producing sub-basins.

Author Contributions: Conceptualization, M.B.L.S., T.C.T.P., and A.L.M.M.; methodology, M.B.L.S. and T.C.T.P.; software, L.F.S.F. and R.C.A.C.; validation, T.C.T.P., L.F.S.F., and F.A.L.P.; formal analysis, F.A.L.P., T.C.T.P., and M.Z.; investigation, M.B.L.S. and T.C.T.P.; resources, A.L.M.M. and T.C.T.P.; data curation, M.Z., L.F.S.F., and T.C.T.P.; writing—original draft preparation, M.B.L.S., M.Z., and T.C.T.P.; writing—review and editing, F.A.L.P. and M.C.L.; visualization, F.A.L.P. and M.C.L.; supervision, T.C.T.P. and F.A.L.P.; project administration, T.C.T.P. and M.B.L.S.; funding acquisition, A.L.M.M. and T.C.T.P. All authors have read and agreed to the published version of the manuscript.

Funding: The present study was carried out within the framework of the Post Graduation Research Programme of Coordenação de Aperfeiçoamento de Pessoal de Nível Superior (CAPES); Conselho Nacional de Desenvolvimento Científico e Tecnológico (CNPq); Agência do Ministério da Ciência, Tecnologia, Inovações e Comunicações (MCTIC); and Land Use Policy Brazilian Group (PolUS). For the author integrated in the CITAB Research Centre, the research was further financed by National Funds of FCT—Portuguese Foundation for Science and Technology, under the project UIDB/AGR/04033/2020. For the author integrated in the CQVR, the research was further financed by National Funds of FCT—Portuguese Foundation for Science and Technology, under the project UIDB/QUI/00616/2020.

Acknowledgments: The authors would like to thank: the Coordination of Improvement of Higher Education Personnel (CAPES) for the scholarship; the National Council for Scientific and Technological Development (CNPq) for Research funds; the Agronomy (Soil Science) Post-Graduation Program from the School of Agricultural and Veterinary Sciences, São Paulo State University (UNESP); The Land Use Policy Research Group—PolUS. Mariana Bárbara Lopes Simedo (M.B.L.S.), Teresa Cristina Tarlé Pissarra (T.C.T.P.), Antonio Lucio Mello Martins (A.L.M.M.); Maria Conceição Lopes (M.C.L.), Renata Cristina Araújo Costa (R.C.A.C.), Marcelo Zanata (M.Z.), Luís Filipe Sanches Fernandes (L.F.S.F.), and Fernando António Leal Pacheco (F.A.L.P.).

Conflicts of Interest: The authors declare no conflict of interest. The funders had no role in the design of the study; in the collection, analyses, or interpretation of data; in the writing of the manuscript, or in the decision to publish the results.

References

1. Academia Brasileira de Ciências-ABC. *Recursos Hídricos no Brasil: Problemas, Desafios e Estratégias*; Academia Brasileira de Ciências-ABC: Rio de Janeiro, Brazil, 2014; ISBN 9788585761363.
2. United Nations Educational Scientific and Cultural Organization-UNESCO. *Water Security and the Sustainable Development Goals*; Global Water Security Issues (GWSI) Series, Series I; UNESCO Publishing: Paris, France, 2019; ISBN 9789231003233.
3. Zogheib, C.; Ochoa-Tocachi, B.F.; Paul, J.D.; Hannah, D.M.; Clark, J.; Buytaert, W. Exploring a water data, evidence, and governance theory. *Water Secur.* **2018**, *4*, 19–25. [CrossRef]
4. Keys, P.W.; Porkka, M.; Wang-Erlandsson, L.; Fetzer, I.; Gleeson, T.; Gordon, L.J. Invisible water security: Moisture recycling and water resilience. *Water Secur.* **2019**, *8*, 100046. [CrossRef] [PubMed]
5. Van Oel, P.; Chukalla, A.; Vos, J.; Hellegers, P. Using indicators to inform the sustainable governance of water-for-food systems. *Curr. Opin. Environ. Sustain.* **2019**, *40*, 55–62. [CrossRef]
6. Licciardello, F.; Piergiovanni, L. Packaging and food sustainability. In *The Interaction of Food Industry and Environment*; Galanakis, C., Ed.; Academic Press: Cambridge, MA, USA, 2020; pp. 191–222, ISBN 978-0-12-816449-5.
7. Shiklomanov, I.A. *World Water Resources: A New Appraisal and Assessment for the 21st Century: A Summary of the Monograph World Water Resources*; UNESCO International Hydrological Programme, UNESCO-IHP: Paris, France, 1998; 37p.
8. Agência Nacional de Águas (Brazilian National Water Agency)-ANA. Quantidade de água. Available online: https://www.ana.gov.br/panorama-das-aguas/quantidade-da-agua (accessed on 15 January 2020).

9. Departamento de Águas e Energia Elétrica (Department of Water and Electricity)-DAEE Água. Available online: http://www.daee.sp.gov.br/index.php?option=com_content&view=article&id=50&Itemid=29 (accessed on 16 January 2020).
10. United Nations Educational Scientifc and Cultural Organization-UNESCO. *Relatório Mundial das Nações Unidas Sobre Desenvolvimento dos Recursos Hídricos 2016: Água e Emprego, Resumo Executivo*; United Nations Educational Scientifc and Cultural Organization-UNESCO: Paris, France, 2016.
11. Food and Agriculture Organization of the United Nations-FAO. *The State of Food and Agriculture 2019. Moving Forward on Food Loss and Waste Reduction*; Food and Agriculture Organization of the United Nations-FAO: Rome, Italy, 2019; ISBN 9781315764788.
12. United Nations Water-UN-Water. *Water Security & the Global Water Agenda: A UN-Water Analytical Brief*; United Nations Universit, Institute for Water, Environment & Health (UNU-INWEH): Hamilton, ON, Canada, 2013; ISBN 9788578110796.
13. Jepson, W.; Budds, J.; Eichelberger, L.; Harris, L.; Norman, E.; O'Reilly, K.; Pearson, A.; Shah, S.; Shinn, J.; Staddon, C.; et al. Advancing human capabilities for water security: A relational approach. *Water Secur.* **2017**, *1*, 46–52. [CrossRef]
14. Tucci, C.; Chagas, M. Segurança hídrica: Conceitos e estratégia para Minas Gerais. *REGA Porto Alegre* **2018**, *14*, 1–12. [CrossRef]
15. Gunda, T.; Hess, D.; Hornberger, G.M.; Worland, S. Water security in practice: The quantity-quality-society nexus. *Water Secur.* **2019**, *6*, 100022. [CrossRef]
16. Garrick, D.; Iseman, T.; Gilson, G.; Brozovic, N.; O'Donnell, E.; Matthews, N.; Miralles-Wilhelm, F.; Wight, C.; Young, W. Scalable solutions to freshwater scarcity: Advancing theories of change to incentivise sustainable water use. *Water Secur.* **2020**, *9*, 100055. [CrossRef]
17. Vieira, I.F.B.; Rolim Neto, F.C.; Carvalho, M.N.; Caldas, A.M.; Costa, R.C.A.; Silva, K.S.D.; Parahyba, R.D.B.V.; Pacheco, F.A.L.; Fernandes, L.F.S.; Pissarra, T.C.T. Water Security Assessment of Groundwater Quality in an Anthropized Rural Area from the Atlantic Forest Biome in Brazil. *Water* **2020**, *12*, 623. [CrossRef]
18. Tundisi, J.G.; Tundisi, T.M. As múltiplas dimensões da crise hídrica. *Rev. USP* **2015**, *106*, 21–30. [CrossRef]
19. Purvis, L.; Dinar, A. Are intra- and inter-basin water transfers a sustainable policy intervention for addressing water scarcity? *Water Secur.* **2020**, *9*, 100058. [CrossRef]
20. Flörke, M.; Schneider, C.; McDonald, R.I. Water competition between cities and agriculture driven by climate change and urban growth. *Nat. Sustain.* **2018**, *1*, 51–58. [CrossRef]
21. Marengo, J.A.; Alves, L.M. Crise Hídrica em São Paulo em 2014: Seca e Desmatamento. *GEOUSP Espaço e Tempo* **2015**, *19*, 485. [CrossRef]
22. Soriano, É.; Londe, L.D.R.; Di Gregorio, L.T.; Coutinho, M.P.; Santos, L.B.L. Crise hídrica em São Paulo sob o ponto de vista dos desastres. *Ambient. Soc.* **2016**, *19*, 21–42. [CrossRef]
23. Empinotti, V.L.; Budds, J.; Aversa, M. Governance and water security: The role of the water institutional framework in the 2013–15 water crisis in São Paulo, Brazil. *Geoforum* **2019**, *98*, 46–54. [CrossRef]
24. Tundisi, J.G. Governança da água. *Rev. Univ. Fed. Minas Gerais* **2013**, *20*, 223–235. [CrossRef]
25. Simedo, M.B.L.; Martins, A.L.M.; Pissarra, T.C.T.; Lopes, M.C.; Costa, R.C.A.; Valle-Junior, R.F.; Campanelli, L.C.; Rojas, N.E.T.; Finoto, E.L. Effect of watershed land use on water quality: A case study in Córrego da Olaria Basin, São Paulo State, Brazil. *Braz. J. Biol.* **2018**, *78*, 625–635. [CrossRef]
26. Pissarra, T.C.T.; Valera, C.A.; Costa, R.C.A.; Siqueira, H.E.; Martins Filho, M.V.; Valle Júnior, R.F.D.; Sanches Fernandes, L.F.; Pacheco, F.A.L. A Regression Model of Stream Water Quality Based on Interactions between Landscape Composition and Riparian Buffer Width in Small Catchments. *Water* **2019**, *11*, 1757. [CrossRef]
27. Valera, C.; Pissarra, T.; Filho, M.; Valle Júnior, R.; Oliveira, C.; Moura, J.; Sanches Fernandes, L.; Pacheco, F. The Buffer Capacity of Riparian Vegetation to Control Water Quality in Anthropogenic Catchments from a Legally Protected Area: A Critical View over the Brazilian New Forest Code. *Water* **2019**, *11*, 549. [CrossRef]
28. Ulibarri, N.; Escobedo Garcia, N. Comparing Complexity in Watershed Governance: The Case of California. *Water* **2020**, *12*, 766. [CrossRef]
29. Rodrigues, V.; Sansígolo, C.; Cicco, L.; Viana, S.; Coneglian, A.; Haas, J. Avaliação do fluxo de água dos canais nas ruas de Freiburg-Alemanha. *Rev. Científica Eletrôn. Engenharia Florestal Garça* **2012**, *19*, 13–22.

30. Marmontel, C.V.F.; Rodrigues, V.A. Parâmetros Indicativos para Qualidade da Água em Nascentes com Diferentes Coberturas de Terra e Conservação da Vegetação Ciliar Indicative Parameters for Water Quality in Water Springs with Different Land Cover and Conservation of Riparian Vegetation. *Floresta e Ambiente* **2015**, *22*, 171–181. [CrossRef]
31. Debastiani, A.B.; Silva, R.D.A.D.; Rafaeli Neto, S.L. Eficácia da arquitetura MLP em modo closed-loop para simulação de um Sistema Hidrológico. *RBRH* **2016**, *21*, 821–831. [CrossRef]
32. McNally, A.; Verdin, K.; Harrison, L.; Getirana, A.; Jacob, J.; Shukla, S.; Arsenault, K.; Peters-Lidard, C.; Verdin, J.P. Acute Water-Scarcity Monitoring for Africa. *Water* **2019**, *11*, 1968. [CrossRef]
33. Lindhe, A.; Rosén, L.; Johansson, P.O.; Norberg, T. Dynamic water balance modelling for risk assessment and decision support on MAR potential in Botswana. *Water* **2020**, *12*, 721. [CrossRef]
34. Tariq, M.A.U.R.; van de Giesen, N.; Janjua, S.; Shahid, M.L.U.R.; Farooq, R. An engineering perspective of water sharing issues in Pakistan. *Water* **2020**, *12*, 477. [CrossRef]
35. Citrini, A.; Camera, C.; Beretta, G. Pietro Nossana Spring (Northern Italy) under Climate Change: Projections of Future Discharge Rates and Water Availability. *Water* **2020**, *12*, 387. [CrossRef]
36. Honda, E.A.; Durigan, G. A restauração de ecossistemas e a produção de água. *Hoehnea* **2017**, *44*, 315–327. [CrossRef]
37. Zhang, S.; Li, W.; Lei, X.; Ding, X.; Zhang, T. Implementation methods and applications of flow visualization in a watershed simulation platform. *Adv. Eng. Softw.* **2017**, *112*, 66–75. [CrossRef]
38. Caldas, A.; Pissarra, T.; Costa, R.; Neto, F.; Zanata, M.; Parahyba, R.; Sanches Fernandes, L.; Pacheco, F. Flood Vulnerability, Environmental Land Use Conflicts, and Conservation of Soil and Water: A Study in the Batatais SP Municipality, Brazil. *Water* **2018**, *10*, 1357. [CrossRef]
39. Park, Y.; Kim, Y.; Park, S.K.; Shin, W.J.; Lee, K.S. Water quality impacts of irrigation return flow on stream and groundwater in an intensive agricultural watershed. *Sci. Total Environ.* **2018**, *630*, 859–868. [CrossRef]
40. Choto, M.; Fetene, A. Impacts of land use/land cover change on stream flow and sediment yield of Gojeb watershed, Omo-Gibe basin, Ethiopia. *Remote Sens. Appl. Soc. Environ.* **2019**, *14*, 84–99. [CrossRef]
41. Miraji, M.; Liu, J.; Zheng, C. The Impacts of Water Demand and Its Implications for Future Surface Water Resource Management. *Water* **2019**, *11*, 2–11.
42. Santos, R.M.B.; Sanches Fernandes, L.F.; Vitor Cortes, R.M.; Leal Pacheco, F.A. Hydrologic Impacts of Land Use Changes in the Sabor River Basin: A Historical View and Future Perspectives. *Water* **2019**, *11*, 1464. [CrossRef]
43. Ul Hasson, S.; Saeed, F.; Böhner, J.; Schleussner, C.F. Water availability in Pakistan from Hindukush–Karakoram–Himalayan watersheds at 1.5 °C and 2 °C Paris Agreement targets. *Adv. Water Resour.* **2019**, *131*, 103365. [CrossRef]
44. Hernández-Guzmán, R.; Ruiz-Luna, A.; Cervantes-Escobar, A. Environmental flow assessment for rivers feeding a coastal wetland complex in the Pacific coast of northwest Mexico. *Water Environ. J.* **2019**, *33*, 536–546. [CrossRef]
45. Thakur, J.K.; Khanal, K.; Poudyal, K. Land cover changes for enhancing water availability in watersheds of tanahun and kaski, Nepal. *J. Water Clim. Chang.* **2019**, *10*, 431–448. [CrossRef]
46. Bessa Santos, R.M.; Sanches Fernandes, L.F.; Vitor Cortes, R.M.; Leal Pacheco, F.A. Development of a Hydrologic and Water Allocation Model to Assess Water Availability in the Sabor River Basin (Portugal). *Int. J. Environ. Res. Public Health* **2019**, *16*, 2419. [CrossRef]
47. Andrade, E.M.; Guerreiro, M.J.S.; Palácio, H.A.Q.; Campos, D.A. Ecohydrology in a Brazilian tropical dry forest: Thinned vegetation impact on hydrological functions and ecosystem services. *J. Hydrol. Reg. Stud.* **2020**, *27*, 100649. [CrossRef]
48. Chen, H.; Fleskens, L.; Baartman, J.; Wang, F.; Moolenaar, S.; Ritsema, C. Impacts of land use change and climatic effects on streamflow in the Chinese Loess Plateau: A meta-analysis. *Sci. Total Environ.* **2020**, *703*, 134989. [CrossRef]
49. Emerton, L.; Bos, E. *Value: Counting Ecosystems As an Economic Part of Water Infrastructure*; IUCN: Gland, Switzerland; Cambridge, UK, 2004; ISBN 283170720X.
50. Ministério do Meio Ambiente (Ministry of the Environment)-MMA. *Pagamento por Serviços Ambientais na Mata Atlântica: Liçoes Aprendidas e Desafios*, 2th ed.; Guedes, F.B., Seehusen, S.E., Eds.; Ministério do Meio Ambiente-MMA: Brasília, Brazil, 2011; ISBN 9788577381579.

51. Empresa Brasileira de Pesquisa Agropecuária (Brazilian Agricultural Research Corporation)-Embrapa. *Marco Referencial em Serviços Ecossitêmicos*, 1st ed.; Ferraz, R.P.D., Prado, R.B., Parron, L.M., Campanha, M.M., Eds.; Embrapa Soils: Brasília, Brazil, 2019; ISBN 9788570359094.
52. Tundisi, J.G.; Tundisi, T.M. Potencial impacts of changes in the Forest Law in relation to water resources. *Biota Neotrop.* **2010**, *10*, 67–75. [CrossRef]
53. Santos, F.L.D.; Silvano, R.A.M. Aplicabilidade, potenciais e desafios dos Pagamentos por Serviços Ambientais para conservação da água no sul do Brasil. *Desenvolvimento e Meio Ambiente* **2016**, *38*, 481–498. [CrossRef]
54. Valera, C.A. Avaliação do novo Código Florestal: As Áreas de Preservação Permanente—APPs, e a Conservação da Qualidade do solo e da água Superficial. Ph.D. Thesis, Universidade Estadual Paulista, Jaboticabal-São Paulo, Brazil, 2017.
55. Periotto, N.A.; Tundisi, J.G. A characterization of ecosystem services, drivers and values of two watersheds in São Paulo state, Brazil. *Braz. J. Biol.* **2018**, *78*, 397–407. [CrossRef] [PubMed]
56. Naeem, S.; Ingram, J.C.; Varga, A.; Agardy, T.; Barten, P.; Bennett, G.; Bloomgarden, E.; Bremer, L.L.; Burkill, P.; Cattau, M.; et al. Get the science right when paying for nature's services: Few projects adequately address design and evaluation. *Science* **2015**, *347*, 1206–1207. [CrossRef] [PubMed]
57. Siqueira, H.E. Identificação de Áreas para Conservação do solo e da água na área de Proteção Ambiental do rio Uberaba cm Geoprocessamento. Ph.D. Thesis, Universidade Estadual Paulista, Jaboticabal-São Paulo, Brazil, 2019.
58. Chu, L.; Quentin Grafton, R.; Keenan, R. Increasing Conservation Efficiency While Maintaining Distributive Goals With the Payment for Environmental Services. *Ecol. Econ.* **2019**, *156*, 202–210. [CrossRef]
59. Alvares, C.A.; Stape, J.L.; Sentelhas, P.C.; De Moraes Gonçalves, J.L.; Sparovek, G. Köppen's climate classification map for Brazil. *Meteorol. Z.* **2013**, *22*, 711–728. [CrossRef]
60. Empresa Brasileira de Pesquisa Agropecuária (Brazilian Agricultural Research Corporation)-Embrapa. Banco de Dados Climáticos do Brasil. Available online: https://www.cnpm.embrapa.br/projetos/bdclima/balanco/resultados/sp/419/balanco.html (accessed on 18 January 2020).
61. Rossi, M. *Mapa Pedológico do Estado de São Paulo: Revisado e Ampliado*; Instituto Florestal (Florestal Institute)-IF: São Paulo, Brazil, 2017; Volume 1, ISBN 3239660180.
62. Instituto Brasileiro de Geografia e Estatística (Brazilian Institute of Geography and Statistics)-IBGE. *Manual Técnico da Vegetação Brasileira*, 2th ed.; Instituto Brasileiro de Geografia e Estatística (Brazilian Institute of Geography and Statistics)-IBGE: Rio de Janeiro, Brazil, 2012; ISBN 9788524042225.
63. Instituto Brasileiro de Geografia e Estatística (Brazilian Institute of Geography and Statistics)-IBGE. *Censo Agropecuário 2017*; Instituto Brasileiro de Geografia e Estatística (Brazilian Institute of Geography and Statistics)-IBGE: São Paulo, Brazil, 2017.
64. Vieira, S.R.; Martins, A.L.M.; Silveira, L.C.P. *Relatório de Implantação do Projeto de Recuperação Ambiental da Estação Experimental de Agronomia de Pindorama, SP, Pindorama*; Polo Regional Centro Norte-APTA: Pindorama-São Paulo, Brazil, 1999; 13p.
65. Abdo, M.T.V.N.; Vieira, S.R.; Martins, A.L.M.; Silveira, L.C.P. *Estabilização de Uma Voçoroca no Polo Regional Centro Norte-APTA Pindorama-Sp*; Anais do III Congresso Latino Americano de Ecologia: São Lourenço, Minas Gerais, Brazil, 2009.
66. Martins, A.L.M.; Lopes, M.C.; Pissarra, T.C.T.; Abdo, M.T.V.N.; Valaretto, R.S.; Bonatti, M.B.L. Monitoramento de vazão e avaliação da interferência de sistemas agroflorestais na qualidade de recursos hídricos na microbacia "Córrego Da Olaria"-Polo Regional Centro Norte, Pindorama, SP. *Pesqui. Tecnol.* **2012**, *9*, 1–9.
67. Abdo, M.T.V.N.; Martins, A.L.M.; Finoto, E.L.; Fabri, E.G.; Pissarra, T.C.T.; Bieras, A.C.; Lopes, M.C. Implantação de sistema agroflorestal com seringueira, urucum e acerola sob diferentes manejos. *Pesqui. Tecnol.* **2012**, *9*, 1–15.
68. Strahler, A.N. Quantitative analysis of watershed geomorphology. *Eos Trans. Am. Geophys. Union* **1957**, *38*, 913–920. [CrossRef]
69. Lima, W.D.P. *Hidrologia Florestal Aplicada ao Manejo de Bacias Hidrográficas*, 2th ed.; Escola Superior de Agricultura "Luiz de Queiroz" Departamento de Ciências Florestais Piracicaba–São Paulo: Piracicaba, Brazil, 2008.
70. Tazioli, A. Méthodes expérimentales pour mesurer le débit des cours d'eau: Comparaison entre les traceurs artificiels et le courantomètre. *Hydrol. Sci. J.* **2011**, *56*, 1314–1324. [CrossRef]

71. Agência Nacional de Águas (Brazilian National Water Agency)-ANA. *Medição de Descarga Líquida em Grandes Rios: Manual Técnico*, 2th ed.; Agência Nacional de Águas (Brazilian National Water Agency)-ANA: Brasília, Brazil, 2014; ISBN 978-85-8210-026-4.
72. Valera, C.A.; Pissarra, T.C.T.; Martins Filho, M.V.; Valle Junior, R.F.; Sanches Fernandes, L.F.; Pacheco, F.A.L. A legal framework with scientific basis for applying the 'polluter pays principle' to soil conservation in rural watersheds in Brazil. *Land Use Policy* **2017**, *66*, 61–71. [CrossRef]
73. Araújo Costa, R.C.; Pereira, G.T.; Tarlé Pissarra, T.C.; Silva Siqueira, D.; Sanches Fernandes, L.F.; Vasconcelos, V.; Fernandes, L.A.; Pacheco, F.A.L. Land capability of multiple-landform watersheds with environmental land use conflicts. *Land Use Policy* **2019**, *81*, 689–704. [CrossRef]
74. Agência Nacional de Águas (Brazilian National Water Agency)-ANA. *Conjuntura dos Recursos Hídricos no Brasil 2017: Relatório Pleno/Agência Nacional de Águas*; Agência Nacional de Águas (Brazilian National Water Agency)-ANA: Brasília, Brazil, 2017.
75. Rodrigues, V.A. Processos Hidrológicos e Sustentabilidade da água em Microbacias com Pinus Halepensis Mill Albacete-Espanha. Postdoctoral Thesis, Universidade Estadual Paulista (UNESP), Botucatu-São Paulo, Brasil, Universidad de Castilla-La Mancha, Albacete, Spain, 2016.
76. Krishnaswamy, J.; Bonell, M.; Venkatesh, B.; Purandara, B.K.; Rakesh, K.N.; Lele, S.; Kiran, M.C.; Reddy, V.; Badiger, S. The groundwater recharge response and hydrologic services of tropical humid forest ecosystems to use and reforestation: Support for the "infiltration-evapotranspiration trade-off hypothesis". *J. Hydrol.* **2013**, *498*, 191–209. [CrossRef]
77. Aguilar, F.X.; Obeng, E.A.; Cai, Z. Water quality improvements elicit consistent willingness-to-pay for the enhancement of forested watershed ecosystem services. *Ecosyst. Serv.* **2018**, *30*, 158–171. [CrossRef]
78. Ferreira, P.; van Soesbergen, A.; Mulligan, M.; Freitas, M.; Vale, M.M. Can forests buffer negative impacts of land-use and climate changes on water ecosystem services? The case of a Brazilian megalopolis. *Sci. Total Environ.* **2019**, *685*, 248–258. [CrossRef] [PubMed]
79. Bremer, L.L.; Wada, C.A.; Medoff, S.; Page, J.; Falinski, K.; Burnett, K.M. Contributions of native forest protection to local water supplies in East Maui. *Sci. Total Environ.* **2019**, *688*, 1422–1432. [CrossRef]
80. Rodrigues, V.; Estrany, J.; Ranzini, M.; de Cicco, V.; Martín-Benito, J.M.T.; Hedo, J.; Lucas-Borja, M.E. Effects of land use and seasonality on stream water quality in a small tropical catchment: The headwater of Córrego Água Limpa, São Paulo (Brazil). *Sci. Total Environ.* **2018**, *622*, 1553–1561. [CrossRef]
81. Vanzela, L.S.; Hernandez, F.B.T.; Franco, R.A.M. Influência do uso e ocupação do solo nos recursos hídricos. *Rev. Bras. Eng. Agrícola e Ambient.* **2010**, *14*, 55–64. [CrossRef]
82. Zakia, M.J.B. Identificação e Caracterização da zona Ripária em uma Microbacia Experimental: Implicações no Manejo de Bacias Hidrográficas e na Recomposição de Florestas. Ph.D. Thesis, Universidade de São Paulo, São Carlos-São Paulo, Brazil, 1998.
83. Tucci, C.E.M.; Mendes, C.A. *Avaliação Ambiental Integrada de Bacia Hidrográfica*, 2th ed.; Ministério do Meio Ambiente (Ministry of the Environment)-MMA: Brasília, Brazil, 2006; ISBN 9788578110796.
84. Oliveira, L.F.C.D.; Fioreze, A.P. Estimativas de vazões mínimas mediante dados pluviométricos na Bacia Hidrográfica do Ribeirão Santa Bárbara, Goiás. *Rev. Bras. Eng. Agrícola e Ambient.* **2011**, *15*, 9–15. [CrossRef]
85. Cicco, L.S.D. Evolução da Regeneração Natural de Floresta Ombrófila Densa Alto-Montana e a Produção de água em Microbacia Experimental, Cunha-SP. Masters Thesis, Universidade Estadual Paulista (UNESP), Botucatu-São Paulo, Brazil, 2013.
86. Rodrigues, V.A. Análise dos processos hidrológicos em modelo didático de microbacias. *Rev. Científica Eletrônica Eng. Florest.* **2011**, *17*, 1–15.
87. Ren, H.Q.; Yuan, X.Z.; Liu, H.; Zhang, Y.W.; Zhou, S.B. The effects of environment factors on community structure of benthic invertebrate in rivers. *Ecol. Soc. China* **2015**, *35*, 3148–3156.
88. Zhang, X.K.; Fan, J.; Cheng, G.W. Modelling the effects of land-use change on runoff and sediment yield in the Weicheng River watershed, Southwest China. *J. Mt. Sci.* **2015**, *12*, 434–445. [CrossRef]
89. Guidotti, V.; Ferraz, S.F.D.B.; Pinto, L.F.G.; Sparovek, G.; Taniwaki, R.H.; Garcia, L.G.; Brancalion, P.H.S. Changes in Brazil's Forest Code can erode the potential of riparian buffers to supply watershed services. *Land Use Policy* **2020**, *94*, 104511. [CrossRef]
90. Likens, G.E.; Bormann, F.H. Linkages between Terrestrial and Aquatic Ecosystems. *Bioscience* **1974**, *24*, 447–456. [CrossRef]
91. Tundisi, J.G.; Tundisi, T.M. *Limnologia*, 1st ed.; Oficina de Textos: São Paulo, Brazil, 2008.

92. Diário Oficial [da] República Federativa do Brasil, Poder Executivo. *Lei nº 12.651, de 25 de maio de 2012. Dispõe Sobre a proteção da Vegetação Nativa;* Diário Oficial [da] República Federativa do Brasil, Poder Executivo: Brasilia, Brazil, 2012; p. 38.
93. Assembleia Legislativa do Estado de São Paulo, Poder Executivo. *Lei Nº 13.798, de 9 de Novembro de 2009. Institui a Política Estadual de Mudanças Climáticas-PEMC O;* Assembleia Legislativa do Estado de São Paulo, Poder Executivo: São Paulo, Brazil, 2009.
94. Diário Oficial do Estado de São Paulo, Poder Executivo, Brasil. *Decreto No 55.947, DE 24 de Junho de 2010. Regulamenta a Lei no 13.798, de 9 de Novembro de 2009, que Dispõe Sobre a Política Estadual de Mudanças Climáticas;* Diário Oficial do Estado de São Paulo, Poder Executivo, Brasil: São Paulo, Brazil, 2010; pp. 1–28.
95. Richards, R.C.; Rerolle, J.; Aronson, J.; Pereira, P.H.; Gonçalves, H.; Brancalion, P.H.S. Governing a pioneer program on payment for watershed services: Stakeholder involvement, legal frameworks and early lessons from the Atlantic forest of Brazil. *Ecosyst. Serv.* **2015**, *16*, 23–32. [CrossRef]

www.ingramcontent.com/pod-product-compliance
Lightning Source LLC
LaVergne TN
LVHW070123170726
843515LV00007B/1740

* 9 7 8 3 0 3 6 5 0 2 6 6 3 *

MDPI
St. Alban-Anlage 66
4052 Basel
Switzerland
Tel. +41 61 683 77 34
Fax +41 61 302 89 18
www.mdpi.com

Water Editorial Office
E-mail: water@mdpi.com
www.mdpi.com/journal/water